March 10, 2006

Wooden Rigs—**Iron Men**

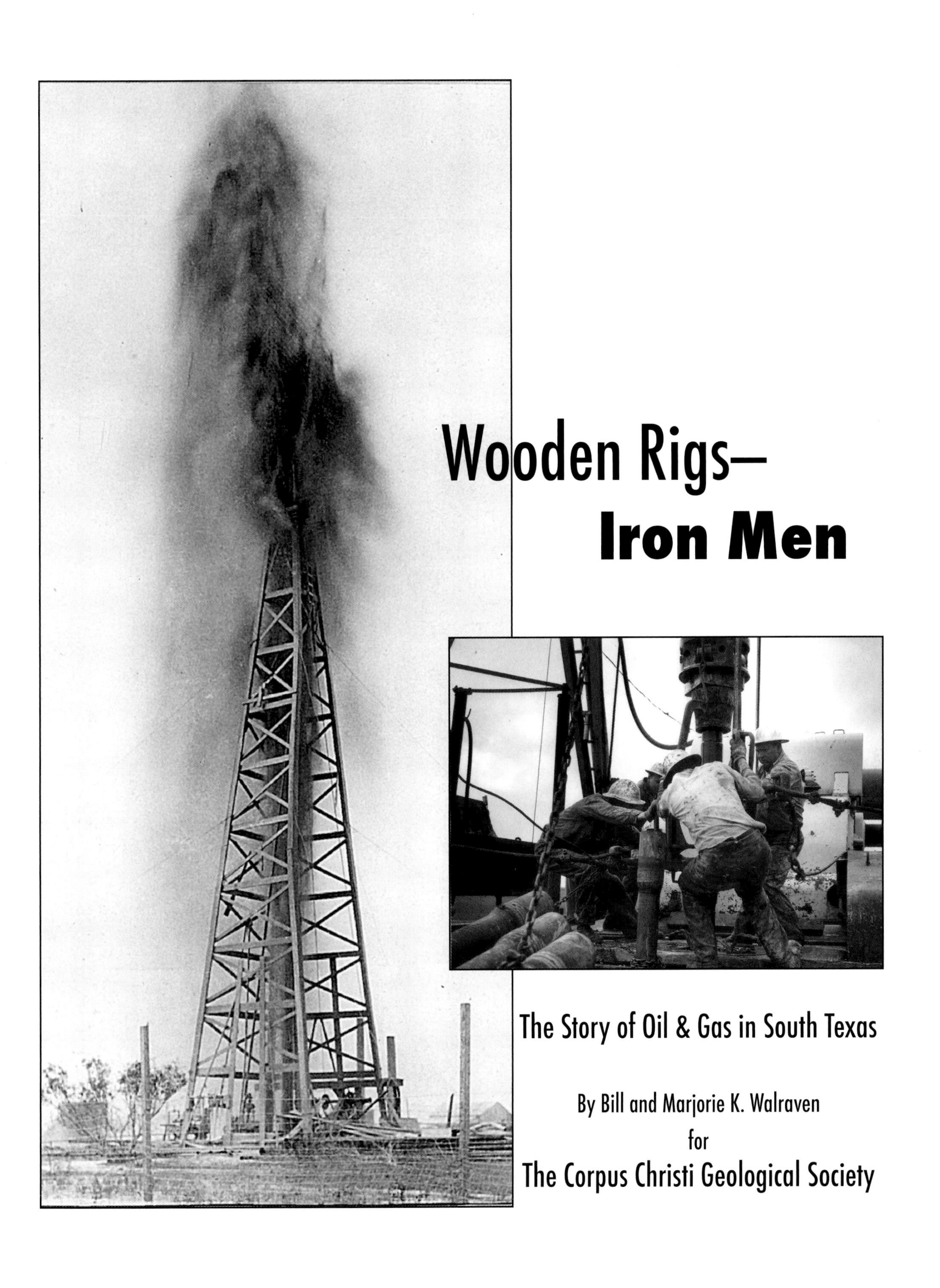

Wooden Rigs— **Iron Men**

The Story of Oil & Gas in South Texas

By Bill and Marjorie K. Walraven
for
The Corpus Christi Geological Society

Imprint of Javelina Press
P. O. Box 60181
Corpus Christi, Texas 78466
361-992-8031

ISBN 0-9646325-6-X
Library of Congress Control Number 2005901543

Printed in the United States of America

The oil business . . . over the past one
hundred years has been of great benefit
to South Texas. The land, the port, which
came along at the same time as the
oil business—it was a long and happy
marriage.

—Richard N. 'Dick' Conolly, Sr.

Introduction

You likely would not find a poet like Samuel Taylor Coleridge, Percy Bysshe Shelley, or William Wordsworth among them, but geologists are considered by some to be romantics. And, surprisingly, those Romantic Poets were among the scientists of their day, the early Nineteenth Century, at the beginning of modern science.

Shelley read scientific works on astronomy and incorporated them into his poetry. Astronomer Sir William Rowan Hamilton was a close friend of Wordsworth.

Theory of the Earth, written by James Hutton in 1798 and reported to have laid the foundation for modern geology, influenced Wordsworth's *Lyrical Ballads*, published the same year.

William Smith, said to have been the "father of modern geology," began his work in 1799. Both Shelley and Wordsworth refer to his theories. Pioneer chemist Sir Humphry Davy lined up a course of study in chemistry for both Woodsworth and Shelley. All three poets felt that the future of the world was to be bound up with the progress of science.

Wordsworth described his feelings: "Poetry is the breath and finer spirit of all knowledge; it is the impassioned expression which is the countenance of all Science.... The poet," he said, "will be ready to follow the steps of the man of science . . . carrying sensation into the midst of the objects of science itself."

In a TIPRO oral history interview conducted in 1993 by the Center for American History at the University of Texas at Austin, geologist Patrick J. F. Gratton advanced the theory that geologists are the modern day romantics.

He said he could pick out the exploration geologists at a social gathering of geologists, engineers, landmen, and others by things they talk about and their attitude in general.

This is not to run engineers and landmen down or say they don't have scientific capability, he said. "They take care of a lot of

This cloud-wrapped drilling rig might support the theory that geologists are the modern-day Romantics. Geologist Patrick J. F. Gratton said that to be effective a geologist must be an artist and a scientist.

details but don't know the value of anything. They just know the price of it."

The landmen's strength is in dealing with people, and you don't get leases unless you are good with people, he said.

"Then you have the geophysicists, who are probably just the alter ego of landmen," he said, "because they are extremely technically oriented and that is about all they are interested in.

"The petroleum engineer is similar to the geophysicist. He is detail conscious but not as imaginative. He is technically oriented, and a good one is very capable and responsible for finding a lot of oil and gas on [his] own," he said. "There are a lot of different types of them...some are trained and working particularly in drilling wells and how to do it mechanically correct. Some are in the area of evaluating what you find, called reservoir engineers. Others are good on completion work—how to get what you think is there out of the ground. So there are a lot of subdisciplines, I should say.

"But geologists in general to be effective must be an artist and a scientist," he said. "None of the other things I have described requires dualism. In a way they are conflicting because frequently people who go into geology have to be strong in physics and math and chemistry and at least acquainted with biology. These things take out an awful lot of the people who are artistically inclined.

"But at the same time people in geology, which is probably the least—maybe anthropology would fight it out for the least precise of the scientific disciplines—who are effective at it need a lot of imagination, which is something an artist considers to be almost his monopoly."

He explained that he was talking about having the imaginative capacity to visualize not something that is not there, which the artist does when it is a new concept, but to do the next step down from that.

"And that is to visualize what might be there and hopefully is there, that Mother Nature left you just a few clues," he said. "So it is like the artist who is going to draw or paint a more or less representational work with water colors. And we are trying to do an accurate job. But we are going to let her imagination go a little bit. But she knows she has to have a tree in there and a hill there and a stream. But she does not know where the stream is for sure except that the tree cannot grow unless it is near the stream and this kind of thing. So she does it all and is trying to recreate this thing."

That is what a geologist does, he said. "He has to be able to kind of imagine from a few clues. So it is kind of like a detective function, but he relies on science and imagination, and it is hard to bring the two together.

"So you have a lot of geologists who have characteristics like petroleum engineers, and you find a lot of geologists who have characteristics of geophysicists."

But the finest geologists, according to Gratton, "are the kind who can hold the whole bundle."

The Romantic poets would understand imagination and advancements in science that they predicted. They would be proud that a man looking at holes in the ground could be a Romantic.

Acknowledgments

We turned to many sources for information, photographs and/or leads to find subjects for interviews for this book. The greatest of these, of course, was the membership of the Corpus Christi Geological Society. We interviewed a number of the current members, and many of the older ones told their stories on videos for possible inclusion in a TV documentary being prepared by Frank Van Heughten's Quadrant Productions. The society also provided us with a collection of its yearbooks.

One member—William Carlton "Tubby" Weaver—stands out as a super source. With remarkable recall for a man in his nineties, Tubby gave us anecdotes from his long lifetime in the oil industry and information about the literally thousands of oil people he knew. He also arranged for us to travel to other South Texas towns and interview his friends. He was not only a source but also an inspiration.

Our special thanks go also to the library, archives, and museum staffs who were so cooperative with this project. Among these was the Corpus Christi Museum of Science and History, which houses most of the collection of longtime local photographer Frederick "Doc" McGregor. Curator Patricia Murphy and her staff located hundreds of oil-related photographs for us. Dr. Tom Kreneck, archivist, and Jan Weaver, his assistant, at the Mary and Jeff Bell Library, Texas A&M University-Corpus Christi, and Cecilia Aros Hunter, archivist at Texas A&M University-Kingsville, were invaluable in supplying photographs and information.

At the Center for American History at The University of Texas at Austin, we discovered a treasure trove of interviews conducted by the Texas Independent Producers and Royalty Owners Association. Much of the information in those files is from oilmen no longer living. The DeGolyer Library at Southern Methodist University, the Fondren Library at Rice University, and the Del Mar College Library in Corpus Christi also provided sources.

Carol Voight helped us with a wealth of historical material from the Central Texas Oil Patch Museum at Luling. Other valuable information came from the library of *The Corpus Christi Caller-Times*, Margaret Neu, librarian; the *San Antonio Express-News* library, courtesy of Robert Rivard, editor; the Corpus Christi Public Library, Margaret Rose and Laura Garcia, local history librarians; the Institute of Texan Cultures of the University of Texas at San Antonio, Tom Shelton, photo archivist; and the public libraries of Alice, San Antonio, Houston, Dallas, Sinton, and Refugio. Robert Marlin and Julie Grob of the University of Houston

Library gave special assistance in helping us acquire photographs of Hugh Roy Cullen. The district and state offices of the Texas Railroad Commission also provided vital information.

A number of company archives were opened to us, including the Exxon-Mobil collection at Las Colinas near Dallas by public relations director Mike Long; the library of Greene & Associates of Dallas, Steanson Parks; and the O.W. Killam files and the Killam headquarters in Laredo. Corpus Christi Oil and Gas provided us with company history. Rich Tuttle and Sharon Walker of Flint Hills helped us obtain information about that refinery, and Carl Newlin provided photographs and background information on Suntide. The *Oil & Gas Journal* and *Business Week* kindly gave us permission to use photographs from those publications.

Family members provided information and pictures. Bill Tinney gave us documents and photographs relating to Piedra Pintas and the Tinney family. Bob Fullbright of Freer provided photographs and information about his grandfather, Robert Hinnant, owner of the land of the historic Killam discovery, considered by some the birth of the South Texas oil industry. Jack Graham provided photos and stories about his legendary father, Tom Graham. Ralph Storm furnished pictures of Storm and Glasscock operations, and members of the Glasscock family gave us permission to use a photograph of Gus Glasscock. Sammy Gold, the leading commercial photographer in Corpus Christi for many years, made a number of his photographs available.

Cecilia Venable was one of our most valuable assistants. Her work in inventory and photo filing at both the public and the A&M-CC libraries helped our efforts in both places. She had worked for several oil operators and held membership in the Desk and Derrick Club. If that were not enough, she put us in touch with her husband, Jake, who gave us information on all phases of the oil business from roughnecking to running the district office for high-tech drilling support companies. And she volunteered to copyread this publication.

Last but far from least, Ray Govett, Ph.D. consulting geologist and engineer, compiled the extensive list of South Texas field discoveries that is included and rode herd on this whole production as chairman of the Corpus Christi Geological Society History Committee.

Bill and Marjorie K. Walraven

Contents

The Quest

South Texas was a wild, wide, virgin country waiting for the strong of heart. It was an attractive paradise for the land hungry and those with enough toughness to come and take it.

The grassy plains and rolling hills provided buffalo, deer, and other wild game—a bountiful living for the Native Americans. Anyone who came would have to fight for the land. The Spaniards came but failed to vanquish the Indians with either the gun or the Cross.

Mexicans, unable to fully conquer the war parties that swept down on them, invited their Anglo neighbors and Irish immigrants to join them to give permanence to the settlements. The native-born Tejanos, adventurous Americans, and other immigrants who came in waves won a revolution and set up a Republic. South Texas remained largely uninhabited.

Many colonists received generous land grants. Most of these came into the possession of enterprising large landowners and tough ranchers. The land was head high in grass. There were no thorny brasadas that would later infest it. The brush grew when ranches were fenced and the plains were no longer cleared by fire as the Indians had done for centuries.

But the promised bounty of the land would not be realized until the turn of the new century, when a man appeared to bring new prosperity. He was the man of iron—the man who would change the face of the land. He was the Wildcatter.

Dr. William E. Hewit, a dentist who came to Texas from Nebraska, surveys the Pratt-Hewit leases. Dr. Hewit's grandsons, John, Richard, and George Hawn, became prominent Corpus Christi oilmen.

Waterway leases like this one are considered on state property in Texas. Under this policy, billions of dollars have been generated to help finance public education in the state.

SouThex
DevelopMENT Co
#1

(Opposite page) An early South Texas Development Company rig. Note the women on the rig floor. (Left)) A touring car provided transportation for early oilmen. (Below) Mules used fresnos to scoop out pits. (Bottom) The DeeWeek crew of Luling

(Above) Oil-field fires were dangerous at times not only to workers but also to homeowners. (Right) An early-day rotary drilling crew at work

(Left) A blowout at Old Seagraves created a mountain of slag. This was the first of many, many blowouts in the area. (Below left) Wooden derrricks dot early Saxet Field on the west side of Corpus Christi. (Below) Before the Flying Red Horse, the Magnolia Company was represented by the flower. This antique pump is in the Central Texas Oil Patch Museum in Luling.

(Below left) Some early drilling rigs, like this one from the 1930s, were conveniently located near roads. Often, however, early day oil-field workers faced primitive conditions, including outdoor 'plumbing' (left) and mule-drawn transportation (right). Sometimes the only place they had to sleep was a tool shed, commonly known as the 'doghouse,' like the one below.

(Right) Escape capsules like this one were used to save offshore workers in cases of emergencies on the rigs. (Opposite) Suntide Refinery in the 1950s (Below) The platform of Mr. Gus II, owned by Corpus Christi's Glasscock Drilling Company off the coast of Louisiana in 1957, resembled a Spanish caravel.

(Above) A well dressed rig visitor finds a cooling-off spot on a hot summer day. (Right) A tractor-truck hauls off debris from a collapsed rig.

(Left) Blowouts sometimes left craters like this one in Saxet Field. (Opposite and below) Roaring flames and clouds of smoke, sights that were too familiar in South Texas oil fields, sometimes drew huge audiences.

Throughout the twentieth century and into the twenty-first, oil-field rigs, refineries, recycling plants and other forms of industry construction altered the Texas landscape and provided billions of dollars to the state's economy.

Falling rigs and well fires were among the dangers workers faced. Ineptly drilled wells charged shallow sands, causing violent blowouts.

(Right) William Carlton 'Tubby' Weaver was a geologist and independent operator, but his proudest achievement was his service as a bull scout. (Below) Ingleside's Kiewit firm adds modules to BP's Thunder Horse, the world's largest offshore platform, brought from South Korea in September 2004 to be anchored in the Gulf of Mexico. (Bottom) 'Jake' Venable points out a PZIG (pay zone inclination gamma) tool at Corpus Christi's Pathfinder yard.

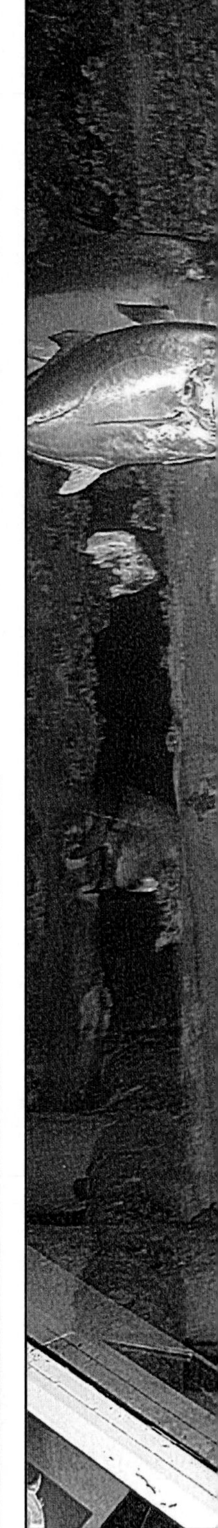

(Above) A plaque at Galveston's Ocean Star Offshore Energy Center recognizes Corpus Christi oilman James C. 'Jimmie' Storm as a member of the Offshore Pioneers Hall of Fame.(Left) Instead of harming marine life, as many had feared, offshore drilling platforms serve as reefs for the development of plant and animal marine life. This exhibit at the Texas State Aquarium in Corpus Christi illustrates this fact for young and old.

Burning wells, blowouts and muddy roads, transportation difficulties, and other problems did not deter the pioneers of the South Texas oil fields from their quest.

GREY WOLF
RIG 520

Foreword

Between these covers is the story of the development of oil and natural gas in the Texas Gulf Coast over the last one hundred-odd years. The storytellers are the Geologists, Geophysicists, and Oil Men of South Texas. The extent of their experience varies from seventy-odd years to just the last few. The timing is perfect because memories fade with age and perspective is immature without adequate time.

The discovery of oil and gas does not come easily. A Land Surveyor must first establish an accurate and permanent point-of-beginning before his work can progress. It seems proper to do the same here.

Since there must be a discovery well, and since it was a wildcatter who inspired it in the first place, it seems proper to use him as the point-of-beginning. After discovery, especially during the first half of the 20th century, conditions were often suitable for a "boom" there. Within this frenzied mass of people, there were those who drilled the development wells and prepared the leases for production. Most found their rewards by servicing the developers, but a few were seeking not jobs but "opportunities" that were often improper if not outright illegal.

A new-field wildcat is a well drilled for oil or gas in an area not previously productive. The men who were first convinced that oil would be discovered at a particular spot and managed to get an exploratory well drilled there were—and are—called "Wildcatters." Apparently this term originated to describe obsessive, reckless tenacity associated with wildcats and the remote areas they inhabit.

The general concept that a typical Wildcatter was, or is, a rather impractical visionary who builds his future out of rainbows is not fully accurate. Wildcatters generally are very sensible people who give due consideration to the experiences of others. Some studied the geological aspects of oil accumulations, and most sought advice from experienced people, though many discoveries have been made by the Wildcatter using unsound reasoning, including the use of doodlebugs. And the final investors must have thought they could make a reasonable decision, or they would not have invested. Witness Andrew Mellon pledging $300,000 toward the drilling of the Lucas No. 1 that blew in at Spindletop, Texas, in January 1901.

Those Wildcatters who could afford to often employed or associated themselves with Geologists. After a discovery subsequent wells are classified as "development" wells. Now there is a business to operate, and the Geologist puts on his exploitation "hat." Up until forty or fifty years ago, the Geologist carried

(Opposite) A Grey Wolf rig near Driscoll in the summer of 2002

nearly all the burden of development. Seismology has developed amazingly and now provides the primary technology in exploring for oil and gas. Now the Seismologists, Geologists, and Petroleum Engineers must work together to efficiently develop a field.

The Production Department was built as the need arose. We should never forget these faithful and sturdy men who got the oil into the tanks and the gas into the pipelines. They were as important as any.

Usually there were several operators in each field who shared in, or just benefitted from, a discovery well. Established operators simply developed their own leases, while speculators who owned leases sold them or made a deal with an operator to develop them.

In the early days of the industry, regulations were minimal, and investors were attracted by a 27 1/2-% depletion allowance, expensing of intangible drilling and dry-hole costs, and the lottery-like odds. Everyone who wished to could participate. There were only the unwritten rules of the marketplace to obey.

Workers came from the ranks of those accustomed to hard work. So many roughnecks came from farms that they were customarily referred to as "weevils," after the boll weevils that often infest cotton fields. Each newcomer was told that there were only two can'ts in the oilfield: "If you (1) can't get the job done, you (2) can't stay." There was discipline up through the ranks, too, that required honesty and integrity and a good understanding of what they were trying to do.

The productiveness and success of these people were astounding. Give credit first to the existence of many oil fields. A moment of reverence is in order here as one ponders this fact. Next, the proliferation of the internal combustion engine and general industrial growth provided the demand for oil and gas. One superlative is a thought by Rush Limbaugh: "As I've explained to you, oil is the fuel that powers the engine of freedom."*

The observations collected out of the memories of the subjects of this book are essential to understanding this great 80-odd year oil boom. They deserve to be honored for compiling this record.

Call it what you will, a great boom, a magnificent adventure, or a "fulfilling life." It was all these, characterized by the exhilaration of the chase, the long-odds wager, the vision of wealth unlimited. The privilege of creating our individual goals and being allowed the freedom to try to achieve them in our own way were gifts beyond compare. Our gratitude for the opportunities and freedom to pursue them knows no bounds.

All this changed in the early and middle 1980s, as investment money fled to the computer/communications, electronic/high-technology, financial manipulation enterprises. The emphasis, to a significant extent, changed from earning money to making money. Nevertheless, the surviving participants in the Great Oil Boom are convinced that theirs were the golden days.

O. G. McClain
Houston, Texas
November 2, 2003

* The Limbaugh Letter, October 2003, p. 14.

Industrial plants along the harbor of the Port of Corpus Christi, like the one shown in the background, were relatively new when this tanker set sail in 1937.

The Early Years

Spindletop, the famous East Texas discovery of 1901, had an intoxicating effect on South Texas citizens at the beginning of the twentieth century. They felt there were vast riches under the grasslands of the Coastal Plain or maybe out there in the brush lands.

There were tantalizing clues—cowmen's complaints about oil befouling their water wells, seepages here and there, and reports of cowboys heating their coffee over flames lapping up from cracks in the ground.

For years blobs of oil had floated ashore from the Gulf of Mexico. Mixing in sand and hardening in the sun, natural asphalt paving stones were used as sidewalks in Tarpon, later known as Port Aransas.

The searchers came, dug shallow holes, and met with some success, but they were interested in "gushers." They were looking for salt domes. They wanted oil to blow out the top of the derrick and flood the countryside.

As early as 1901, Sharp and Johnson of Corsicana drilled a Duval County well that produced 200 barrels of oil a day. Unfortunately, it was completely overlooked in the barrage of publicity created by Spindletop. Other discoveries in Bexar County and the Rio Grande Valley just didn't seem important.

So South Texas would continue to be the poor cousin of the rest of the state. When things started looking up, gushers were being found in Corsicana, Electra, Ranger, and other places.

But the day was coming. Ironically, a major hurricane in 1919 would bring a long-sought deepwater port to Corpus Christi. Oilmen would lead the way in developing the port and bringing the industries that would ensure the prosperity of South Texas.

Accidents on the job were not uncommon in the early days, and safety was not much of a consideration. Note the cigarette.

Blowouts were so frequent around Corpus Christi that some operators were reluctant to bring their rigs into the area.

((Below) Wagonloads of casing went through Corpus Christi on their way to King Ranch wells. Captain Richard King wanted water but would also have welcomed oil. (Bottom) Early development at White Point (Right) An early blowout in Saxet Field. Note the wooden rig.

South Texas well predated Drake's

God made petroleum long before he thought of creating man. Of course, He had to tell Noah how to use it. And over the millennia men have been invoking the Deity to help them find it.

Noah needed a lot of engineering advice to build a vessel that would hold an entire zoo of animals and the few humanoids who would survive the great flood. It was to be 450 feet long, 75 feet wide, and 45 feet deep, with three decks, stalls throughout, and a skylight around the ship below the roof. It was to be made of resinous wood, sealed inside and out with tar or pitch. Apparently Noah could have used power tools, because it took him 120 years to finish the job.

Many religions record the great flood. James A. Clark in his *Chronological History of the Petroleum and Natural Gas Industries* places it in the Tigris-Euphrates Valley at 6000 B. C. Some geologists believe the Mediterranean Sea was greatly swollen by an ice age melt and broke through a natural barrier, flooding that part of the world.

After Noah made use of petroleum to seal his great ark, it wasn't long before other civilizations discovered its value. Egyptians were cooking with fuel by 4000 B.C., asphalt was used in 3000 B.C., and the Chinese used natural gas in 900 B.C. Oil seepages have been known through the ages.

Apparently in 600 A. D. the Japanese dug by hand the first wells specifically for oil. They found natural gas a few years later. Greeks forced ram-filled flaming naphtha into the hulls of enemy ships. The Romans used flaming oil to defeat the fleet of the Saracens. The Chinese drilled to depths of 3500 feet in 1100 A. D. People busily collected oil from seepage wells in Baku in 1272.

Some time in the fourteenth century the word "petroleum" was used in the court of English King Edward III. The entry in his Wardrobe Account read "Delivered to the King in his chamber at Calais: 8 lbs. of petroleum."[1] Columbus picked up samples of asphalt in Trinidad in 1510. A few years later Spanish explorers acquired some know-how from Native Americans, who waterproofed their canoes with asphalt or pitch along the upper Texas coast. In the next century tar pits and oil springs were

Spanish explorers acquired some know-how from Native Americans, who waterproofed their canoes with asphalt or pitch along the upper Texas coast.

Early prophet was right

William DeRyee, who came to Texas in 1856, maintained there were large oil fields in the vicinity of Corpus Christi.

Born in Bavaria, he studied philosophy, botany, meteorology, and geology. In 1848 he participated in the overthrow of the king. After coming to America, he represented several copper mines in Tennessee. He invented an early photographic process, photographed the Texas Legislature, and made two famous photographs of Sam Houston.

During the Civil War he led the development of nitrates for explosives manufactured for the Confederacy. In 1866 he opened a drug store in Corpus Christi. With a lifelong interest in chemistry and geology, he held classes for young men of the community on those subjects. In 1888 he published a paper, "Economic Geology of Webb County" in the *Geological and Scientific Bulletin*.

He died in 1903, well before his theories about South Texas oil could be proved.

New Handbook of Texas, 2: 604–605

discovered all over the world, and oils were valued for their medicinal powers and for illumination.

Gas lighting became more common in the early 1800s. In 1815 Englishman Samuel Clegg invented the first gas meter. He laid out 6-inch gas mains. House pipes were made of old musket barrels connected from barrel to breach.[2]

On August 27, 1859, the famous Titusville, Pa., discovery by Edwin L. Drake flowed ten barrels of oil from 69 feet, the first North American well deliberately drilled to find oil. On October 7 the well caught fire, thereby becoming the first oil well fire on record. Reworked, the well became a historical landmark in petroleum, accelerating the search for oil all over the country.

In spite of Drake's fame, he was probably upstaged down in South Texas by a Tejano rancher, Juan Lopez Saenz, a native of Mier, Mexico. Saenz moved from Rio Grande City, Texas, to Duval County in 1854. He gave the settlement its name, Piedras Pintas—Painted Rocks—although for a while it was known as Noleda, which some oil pioneers chose to call the station where the Tex-Mex Railroad later dropped their mail. It was a wild, untamed frontier. During an Indian uprising, Saenz sent his family back to Rio Grande City for their safety.

Capt. E. R. Tarver was quoted in the *Laredo Times* in 1901, "Piedras Pintas takes its name from several large masses of colored rock which gave the appearance of having been at some time in the past thrown to the surface by some great internal force of nature. These rocks a few years ago were blown up and quarried and carried to (the) Aransas Pass and put in the jetties there."

As he was digging a well shortly after his arrival at his new home, Saenz struck oil—some four years before Drake's Pennsylvania discovery. He was not deliberately drilling for oil—he needed water in an arid land. His crude rig, with a drop tool suspended by a manila rope and powered by mules led by handlers, encountered oil at 30 feet. He cut a mesquite plug, covered it with burlap, and jammed it into the hole, stopping the oil flow. Then he continued drilling until water was found.[3] Some wag has dubbed this Texas's first dual completion.

The well was used primarily for water, making Piedras Pintas an oasis for travelers, who could head for Laredo or other border parts refreshed, although the water was potable only after it stood until the thin sheen of oil evaporated from the surface. And if they wished, they could use Saenz's oil to grease the squeaky wheels of their carts or wagons. Such a marketable item qualified the Saenz well as a commercial venture.

As for working parts, the oil-retrieval system included a pioneer roustabout, a small boy who was lowered by rope to remove the plug so it could be used as an oil swab for the cart wheel and also, on occasion, "for greasing scratches and boils on the bodies of the freighters and their families." Through the years Señor Sanez must have introduced a number of small boys to the oil business.

In 1867 Saenz drilled another well, using a gin pole as a derrick. The laborers noticed a strange odor at 30 feet, and one of the men passed out. The other men scattered, and Saenz had no easy time rounding them up and getting them back to work. This well was said to have found oil sand around 60 feet and a fair flow of oil at 310 feet and to have produced 23 gravity Pennsylvania-type paraffin base crude for six or seven years.[4]

In 1870 William Archibald Tinney, a native of Illinois, moved from Corpus Christi to Piedras Pintas with his brother, Samuel H. Tinney, who opened a store there. William Tinney said he tried without success to get Saenz to hire an expert to test his well in 1872. His efforts were interrupted when some seventeen bandits attacked after they heard there was four thousand dollars in town to be used for the purchase of sheep. They were met with blazing rifle and shotgun fire and the money and sheep saved. Again in 1876 nine bandits attacked the town with the same results. Two of the bandits fell, "leaving the Glovers [owners of another store] and Tinneys with two saddled horses for their work."

In 1875 William Tinney married Saenz's daughter, Candelaria Saenz, and continued spending a good part of his time thinking about his father-in-law's well.

Saenz died in 1883 during a trip to Mexico. William Tinney's efforts to get a well drilled continued, only to be met with years of frustration. He made a lease in 1890 and another in 1893. There was no shortage of

William Tinney (right) in his clerk's office. The other two men are probably drillers. After a number of failed ventures, Tinney found profits in leasing his land.

backers, but as none could raise the money, nothing came of their efforts. Even before Pattillo Higgins started trying to sell his salt-dome theory at Spindletop, William Tinney tried to get somebody interested in drilling a test well in the salt dome of Piedras Pintas.

For years, nobody listened. Then, in 1900, he and Col. S. R. Peters, John C. Negerle, and Charles Negerle engaged Jack Fraley, a Pennsylvania driller, to try his luck. They drilled four holes. One produced sulfur and salt water, the second, a small amount of good oil. The third stopped because of gas flow that burned "night and day for more than a year, throwing a flame three feet high." The last produced sulfur and salt water.

Later that same year Tinney and his partners leased land to A. C. Hall of Corsicana and Henry Keller, a one-time mayor of Corpus Christi. Keller, who was then representing a Corsicana well and prospecting company, knew of Saenz's discovery. He reported that many years earlier "a Mexican dug a well about 3 1/2 by 6 feet for water. When he got down to 49 feet, oil began to flow through a crevice in the rock. He plugged the crevice with wood and rags. We had occasion to clean out this well so that we could use the water for our boiler. We pulled out the remains of the plug and oil began to run again. Fifty yards south of our well, you can dig a foot down and the gas that escapes will burn."

"Then," Tinney said, "came the Producers Oil Company of Houston and Well & Prospecting Company of Corsicana."

(Opposite) A 1900 Piedras Pintas well. William Tinney is standing on top of the derrick, and seated next to him is Jack Fraley, a Pennsylvania driller brought to Texas to search for the Piedras Pintas oil. (Below) In the later years of his life, Pattillo Higgins lived in this home on Drexel Avenue on the south side of San Antonio.

Tom Wren, a Tinney friend and San Antonio oilman, was quoted in the 1913 *Caller* as saying that Sharpe and Johnson of Corsicana, who had been involved in early Spindletop work, drilled a 200-barrel oiler at 1400 feet at Piedras Pintas in 1901. Wren added, "But Spindletop came in about then and being so big, got all the publicity."

According to the *Corpus Christi Caller,* it was Producers Oil Company that made the first substantial Piedras Pintas production, with a well completed at 500 feet. "In June 1902 Knight, McKnight and Lawson drilled 300 yards due west of the 1901 well. They formed the Piedras Pintas Oil Co. [and] on July 4, 1902 entertained the community at a barbecue," probably the first such celebration in South Texas.[5]

In 1906 John D. Cleary and T. J. Lawson developed the Mary Owens Field, and in 1912 they sold their holdings to the Mary Owens Oil Company. In 1913 the San Diego Oil and Gas Company, headed by J. B. Wells, had five producers on the Tinney land.

"The Mary Owens Oil Company had eight wells on the Tinney land west of the creek while in the Noleda Field on the east side of the creek and on the William Peters land were five producers owned by A. L. Jack. Total production was about 100 barrels daily, or about five barrels from each well."

The gun in the boot of Pat Tinney, (on right) William's son, shows that the early oilmen still lived on the old frontier. It was positioned for quick access on horseback.

Eli Merriman, editor of the *Caller*, wrote that this oil was pumped from an average depth of 300 feet by "individual gasoline engines" and was selling for two dollars a barrel.

In the next thirty years more than one hundred wildcats were drilled around Piedras Pintas. Some found good production, but the field would not be a big producer until 1945, when H. H. Howell opened major production with a multipay discovery.[6]

As for the town of Piedras Pintas, in 1880 it was a community of some six hundred, with a bull ring, racetrack, schoolhouse, stores, a bar, and a billiard room. Saenz signed its death warrant that year when he refused to grant a right-of-way to the Corpus Christi, San Diego & Rio Grande Railroad, later the Texas-Mexican. Placido Benavides, nephew of the Texas Revolutionary hero of the same name, gave railroad builder Uriah Lott 160 acres, and the town of Benavides was born. Years later nothing remained of Piedras Pintas except the cemetery and one lone telegraph pole. The others were used as fence posts.

Fortunately, Tinney had success in other endeavors. In 1887 he opened a country store, and in 1890 he was elected district and county clerk and moved to San Diego. He was philosophical about his efforts in the oil business.

"I have been in various and many different leases," he said before his death in 1924. "I have been working or pretending to do so, more in selling leases, etc., than working." His view of his plunge into the oil business was summed up in some verses from a poem found in his effects:

"We wish we had someone to love us—
A banker to make us a loan.
This oil game's about got us busted,
And caused us to weep and to moan."
. . . .
Tomorrow we go to the rancho
Good-bye to the well of our dreams
In that cold ranch house we will supper
On nothing but goat meat and beans.

However, the faith he had expressed in South Texas oil proved justified. Before World War I he had described to a *Caller* reporter how the sheep industry had given way to cattle when brush took over the country, but he had added, "... Oil no doubt eventually will be the means of compensating for those who have struggled for so many years."[7] He was right, although it would be a good many years before his prediction would come true.

While Saenz discovered his well by accident, another early South Texas driller was actively searching for oil. He was Lyne Taliaferro Barret, who had been fascinated by oil seepages when he was a child. After he was discharged from the Confederate Army, he was convinced oil could be found in the region of Nacogdoches, and he set about buying leases to that end. For years he sought financial backing and tried to acquire drilling machinery. In September of 1866 his well came in at 10 barrels a day. Total depth was 160 feet. Historians have found that well to be the first producing oil well in Texas.

Possibly Barret was associated briefly with Dick Dowling, Confederate hero at the Battle of Sabine Pass, who shared his dream of finding oil. Dowling's foresight was apparent, as he became the first large-scale oil lease operator, this before a single well had been drilled. He joined two Irish partners, John Fennerty and John Riordan, "to rent, lease and purchase lands containing minerals, coal or gas, for the purpose of developing such properties."

In 1857 Dowling married Anne Elizabeth Odlum, daughter of Benjamin Odlum, colonist in Old San Patricio. Unfortunately, Dowling died in a yellow fever epidemic in 1866. Lands his group had leased now contain oil fields all over East Texas.[8]

Barret had begun his quest in 1859, the same year Drake completed his Titusville well. After his well came in, Barret traveled to Pennsylvania to seek financial backing and machinery to continue his efforts. He returned to Texas with an oil operator, John F. Carll, and five thousand dollars worth of machinery. They began to drill a second well in the area but found no oil, and Carll abandoned the effort at 80 feet.

By this time production in Pennsylvania had reached 3,000,000 barrels a year, and prices had dropped to $1.35 a barrel. Barret, whose wife had inherited large tracts of Texas land, "tried to interest Pennsylvania

Lyne Taliaferro Barret was convinced oil could be found in the region of Nacogdoches, and he set about buying leases to that end. . . . In September of 1866 his well came in at 10 barrels a day. Total depth was 160 feet. Historians have found that well to be the first producing oil well in Texas.

oil operators in taking land certificates for their machinery and services. It was no use."

In addition to drilling the first commercial oil well in Texas, Barret employed the principal of rotary drilling with an augur clamped on a pipe, rotated by a small steam engine. Optimistic about the future of oil in the state, he believed there were huge lakes deep under the East Texas surface.

In 1865 he wrote in a letter, "The great excitement of this age is oil. . . . This region of Texas will be wild upon the subject in a few months. . . . None doubt that oil will be found and the excitement will not wait for the slow process of experiments, and if we are prepared for excitement, we will make our fortune. What is the use of toiling and struggling with aching brains and weary hands for bread when gold so temptingly invites you to reach out and clutch it."[9]

Barret was just slightly ahead of his time. A boom of sorts came to the Oil Springs area of Nacogdoches County twenty years later. In 1886 B. F. Hitchcock of Nacogdoches and Edgar H. Farrar of New Orleans organized the Petroleum Prospecting Company. They brought men and cable-tool machinery from Pennsylvania and drilled near Oil Springs in Nacogdoches County. With casing set at 70 feet, the well began to flow

A rotary table on a wooden rig

about 300 barrels a day. Oil rushed up, providing Texas's first gusher. It poured on the ground and the flow stopped two days later, but the state had its first oil boom.

With a new partner, J. E. Pierce of New Orleans, Hitchcock drilled some forty wells. Thirty of them were producers, with oil brought to the surface in buckets. They laid a 14 1/2-mile pipeline from the field to Nacogdoches, the first pipeline in Texas. Another company with headquarters in New Orleans, the Lubricating Oil Company, came to the area. It set up a plant similar to Hitchcock's and established a crude refinery, the first refinery in the state. By 1890 activity in the area decreased. The pipeline was taken up and used in the Nacogdoches water system.[10]

In 1886, just as the play was ending in East Texas, rancher George Dullnig dug a water well six miles southeast of San Antonio. His drilling crew was made up of ranch hands who had no oil experience. At 235 feet there was a flow of thick oil. This same viscous oil flowed from a second well at 300 feet. As the oil was found to be a good lubricant in its natural state, Dullnig sold it, charging twenty cents a gallon for oil in a barrel, thirty cents a gallon for five gallon cans, and thirty-five cents a gallon for lesser amounts. A third Dullnig well was drilled to 900 feet but found no oil below the oil producing sand. A fourth well produced natural gas, which he piped to his ranch house.

Because of Dullnig's wells, Texas appeared for the first time in statistical tables as a producer of oil and gas. The 1889 reports of the United States Geological Survey and the census of mineral industries in the United States showed a production of 48 barrels of oil from the Dullnig wells, plus $1,728 worth of natural gas. Dullnig arrived at that figure by estimating the amount of coal he would have used in firing his boilers. The gas well was still furnishing fuel to the ranch as late as 1939.[11]

Following Dullnig's success, other marginal wells—one on a farm northeast of San Antonio, another on a ranch south of town at Sutherland Springs in Wilson County, and a third south of Dunlay in Medina County—were sunk in 1890. Oil was also found at 250 feet in a water well south of Waco.[12]

Production from these wells was insignificant and completely overshadowed by discoveries in 1887 at Corsicana, which many experts consider the true birthplace of the Texas oil industry.

In those early days the discovery of oil in Texas was usually, like Saenz's, a result of a search for water. Texans have always had problems finding water for crops, cattle, and drinking. One reason Texas frontiersmen drank whisky so freely is that it helped them face the droughts, heat, cold, bugs, and people. But a more practical reason was that it enabled them to drink bad water without taking to bed with a deadly fever. They were constantly looking for clear drinking water for themselves and their animals. This resulted in a lot of drilling.

. . . like a sailor

Many of the early oil-field workers who earned their apprenticeship in the Pennsylvania fields would find their way to Texas, where their expertise was needed. They brought more than drilling knowledge: They helped create the oil-patch lingo.

Many were sailors who knew machinery, the moving, hoisting, cable and rope handling, and other skills that would be of use on a derrick floor in an entirely new industry. They brought their nautical terms: rig or rigging, rigging up, mast, catwalk, cathead, hatch, crow's nest, and others. A written record, the ship's log, became "the log" in drilling nomenclature.

Lalia Phipps Boone, *The Petroleum Dictionary*, 7–8

South Texas rancher Denis O'Connor was furious when a water-well driller he hired kept finding oil instead of water. It was hot and dry, and his cattle were thirsty. The driller, who was familiar with the new uses for petroleum, thought Denis would be happy with his discoveries.

"If you can't find me some water," O'Connor roared, "I'm gonna find somebody who can."[13]

The city fathers at Corsicana had the same view. In 1894 they contracted for three deepwater wells. On June 9 oil was encountered at 1027 feet. City officials were annoyed and ordered the search for artesian water to continue. At 2470 feet more oil was encountered, and the well site was well soaked with oil. Not only did the city not have water, it also lost derricks over three completed wells when careless onlookers, who had no experience with oil, accidentally ignited them.

Corsicana businessmen had a different view. They formed the Corsicana Oil Development Company and began leasing like crazy. A committee went to Pennsylvania and offered J. M. Guffey and John H. Galey, both experienced in exploration and drilling in Pennsylvania, this deal: one-half of their Corsicana leases to drill five wells. The Guffey-Galey wells were not prolific—20 to 25 barrels. But the following year Corsicana was engulfed by a wave of operators, workers, investors, and con men. In 1896 total oil production in Texas, most of it from Corsicana, was 1,450 barrels. The first oil in the United States shipped from west of the Mississippi River came from Corsicana.

In short, Texas was introduced to the oil boom, though it was far tamer than future ones. In 1897 Texas production soared to 66,000 barrels, again most of it in Corsicana, where fifty-seven wells were drilled that year. New oil companies formed. The entire town was crowded with derricks in 1898, as 291 producers were completed.

As storage space and the market for crude were limited, there was a desperate need for a refinery. J. S. Cullinan, a Pennsylvania refiner, built a refinery, pipelines, and storage facilities as J. Cullinan & Company. The price of crude dipped to fifty cents a barrel while the plant was being built but nearly doubled to ninety-eight cents as the refinery went into operation and Cullinan succeeded in efforts to promote new uses for his product.[14]

Corsicana's discoveries would have an indirect effect on South Texas, as oil fever moved south to Luling, Refugio, Corpus Christi, and Laredo. Thousands of men from farms, ranches, and small towns became skilled at the oil-patch trade and passed it on to generations of Texans—and newcomers who would become Texans.

Oil fever, however, brought with it hardships and dangers. When the first gas well was drilled in San Patricio County, it was not a happy occasion. The event occurred in 1893 after the Coleman-Fulton Pasture Company bought a drilling rig capable of drilling to 2000 plus feet. The problem was, like Denis O'Connor, they wanted water, not oil or gas. The

A SAAP train on the Nueces Bay Causeway

Oil fuel for engines

July 12, 1901- R. H. Innes, superintendent of the San Antonio & Aransas Pass Railroad, reported that locomotives of the Santa Fe Line had been adapted for the use of oil, which was more economical and efficient than coal pulling a train at more than 60 miles an hour. About four barrels of oil were equivalent to a ton of coal in the running of an engine.

The San Antonio Express

July 12, 1901-The first large oil barge designed for Gulf of Mexico ports was launched in Beaumont in July 1901 by the Charles Clarke Oil Transportation Co.

"It is built of wood and is equipped with twenty-four oil tanks which have a capacity of 200,000 gallons." Others, with 300,000- and 350,000-gallon capacities were being built.

The Corpus Christi Caller

(Left) An early day well-flow test. Drillers sometimes opened the valves to create a gusher and thereby attract investors, hoping that sights like blowout wreckage (above) would not appear to discourage them.

company desperately needed water in the 1880s as tanks dried up and well after well produced salt water.

James C. Fulton, son of George W. Fulton, a partner in the company, was in charge of the search. In March 1893 he selected a site a mile and a half west of Portland at a place called Doyle Water Hole for his first deep test. At 740 feet they struck water that was fairly good but slightly salty. Fulton decided to go deeper in hopes of finding artesian water. At 2000 feet they found mud and water accompanied by a "strange wind." Puzzled, Fulton leaned over the well and struck a match. In a flash the gas flared and his thick beard was seared off.

Clyde T. Reed, a correspondent for *The Corpus Christi Caller-Times,* wrote later that Fulton's well "could be considered the first blowout in South Texas."[15]

Boiling mud and escaping gas erupt from a White Point crater in which the derrick and machinery have sunk to unknown depths.

Since the well had a low gas pressure and it choked off with dirt, only Fulton's pride was injured. When drilling for oil began a few years later, this more sinister scene was far too familiar:

The ground would begin to rumble and shake. The rig would begin to dance about; the crews would jump down and sprint as far as they could before the pipe came roaring out of the hole like a clap of thunder, followed by a huge rush of mud and rocks propelled by thousands of pounds of pressure from millions of cubic feet of natural gas.

Technology lagged and the pipe, unprotected by poorly applied cement, would blow out. Sometimes a huge crater formed. Rigs were lost, and men and cars burned to a cinder by a gigantic blowtorch.

A test drilled eight and one-half miles east of Tilden in McMullen County blew out on June 25, 1908, and remained uncontrolled for several years. Another of the earliest major blowouts occurred around 1912 at White Point.

The promontory jutting southward in the bay north of the Nueces River is white in appearance, but it did not get its name from the light colored cliffs. Before the Civil War Edward P. White and his wife, Aspasia Blanchette, settled the area. They were ranching there when they were fatally stricken by the catastrophic yellow fever epidemic of 1867. For many years the area was called "White's Point."

The property later became the Rachal Ranch after it was purchased by D. C. Rachal. In 1977 Chris Rachal, a retired Corpus Christi police captain, recalled the blowouts he had seen there while growing up.

"The first one was about 1912 or 1913," he said. "The White Point Oil and Gas Company rig was coming out of the hole when she went. There was no drilling mud in those days to control pressure. The drill pipe came straight up through the derrick. It went several hundred feet into the air. When it came down, the drill pipe stretched out with the bit away from the hole. The rig, except for the boilers, was lost. It burned until it cratered over. Mud and water mixed together to choke off the gas flow."[16]

In 1914 a group of local businessmen and ranchers capitalized the White Point Oil and Gas Company for forty-five thousand dollars. Then the J. M. Guffey Petroleum Company—later the Gulf Oil Corporation—paid them sixty thousand dollars cash and forty thousand dollars in oil for three-fourths of their holdings.

Soon after it took over the lease, the Guffey Company, by then the Gulf Coast Oil and Gas Company, drilled the Guffey No. 1 with the intent of marketing gas in Corpus Christi and San Antonio. On January 2, 1916, the *San Antonio Light* reported that the Guffey No. 1 had broken all records in volume and power.

"What experts consider the biggest strike of gas in the world was made a few miles from [Corpus Christi] Friday afternoon. The flow was found by the Guffey Company in No. 1 well. It was estimated that the capacity of the well is 50 million cubic feet per day . . . The well is beyond all control at present . . . The roar of the escaping gas can be heard easily for seven miles. . . pure gas is rushing from the well in a column about three hundred feet high. . . ."

D. M. Pictor of the Gulf Production Company said on January 3, "You can see the column of gas from Corpus Christi and hear the roar for fifteen miles. A thousand feet away, ordinary conversation is inau-

Area residents turned out to see the 1916 White Point gas well blowout. From left to right are Annie Piehl, Minnie Freier holding Paul Freier, and L. J. Piehl. The man on the right is unknown.

dible, and within 500 feet of the well you can't yell loud enough to hear yourself . . . A stick of mesquite cordwood thrown into the stream went at least 500 feet high—200 feet higher than the gas is visible."

Hopes of controlling it vanished when the gas ignited and the monster well caught fire.

"Residents from counties as far as forty miles from White Point watched the progress of the great fire, seeing by day a thin plume of black against the blue sky, while by night a mighty red glow ignited the heavens. Hundreds of visitors made all-day excursions to White Point to see the greatest fire of local history; when the well at last burned itself out, it left a huge black scar to mark its grave."

'Residents from counties as far as forty miles from White Point watched the progress of the great fire. . . .'

"It shook windows in Sinton fourteen miles away," Rachal remembered, "and it lit up the skies in Sinton and Corpus Christi. The cone was 60 feet high and the crater was 100 feet deep and 200 feet across."

With officials convinced that enormous gas pressure was possible in the field, Gulf drilled another well, but it also ran wild and caught fire.

Another of the early blowouts occurred when the Pratt-Hewit No. 1 F. B. Rooke blew out from 2200 feet. It ran wild for several weeks.[17]

Hundreds of visitors made all-day excursions to White Point to see the greatest fire in local history.

The History of Nueces County, published in 1942, says that "blowouts (sic) from the gas sands were so frequent that they actually became accepted as a necessary evil. It was nothing unusual for residents of the western part of [Corpus Christi] to wake up in the morning and see a stream of salt water, gas, boulders and mud shooting high above the derricks. The debris scattered far and wide, and the black soil disappeared beneath a covering of muck."

There were twenty-one blowouts in Nueces County's Saxet Field alone.[18]

On November 6, 1922, near the beginning of the oil development in the Corpus Christi area, a W. L. Pearson and Saxet Company well on the Dunn Tract blew out at a depth between 2300 and 2400 feet, sending mud, water, sand, and stones more than 200 feet into the air. The force of the explosion threw out shale in lumps varying in size from two to eight and nine inches in diameter. The pressure continued for an hour and then choked off.

On November 15, 1936, two men, fireman R. Carlisle and M. Breeman, were killed when the crown blew out the top of the boiler at the Winn Drilling Company Morgan lease in Saxet West Field. The Frank Harmon No. 1 Paul Benedum in Saxet Field eight miles west of Corpus Christi blew out on December 3, 1936. A nitroglycerin blast snuffed out the blaze.

Another wild well fire in the Saxet Field occurred in 1936 on H. H. Howell's No. 1 John Kucurek. A *Caller-Times* story said, "Mud and rocks were being spewed up and out of sight before the well ignited. It blew for nine hours before catching fire. It was the third well in three weeks to blow out and catch fire."

One man was injured coming down from the derrick, and three small nearby homes were destroyed.[19]

Dr. W. Armstrong Price, who became one of the area's most noted geologists, had recently arrived in Corpus Christi when the White Point Oil and Gas Company Well 7, another famous runaway well, was capped in 1927. The well had been producing 5 million cubic feet of gas a day. When it blew, it spewed 75 million cubic feet a day for 110 days. The blowout proved one of Dr. Price's theories—when water is added to a high-pressure flow of gas, static electricity is produced. Every time water was poured on the well, it caught fire. Rachal remembered that famous well fire fighter Tex Thornton capped the well.

Thornton removed the derrick and other wreckage and called his wife in Amarillo to hurry down with a load of nitroglycerin. She loaded a truck, which had special springs, and set out for the 800-mile trip. She averaged 45 miles an hour. It was a remarkable feat, considering the fact that there were few paved roads. The press covered her trip and residents in towns along the way were apprehensive as she sped through carrying the equivalent of a 2000-pound bomb.

Big newspaper story offered little incentive

A North Texas oilman told *Corpus Christi Caller* oil editor Billy Blake the newspaper was not doing enough to attract oilmen to the area.

""Why," he said, "you have the biggest oil story in Texas out in the Agua Dulce Field in that big blowout that has been burning for almost two years. If you would write a full page story and send it out, the oilmen would flock in here by the hundreds."

Actually, that well had been written up and pictures of it had been published all over the world. It was about the biggest blowout on record—not a great attraction to drillers who looked with trepidation at the highly charged South Texas gas sands that had blown out and swallowed up quite a few rigs.

Jewell Webb of the Baltic Drilling Company told Blake he "trembled all the way from Houston to Corpus Christi thinking about those high pressure sands."

Billy Blake, "The Dope Bucket, *Corpus Christi Caller*, April 28, 1938

A photographer, wearing one of Tex's asbestos suits, entered the flames of the blowout for a close-up shot. The heat was too much for him, and he passed out. Tex pulled him out. Another photographer got good pictures of the rescue.

Dr. Price remembered that a derrick was built over the roaring well.

"The jagged edge of the 6 5/8-inch casing was sticking out of the crater. A pipe slightly bigger was slipped over it. I stood on the ridge with my transit

Continued on page 56

(Right) No one doubted the dangers of the oil-well fire fighter's job. (Below) Blowouts left mounds like this one in the White Point Field.

'Tex' Thornton was Paul Bunyan of oil-well fire fighters

W. A. "Tex" Thornton, famed well fire fighter, took an early interest in explosives. He was only nine when his father, a physician, took him to a July 4th picnic at their home in Oxford, Mississippi. He was fascinated by the fireworks display, and the first time he saw a fireworks stand, he spent his entire allowance on firecrackers and rockets and asked his dad for more.

He decided he was going to get a job blasting away at things. When he was sixteen, such a job was available. Dynamite was just coming into use for removing old tree stumps. He joined a crew as a helper blowing up stumps. Ward Thornton, as he was called then, learned the techniques of handling explosives and went into business for himself.

His father moved to Texas and opened an office in the North Texas town of Goree in Knox County. He also bought a farm there, but his boy wasn't much interested in farming. He was fascinated by a "shooter" on an oil rig as he loaded a torpedo carrying a charge of nitroglycerin, which explodes in the pay sand and breaks and cracks the formation to free the oil so it can flow more rapidly. In 1913 Thornton went to Cleveland, Ohio, and worked for a company that manufactured the torpedoes. There he learned how to handle explosives.

He worked as a driller and shooter after World War I. He worked for a torpedo company at Wichita Falls and was branch manager at Amarillo. There he met and married Sarah Troxell, a young lady who would prove to be unfazed as she handled nitroglycerin. The Panhandle's tight limestone formations required as much as 500 quarts of the liquid explosive for each well. Thornton formed a company—Tex Thornton Torpedo Company in which he manufactured his own nitroglycerin.

There were plenty of clients as new oil fields multiplied. Owners of a burning oil well at Electra turned down Tex's first offer to snuff out the blaze in 1919. They weren't convinced explosives would do the job. Tex offered to do the job free if it didn't work. It worked and started his career as a legendary well fire fighter.

He fashioned a homemade suit made of asbestos. Later Johns-Manville Company designed a complete fireproof outfit of suit, gloves, boots, and helmet. He is credited with being the first to employ such sophisticated equipment. Newspapers began to take notice and Tex was receiving more and more publicity as success followed success, first in the Panhandle, then in the nation. He became something of a folk hero.

Tex did nothing to lessen his legendary image. As a matter of fact, he encouraged it by telling Paul Bunyan-type tall tales about his experiences. He was warm and friendly, a trait that led to his death on June 22, 1949. After thirty years of death-defying feats, he was murdered by two hitchhikers he had befriended.

Mody C. Boatright, *Folklore of the Oil Industry*, 111–114; *New Handbook of Texas*, 6:480

and watched the hole to tell everyone to run if there was any movement," he said. "Lead wool was tamped between the two pipes and gas was released through a valve on the pipe sleeve. Concrete was then poured over the pipes. When it set, the valve was turned off and the well was tame."[20] On July 17, 1930, the No. 1 Strauch in Refugio Townsite, Block 29, blew out. The Houston Oil & Refining Company well was very close to some residences.

"Several houses have been torn down and surrounding wells shut-in. Pressure 2,200 pounds," the *Corpus Christi Caller* reported. No effort was made to approach the well, which had flames flaring upward seventy-five feet and formed a crater twenty feet across. The cause of the ignition was unknown. Thornton was first on the scene to help quell the fire.

Huge rocks were thrown from the hole on September 4, 1936, when W. L. Hinds Clara Driscoll Sevier two miles north of Driscoll blew, but the derrick was not knocked down.[21] The well was plugged and abandoned. On November 19, 1936, at 3 a.m., the Texon Drilling Company No. 2 Clara Driscoll Sevier in the same area blew out.[22]

One Driscoll farm blowout was one of the largest in South Texas. John L. Sullivan was the operator of the well, and Jack Calvert was his

(Below and opposite) The Houston Oil & Refining Company No. 1 Strauck in Refugio Townsite, Block 29, blew out July 17, 1930. 'Tex' Thornton was first on the scene to help put out the fire.

The Texon Drilling Company No. 2 Clara Driscoll Sevier blew out at 3 a.m. on November 19, 1936. This aerial photo was taken the following March.(Top) Aerial photo of a Saxet Heights well fire in March 1933

field superintendent. Late one night an Associated Press reporter found Calvert in an Agua Dulce café and asked him what progress was being made in containing the well. Calvert said, "We're doing fine. We have her on a two-acre choke now and expect to have her on a hundred and sixty in about a week."

That crater spread over six acres and continued to burn for several years. As late as 1970, passengers on planes flying over the site could still see mud bubbling from the crater.[23]

Corpus Christi oilman Ralph Storm told of one driller who probably had more blowouts than anybody because he would never put any baroid in his drilling mud. According to the story, the man's well was kicking, "and they told [him] he should get some baroids and weight up his mud. Called it 'bayroid.' He came back with five sacks. Usually when it starts kicking, you come out with four or five truckloads. He had five sacks, and the well blew out."

One operator in the area had so many blowouts that the oil fraternity named the volcanic hills his "range."[24]

Geologist O. G. McClain described the role of the men who fought oil-well fires.

They come out, he said, "and they are so casual. They don't worry, they're not in any hurry, and you say, 'Why don't you do something?' And they just go about their work very methodically, but at some juncture, in all those cases, they have to risk their lives.

"They have to go into that well, into the spew of gas, and shut a valve. And if it should ignite at that moment, they would be just crisp—charcoal. Don't ever think that a wild well fighter doesn't earn his money!"

McClain told of one occasion in which a roughneck friend, not a professional fire fighter, averted a well fire after a blowout. It occurred in Duval County during McClain's roughnecking days.

"On one well we had a 'water flow'—a blowout at shallow depth," he said. "This one blew water, rocks, and pieces of clay formation up through the crown. It had so much gas that there was a danger that a rock striking the derrick would cause a spark that would ignite the gas and destroy the rig—to say nothing of endangering the work crew.

"While the blowout was at full strength, the catline, a heavy $1\,^1/_2$-inch rope that runs over a sheave near the crown block, became overbalanced and fell down on the floor. In order to move heavy valves, tongs, etc., on the floor, we had to have the catline in place."

His close friend Ed Sellers volunteered to put the catline back. As the other crew members watched, Sellers tied a loop in one end of the rope, threw it over his shoulder, and climbed up the windward side of the derrick. Gas, water, and rocks engulfed the top third or fourth of the structure.

(Below) Sometimes fire-fighting crews were able to take a break from their dangerous job to have lunch.

(Right) An oil-field fire is a terrifying sight. (Below) Sometimes sightseers underestimated the dangers of a blowout. Hulks of burrning cars testify to the intensity of the flames.

Well fires were put out in the early days. 'Then two young men were driving across a field to take photographs of a blowout in Saxet. They didn't know gas was coming out of the ground. The car ignited it, and they were burned to death. After that, the wells were set on fire until they could be controlled.'

—W. Armstrong Price

"Ed climbed right into this mess," McClain said. "His gloves became wet and slippery, and rocks could have easily knocked his handhold loose. Many of us thought it was certain he would fall. He had to reach out at the top, catch the edge of the water table, and swing himself out, up and onto its top."

McClain said he had never felt such a sinking feeling as he did when he saw his friend swing out, but Sellers accomplished the feat.

"He came down the same way he went up," McClain said, "arriving with the typical grin on his face. I was so weak I could hardly stand."[25]

Price said that well fires were put out in the early days. "Then two young men were driving across a field to take photographs of a blowout in Saxet. They didn't know gas was coming out of the ground. The car ignited it, and they were burned to death. After that, the wells were set on fire until they could be controlled."[26]

As there was more gas than markets would support in those days, this seemed a viable solution. Even after statutory proration laws said that gas produced with oil could not be flared, billions of cubic feet still turned the sky red, lighting up the night for miles in all directions. Few felt any regret at the loss of such huge amounts. That would change—but the change would be years in coming.

By the early 1900s the problem was not that the value of gas was unrecognized but that the difficulty of getting it to markets outweighed the value. In 1908 C. E. Williams, Coleman-Fulton superintendent who was still looking for a reliable source of fresh water, contracted for Pat Leary to drill behind the hotel at Gregory. Gas was found at 1350 feet, but again no water at 2000 feet. The well was completed as a small gas well. Gas was piped into the town and became the only source of fuel and light for Gregory. In 1937 the well was still producing gas, but the lack of fresh water in that area caused the hotel and much of the town to move to the new town of Taft, nearby to the west.

White Point was one area of major importance in South Texas natural gas exploration. In 1902 Randolph Robinson found shows in a 400-foot test in what would be the White Point Gas Field. It was abandoned after flowing gas for several days.

A Reiser Gas Field rig. The field, east of Laredo, was discovered by a rancher drilling for water.

Ten years later the blowout of the No. 1 White Point occurred near the Robinson site. The crew encountered a show of gas and water and a slight showing of oil at 1700 feet, and another test was drilled to 2195 feet. After they drilled through a heavy gas sand, it blew out, flowing an estimated 60 million cubic feet of gas daily.[27]

In 1908-1909 a rancher drilling for water discovered the Reiser Gas Field east of Laredo. This discovery, along with the 1908 test well that blew out in McMullen County, "revived interest in Southwest Texas; and in 1911 the Border Gas Company completed a pipe line from the Reiser field to Laredo and began the delivery of gas to consumers there."

In 1912, Rinehart's *South Texas Oil* reported, a water well drilled south of Kingsville in Kleberg County had a showing of gas that led to the drilling of a producing gas well in 1914. Not much attention was paid to it, but the Jennings Gas Field was opened in Zapata County in 1914. Gas was discovered in northern Live Oak County in 1915. In 1918 Grubstake Investment Company had completed its No. l Brown in the Calliham area of McMullen County for 40 million cubic feet of gas from a depth of 880 feet. Several additional wells indicated a considerable gas reserve to the satisfaction of Southern Gas Company, which installed a gas line into San Antonio. The first natural gas from McMullen County reached that city in November 1922.

The White Point natural gas flow was finally harnessed in 1919 and the Nueces Gas Company Incorporated. White Point Oil and Gas took one Gulf well and ran a gas line to Taft in 1919 and to Corpus Christi for a time in 1920.[28]

In spite of the obvious abundance of natural gas in the area, exploration was still largely focused on the search for oil. In 1915 John J. Welder, J. C. Dougherty, and Dave Odem organized an oil company in Sinton. Two thousand of the 2500 shares they offered were sold on the first day.[29]

However, it would be two newcomers who would bring the South Texas oil industry into its own. The first of these was O. W. Killam.

Nancy Heard, widely respected oil editor for *The Corpus Christi Caller-Times* and *The San Antonio Express-News*, wrote in 1965, "Most historians consider the real beginning of commercial oil production for South Texas as April 1921, when the third try drilled by the late O. W. Killam and his partners, a 30-barrel-a-day well, was completed."

With a vision that made him unique, Killam opened the eyes of the nation to the resources of South Texas. He opened oil fields, and when he found no market for his oil, he ran pipelines, built a refinery, readied a railroad shipping point, and exported millions of barrels of oil from Harbor Island, near Port Aransas.

Even though he was a frugal person, he was willing to take risks in uncertain Mirando City; and, with the foresight to lease vast tracts of land to ensure his fortune, he acquired natural gas rights over a huge area when gas was considered virtually worthless. He brought Magnolia Petroleum Company, the first major company, to South Texas, an achievement that earned him more than a million dollars. Thus he fulfilled what had been his main goal in life since he was eighteen, some twelve years before the beginning of the twentieth century.

"My mother asked me what I wanted to do in life. She said if I wanted it bad enough, I could do it," he told Mody C. Boatwright in 1956. "About then there were a number of millionaires in the news—the Vanderbilts, the Morgans, Gates, Carnegie, and some of the steel magnates. I decided, 'I am going to make a million dollars, and I was going to do it before I was twenty-five.' It didn't work out that way. I passed thirty, then forty. I was like a frog in a well, jumping up and falling back."

By 1920 he had accomplished a number of things. He had starred in football at La Grange College in Missouri, graduating in 1896, earned a degree in law at the University of Missouri, sold lumber, and dealt in real estate. He was a member of a delegation from Oklahoma Territory that went to Washington to petition for statehood—the same type of service his brother, Lloyd R. Killam, would perform in 1959 for Hawaii.

He served as a representative in the Oklahoma Legislature, then a term in the state senate, ending in 1919 when Charles N. Haskell, the first governor of the state, suggested that he move to Texas to look for oil.

He still didn't have his million dollars. He was fascinated when he heard of ranchers getting traces of oil when they drilled water wells. He brought his wife and three children to Laredo, which seemed the end of the earth. Then he decided to use his savings and take a plunge into the oil business. He could have literally done just that. When he arrived, water was a foot deep in the streets. In 1919, the year of the great storm on the coast, Laredo had a rainfall total of sixty inches.

O. W. Killam

'My mother asked me what I wanted to do in life. She said if I wanted it bad enough, I could do it. . . . I decided, "I am going to make a million dollars, and I was going to do it before I was 25." It didn't work out that way.'

—O. W. Killam

Killam, T. C. Mann, and L. T. Harned formed the Mirando Oil Company in 1919. Killam had a one-twelfth interest. "I had managed to save twenty or thirty thousand dollars, a long way from a million," he said.

They bought a complete drilling rig for thirteen thousand dollars. "Of course, it wasn't very big . It was a rotary rig with a 45-pound boiler. The drill stem was 4-inch line pipe and pretty weak stuff, too. Wood for the derrick was brought in from Laredo."

They had picked a spot down on the Hinnant ranch in Zapata County because there were some two or three high hills like the hills around Bartlesville, Oklahoma.

"I decided it was a good spot, too," Killam said. "You could see off eight or ten miles, and they looked like some of the domes there in northeast Oklahoma where there was oil production."

The first well had a little showing in it but not enough to make a well. The second failed because of a faulty cementing job. A year later the third test was drilled. Fuel for the boiler was running out at a depth of 1413 feet, 40 feet above the target sand.

Driller Otto Middlebrook of Hebbronville, who had been hired for the job, said, "If you will let me set casing here, we can make it on the fuel we got. But if you're going to go any deeper, you're going to have to get some more fuel down here."

"Well, it didn't make any difference to me," Killam recalled. "I didn't know any better. I told him to go ahead and set it. Then we bailed it to see if the second-hand string of casing had any leaks in it. And we bailed it, we had an oil well. If we had more fuel, we'd have gone on through and we'd never have found that field. It looked like Providence was helping me out."

The 1921 Killam No. 3 Hinnant, in Zapata County, was the discovery well of the Mirando Valley Field. To him, Killam said, it was like a thousand-barrel well. Actually, it was closer to thirty.

(Left and right) Otto Middlebrook and crew before and after the discovery. (Above) They bring in the Killam No. 3 Hinnant,

the discovery well of the Mirando Valley Field. If they had had more fuel, they would have drilled deeper and missed the field.

The discovery did not end his troubles. The Mirando Valley Field introduced the market to a new brand of low gravity crude that the refiners were not familiar with, and there was no market at all for his first 2000 barrels of crude. Yet Killam persisted and the drilling continued. He had some 10,000 acres leased, protecting his flanks.

In September 1921 he bought a section of land and began to lay out the new town of Mirando City on the original grant allotted to Nicolas Mirando. At that time it was a small ranching community that boasted a rail siding for the loading of sheep and cattle. Killam had experience in town building, for he had created Locust Grove, Oklahoma. He had bought the land, subdivided it, and sold the lots. He had owned a number of farms and other real estate in Oklahoma but foresaw a drop in land values, sold all the land, and was looking for something else to do when the Texas opportunity presented itself.

In December 1921, with Colon Schott of Cincinnati, he opened the Schott Field in Webb County. A huge gusher, the Schott No. 2, blew in just south of Mirando City on December 10. This made news all over the United States and started the first real boom in the area. The new town lots were soon selling at premium rates. Supply houses and other allied businesses were moving in.

In F. Michael Black's booklet entitled *Mirando City: A New Town in a New Oil Field*, Luis R. Cadenas wrote in 1972, "Fifty years ago in 1922, some of my brothers and I were working on a farm at Los Ojuelos [near Mirando City] when the first oil gusher was discovered. We heard something like the sound of thunder, due north from where we were—we saw this 'black stuff' up in the air, and we ran all the way, about a quarter of a mile, to this rig. Since we were kids, we had a thrill seeing oil running all through those creeks nearby. We saw men dipping cups into the oil creeks and tasting the oil like they were drinking coffee."

Killam's son, Radcliffe, remembered the thrill of that sight: "Most of the time I went out to the wells with my father. It would take a whole day to make the forty-mile round trip. The Schott No. 2 was a mile away."

He and a friend were looking down on a flock of ducks in a tank, or small lake.

"I looked around, and I saw the thing gushing up over the derrick. It was a gusher. We ran all the way to the location. It was making 2 or 3 million cubic feet of gas and maybe 300 to 400 barrels of oil a day. The well was capped in fourteen to twenty days. When it blew in, they had to dig big holes to put the oil in. They finally built these big concrete tanks that would hold as much as 20,000 barrels."

O. W. Killam started a refinery the next year, in 1923. That was the Misko Refinery, "associated with Texpata Pipe Line Co.," which marketed "Zero Pale diesel fuel and Red Lubricants household fuel."

At this time Killam did some other oil pioneering. He built housing for refinery and other workers so they would not be crowded or forced to live in tents as many of the boom chasers were. And Mirando did not develop into a typical rootin'-tootin' boomtown with uncontrolled vice and violence. It became family oriented, with businesses, schools and churches, and a towering law-enforcement figure who was later to win fame as a Texas Ranger officer, W. W. "Bill" Sterling. An empty boxcar served as the jail, and law and order prevailed.

By hiring Tejano workers, Killam was the first in the industry to rely heavily on Hispanics in his labor force. "If the population is 90 per cent Mexican, then the work force should be 90 per cent Mexican," he said. "I mean citizens of the United States, not imported labor. That's the proportion I have used the last thirty-five years."

The only language requirement, he said, was for foremen to be able to speak both English and Spanish.

(Opposite) Two views of early day Mirando City. (Below right) Killam started his Misko Refinery, which marketed 'Zero Pale diesel fuel and Red Libricants household fuel,' in 1923. (Below) A Schott Oil Company gusher

FOUNDRY
MIRANDO
LUMBER Co
Lumber, Derricks, Cement.
HARDWARE
MIRANDO
TATE PHOTO

NDO LUMBER CO.
LAUREL OIL Co
LAUREL OIL Co
GROCERY
WKM CO
TATE PHOTO

"They [Mexican-Americans, or citizens later referred to as Hispanics] are very loyal and hard working. I have one of them that I think knows more about my business than I know myself and takes care of it better. I have one Mexican driller that's as good as anybody I've ever had. And one is a bookkeeper and a stenographer, who are both fine.

A Killam crew on the rig

"We handle our labor a little more personally than a larger company. We have a loan department in that we loan employees reasonable amounts without interest or expense and deduct it from their pay when they can afford to pay it back. We loan them money to pay hospital bills or doctor bills or any emergencies. The plan is to keep our workers from being victimized by loan sharks. I estimate our losses from this plan have been less than 1 percent."

With pipelines, processing, storage, and shrewd marketing, Killam quickly overcame his earlier marketing difficulties.

"My father also built loading facilities at the rail siding and shipped oil to Port Aransas, where they had storage tanks [at Harbor Island]," Radcliffe Killam said. In one year "he sent over a million barrels of oil to New Jersey by tank car."

O. W. Killam had a unique method of selecting a discovery site. It was to skid the rig sixty feet from the preceding dry hole, a plan that involved having three rigs working at the same time. The wells were shallow and relatively easy and cheap to drill, and his record of successful completions was quite high.

Asked how he located wells, he replied, "Oh, I haven't used any very scientific method. It's kind of a little geology and a little doodlebugology and a little common sense and a few things put together. And in reality I've done most of my own thinking along that line rather than taking other people's advice."

He said he did not depend on geologists because "I've always believed that there would be a direct method of locating oil. . . . While it's not very dependable, it has some virtues; and I think before long it will be a positive proposition, just like locating uranium with a Geiger counter. And I've had a lot of fun with it, and I've had a lot of outdoor exercise, and I've kept up my enthusiasm and good health. And I've made enough to buy the groceries. And what more do you want?

"My definition of doodlebugging is a little instrument that goes up and down and around and around and makes you spend your money drilling holes in the ground. And the law of averages finally hits you with a pool of oil."

He said he knows the practice is "in very bad repute with the real oil people, but I've never been ashamed of my work in it because I've gotten along fine, and I've accomplished most of what I wanted to do in life."

He said if you knew what to look for, you could see an oil field on top of the ground, "but we haven't reached that point yet. If we had the close observation of an Indian tracker, we could probably go out and locate an oil field just from the vegetation, the growth, and the shape of the earth."

It is strange that Killam gave geology such short shrift. Likely he was influenced by geologists who stopped there in 1920 and "said I was here about four million years too soon. I had access to the opinion of a great many geologists going to Mexico. They were all positive there couldn't be any oil in Webb and Zavala counties, or any of this Gulf Coast country for that matter....The country where they said there couldn't be any oil has produced more than a hundred million barrels."

Possibly, as he said, he enjoyed playing with doodlebugs. He once had a Corpus Christi oilman take him up in his plane so he could try a new application for his doodlebug.

"He drilled a well on North Beach," the pilot said. "It was a dry hole."

However, at a geological meeting in Corpus Christi, a geologist asked Killam if he used doodlebugs.

Killam glared at him and ended the question-answer period: "I have a million dollars. How much do you have?"

No doubt he was proud of his reputation as a tough trader. He had signed thousands of acres in long term leases at good prices before the competitors were even in the game. And when dealing in leases and land, he quoted advice his father had given to him: "Never ask a man what he wants. Tell him what you'll give him."

In spite of his statements, however, Killam did use geologists. "The most effective geologist that I've ever dealt with," he said, "was a one-armed quarter Indian, who perhaps worked out more producing areas in this country than all the rest of the fellows put together. And he did most of his geology by climbing up on windmills and studying the country—the surface of the country. And the second field, the Schott Field, it was his work that put me on the exact spot."

That man, who Killam said was not a graduate or licensed geologist, was R. L. Richards. Nancy Heard wrote, "Richards, the son of a geology professor, had little schooling but plenty of books and aptitude for nature. He did much of the geological work on wildcats of Killam, C. L. Witherspoon and other pioneers."

During the Mirando City boom Richards ran an advertisement offering lots and the opportunity to invest in oil wells, giving full geological information to potential buyers.

"He worked for me for a while," Killam said. "And his old maps that he made of this country show some thirty oil fields, or thirty oil pools that have since been developed."

Killam did not know what happened to the maps. He thought they were in the possession of Richards' ex-son-in-law, R. R. "Kirk" Kirkpatrick, a wildcatter who lived in the old Hamilton Hotel in Laredo.[30]

A letter Tom Henny wrote to R. E. Vanderdruff of San Diego, California, on November 17, 1922, tells how Kirkpatrick made his father rich.

Water-witching came to the oil patch

Doodlebugs had their origin in water witching. You cut off a branch, a forked branch, leaving some stem, and you hold it in your hands to your side and grip it, and the stem will go out horizontal as you walk along over the water. The stem will go down, and you squeeze it as hard as you want to, and you can't keep it from going down. And there's water and it was attracting it—that was the origin of it.

Then somebody said, "If it will do that for water, putting a greasy rag on the end of it, oil will probably attract it." Then people began to make variations on that. They'd take a spring and put a bottle of oil and a weight on the bottom of it, and that spring would pump up and down as oil activated it. Or they used a welling rod. You walk along, and when it began to go up and down, it would indicate oil.

Then there were the electrical devices. There never was a case when one of these so-called doodlebugs, if laid on the ground, would operate by itself. It always has to be in contact with the operator. This is a must, and the proper attitude goes with it as an example [of] the subconscious mind and muscle reflexes...You're not conscious of what you are doing.

In this foggy state business, the truth of the matter is that the tighter you grip it, the more positively the end goes down....It's a complete illusion really. It has no basis at all, but a lot of people use them,

One doodlebug enthusiast who claimed to have opened several major fields used his machine somewhat differently. He used it to spur interest in leases. That was more honest than I ever heard another dedicated doodlebugger confess to.

—O.G. McClain

Roy Hinnant (right) of the Hinnant ranching family and an unknown associate out in the field

The most important thing

Bob Fullbright of Freer recalled a tale told by his grandfather, Robert Hinnant.

"When they were drilling those wells, there was an escarpment," he said. "You drive up and you were on the red hill where it's nice and level. Then you fall off and drop a hundred feet, just like that. Well, they were building a road. There were no roads in this country.

"The Killams were there and Ol' Buffalo Wolcott and two or three others. One would say, 'I'll bring the mules and traces.'

"Others said they would bring this and that. Someone asked my grandad, 'Mr. Hinnant, what are you going to bring?'

"'The most important thing,' he responded.

"'What's that?'

"'I'm furnishing this here hill,' he answered."

Author's interview with Bob Fullbright

"There seems to be the rule with all Texas farmers and cattlemen and is readily accounted for, as the oil and gas industry and developments have advanced too fast for them. An illustration of this fact appears in the oil news of this month wherein it is disclosed that J. H. Kirkpatrick, an oldtimer of this city and state and large land owner—owner of a ranch of 17 sections in Webb County, had been endeavoring for several years and until only a few months ago, to 'get rid of it,' and to sell it for a few dollars an acre. The fact that the now famous Mirando Oil Field, discovered less than two years ago, was rapidly extending in his direction, did not appear to interest him, for he still kept offering this large tract at a nominal figure without a thought of having a trained geologist examine it for probable oil structures. It was Kirkpatrick's son (an eastern college graduate [R. R. Kirkpatrick]) who persuaded him about six months ago before starting for Europe to let a drilling test be made on a small acreage lease within the tract. A 200-barrel well of high grade oil was brought in early this month, and since that date, Kirkpatrick has leased a very small portion of his holdings and received over $100,000 in cash as bonus or rents, besides the royalties to be paid. Some men in Texas have oil thrust upon them. But that very fact is our capital so we are not complaining. Signed, Tom Henny."

Later young Kirkpatrick was involved with Richards in other projects. It would seem that he could well afford to live in the Hamilton Hotel.

In 1926 O. W. Killam sold his holdings in the Mirando City, Henne-Winch-Farris Field, his Schott Oil Company, Miami-Laredo Oil Company, and Nueva Oil Company, all in the Mirando District of Jim Hogg, Webb and Zapata counties, to Magnolia Petroleum Company for $1.25 million.

He retained his Misko Refinery, which operated until the early 1930s, the Texpata Pipeline Company, and the Killam Condran and Mirando oil companies.

He had more than his million dollars, but he was far past age twenty-five. He was fifty-six, but there would be many more dry holes and many more discoveries.

He continued wildcatting in the Rio Grande Valley, West Texas, East Texas, Mississippi, Missouri, Louisiana, and Oklahoma. He greatly preferred shallow formations, for he felt deep tests cost too much and took up too much time. One exception was the Pescadito Salt Dome, a deep, deep challenger.

Call it luck, hunches, geological advice, doodlebugs, O. W. Killam was the king of Texas wildcatters. However he found success, he kept at it until his death at eighty-four, on January 1, 1959.

"He lived," his son said wryly, "to get another year on his tax returns."[32]

About the same time as Killam's Zapata County discovery occurred, another newcomer came to Texas, this one to Luling, and less than a year later brought in the first well in the vast Edwards Formation. This newcomer was Edgar Byram Davis, an eccentric New England businessman.

Eighty years after his discovery well signaled an oil boom in the area, Luling definitely remained an oil town. Visitors could not ignore the smell of raw petroleum that permeated the air. If there was dew, the odor was twice as pungent.

Early efforts to find oil in the area had failed. In 1902 wildcatter John A. Otto abandoned his first test in what would later become Salt Flat Field. His second did not hit oil pay, either. It did, however, produce a powerful stream of boiling artesian water. Warm Springs Hospital was

(Above) O. W. Killam continued wildcatting after he sold many of his properties to Magnolia Petroleum Company in 1926. (Left) In the early days oil was stored in cement ground tanks like this one of Killam's.

later erected on the spot. It would be 1922 before Davis's faith and tenacity would bring the opening of production to the area.

Myth has pictured Edgar Davis as an effete clown who made three fortunes and gave most of it away. His biographer, Riley Froh, attributes this to a repetition of "half-truths and exaggerations" by free-lance writers "who emphasize the eccentricities at the expense of history."

Indeed, Davis did not fit the mold of your typical Texas wildcatter. Corpus Christi geologist Carlton "Tubby" Weaver, a youngster in Luling at the time, remembered him as a tall man who weighed some three hundred pounds and was a bit of a "sissy." Of course, a Texas farm boy might have formed such an impression of Davis, a man who was deeply involved in painting, literature, music, drama—a gentleman who spoke several languages and painted, sang, composed music, and played the piano.

At the same time Davis was an entrepreneur who enjoyed tremendous successes, the first as a shrewd businessman who became a millionaire in a New England shoe corporation. With the same work ethic, his brother Oscar also made millions but in a rival shoe firm. The brothers' work ethic was instilled in the New England Calvinist tradition by their parents, Stephen Davis and Julia Anna Copeland, both of whom were eighth-generation descendants of John and Priscilla Alden.

In their home it was "honor to work but dishonor to loaf." When they were younger, Edgar and Oscar, who was three years older, frequently fought, but Edgar won the last fight by throwing his brother over a fence. Oscar later invited him to a dance, and "an entirely different relationship

Pat and Bud Murphy stand in front of Pat's blacksmith shop in Luling. A drilling rig stood behind it.

resulted," one that grew into a bond of devotion and loyalty. As a young man Edgar wanted to go to Harvard and play left guard, but there was no money for him to do so.

He left the shoe business after a relative of the boss received a position he was expecting. Crushed by the event, he suffered a nervous breakdown, an experience that led him to vow that he would never mistreat his employees. Later, true to his word, he gave employees and friends generous financial gifts that consumed much of the $3 million he had accumulated in the shoe business.

In 1908, while on a cruise for his health, he saw that the British were making tremendous profits from rubber and decided that Americans could do the same. He persuaded the U.S. Rubber Company to plant rubber trees on a 90,000-acre tract in Sumatra in the Dutch East Indies. He became a millionaire for the second time and as vice president and director of U. S. Rubber passed up a chance to be president of the company. However, he never lost his interest in the rubber cartel.

He was trying to devise a plan that would eliminate wild fluctuations in rubber prices when one of the U. S. Rubber executives introduced him to Carl Wade, a Fort Worth lawyer who was involved in oil exploration. In the spring of 1919, Wade offered Davis a deal in oil leases near Luling. Edgar was not at all interested in oil, but brother Oscar was. Two of Edgar's assistants, Kelts C. Baker and Arthur Peck, visited Luling on their own and reported favorably on the investment opportunities.

Oscar Davis put up seventy-five thousand dollars for the venture with the stipulation that Edgar manage the Davis holdings in the organization, the Texas Southern Oil and Lease Syndicate, and receive a third of the profits. Edgar sent Baker back to Luling to serve as on-site manager of the investment. Then a fortuitous event happened. Baker, a forester, asked his brother, an oil company employee, for advice, and the brother suggested hiring geologist E. Verne Woolsey to suggest sites for drilling possibilities.

Geologists were scarce at the time because most of them were in the Army. Woolsey, a 1916 graduate of the University of Oklahoma recently discharged from the Army at San Antonio's Fort Sam Houston, arrived in Luling on June 1, 1919.

A mining engineer from Colorado had been hired to make a geological survey. He found anticlines or structures all over Caldwell and Gonzales counties.

"My first week or two in Luling was spent in checking the structures marked by the mining engineer," Woolsey later wrote. "I had to condemn each one of them—they were not there.

"It was in July that I found the surface fault for the Luling Field near Stairtown and on the San Marcos River. About 300 feet down river from the low water bridge on the east side of the river the fault plane stands out in the form of a rock wall running north-east–south-west. I estimated

Edgar B. Davis at age 23

Edgar Davis did not fit the mold of your typical Texas wildcatter. . . . He was deeply involved in painting, literature, music, and drama.

this fault to have a throw of about three hundred feet. Now we know the throw is four hundred fifty feet. There were no fault line fields in Texas at that time that were producing . . . .

"I recommended that oil leases be taken along the line northeast and southwest and this was done. I expected to find good sands in the Navarro and Taylor Marl formation, but there were none. The Edwards was unknown.

"I made a location on the Thompson tract south east of the surface fault and at 1527 there was sufficient gas to shake the floor of the derrick. This well went deeper and was abandoned.

"An interesting item to this report was that Mr. W. F. Pearl, head of the land department [for Davis] was trying to sell spreads in the block to the major companies, but none would buy."

The Luling Field lies within the limits of Woolsey's findings. He had mapped out the fault by pacing the entire area by foot. When his work was finished, he went to work for Atlantic Oil and Refining Company.

The first test was a dry hole. Oscar decided he didn't want to invest another twenty-five thousand dollars but gave Edgar complete interest in what had been spent.

Edgar Davis had some sort of mystical experience and felt a guiding force was leading him to a higher calling, perhaps to be president of the United States. This force seemed to give him the confidence to continue ventures that most other men would have abandoned. He put up the money for the next project, the Southern Oil and Lease No. 1 Thompson.

The Colebank Schulleon derrick crew of Luling

By April 1920 it had become an expensive project. The driller made mistakes as to the depth. Bolts from the rotary table fell into the hole, and fishing for them was taking time. Special pipe had to be ordered from Beaumont. The boiler was ruined because a young helper did not check the water level before firing it up. The well was finally abandoned after oil gushed nearly to the top of the derrick for a few minutes, then turned to salt water.

On March 1, 1921, Davis, now in control of the leases, renamed the firm The United North & South Oil Company, likely to add a touch of conciliation in a time not far removed from the War Between the States.

From 1919 to 1922 he endured difficult times. His brother Oscar died suddenly in March 1922 while Edgar was on a business trip to Europe, and his drilling was producing only dry holes. In spite of financial difficulties, he decided to put down three more wells. The third of these was the Rafael Rios No. 1, on a farm about five miles northwest of Luling and a mile north of the San Marcos River. To complete it, Davis even had to mortgage for four thousand dollars the pumping equipment he used to get river water to his wells.

On August 9, 1922, the Rios No. 1 came in as the discovery well, striking an oil sand at 2161 feet and producing 150 barrels of oil a day.

"I do not know who made the location for the discovery well on the Rios, but I suspect K. C. Baker [Davis's vice president] made it," Woolsey wrote. "This well was drilled by Mr. Drew Mosley when he got into the Edwards Lime."

(Below) A Humble crewman takes a break on his job near Luling.
(Bottom) A grasshopper pump jack

A Luling area gusher comes in. (Right) Jerry Tiller points to the top of Rafael Rios # 1, the discovery well in Old Luling Field.

Strangely enough, the Rios No. 1 discovery elicited no excitement among the major oil companies. Their geologists said numerous dry holes indicated that the Edwards Lime and Austin Chalk "were obviously water, not oil producers."

And Davis was not out of the woods. Deeply in debt, he had run completely out of credit. His employees continued to work, though they had not been paid for months. The Rios No. 1 had been his last hope. He had bills all over town, and nobody would extend him more credit. There was no way a 150-barrel-a-day well would solve his problems.

But he still had hopes. He had managed to hold on to some leases by arranging to sell new production to Magnolia Petroleum Company, which had loaned him twenty-five thousand dollars. One of the leases was for the Merriweather No. 2. On April 14 he had written to Magnolia that he was unable to pay a note due on Saturday, April 15.

That Saturday afternoon he was driven to the well site. "We had entered the motor car to come away, despairing of any immediate action," he wrote, " when I chanced to look up and saw clouds of what looked like black smoke coming out of the hole and shooting up the derrick in increasing force. We hastily got out of the automobile and beheld one of the sights of our lives, for the oil was thrown clear to the top of the derrick, a height of 112 feet. As we were on the leeward of the well, we were soon literally baptised in oil."

Magnolia quickly extended credit on the loan. Another loan allowed him to pay for leases that were expiring that month, and he agreed to sell oil to Magnolia at fifty cents a barrel.

And he continued to seek new challenges. He had gone against scientific advice before, and he did so again. Although geological opinion said the Guadalupe County side of the field would be nonproductive, he maintained that it would outproduce the other half. His first three holes there were dusters, and it seemed that the "experts" were correct.

In *Edgar B. Davis, Wildcatter Extraordinary*, Riley Froh wrote, "Davis placed his faith in the fourth attempt, the Marines No. 1, and optimistically departed Luling for Massachusetts and Christmas with his sisters. Marines No. 1 came in on December 27 [1923], the first of many to come in Guadalupe County; this strike saved all the United North and South leases; and 1923 marked the last year of financial strain with the Luling Field."

On August 9, 1922, the Rafael Rios No. 1 came in as the discovery well, striking an oil sand at 2161 feet and producing 150 barrels of oil a day.

A sign and pump jack mark the site of the R. Rios No. 1.

It is small wonder that the wildcatter believed in divine intervention. He had seen it. Of course, his dogged determination also had something to do with it.

This short item was published in *The Luling Signal,* dated May 14, 1924:

"One of the biggest wells in the entire Luling oil field was completed last Friday afternoon when the United North & South Oil Co. brought in their A. J. Baker No. 7 across the San Marcos River in the Guadalupe County end of the field, the flush production being rated at between 10,000 and 12,000 barrels. The new well is in the sector close to the river and tends to show a connection between the two sides of the field more than ever."

In 1924 United North & South had 113 producing wells in Caldwell and Guadalupe counties.

It was widely rumored that Davis consulted a psychic before choosing a drilling site. In spite of stories to the contrary, he never received drilling advice from Edgar Cayce, who went into a trance and detailed the boundaries of the oil pool. Davis did talk to Cayce's partner, David Kahn, who ran out of money before he could find out if his friend was a real live doodlebug. Froh, Davis's biographer, said it is doubtful that Davis put much stock in such a chance conversation.

Though Davis appeared to operate largely on chance, it must be remembered that his monumental success was due to the early work of geologist Woolsey in defining the fault along which Luling Field would be developed and wells producing an average of 900 barrels a day would appear. And Davis continued to utilize geologists more than he would admit. After the Luling Field became successful, he turned to geologists Ernest W. Brucks as chief geologist and hired George C. Matson and Davis Donoghue as consultants.

Naturally, it was impossible to strike oil in Texas without a boom. And Merriweather No. 2 caught the attention of oilmen all over the country. Drillers, tool pushers, roughnecks, and others looking for honest work flowed in. The work was hard, dirty, and dangerous, but thousands of farm boys took to it for paychecks that looked mighty good. The population of Luling jumped from 1,500 to 10,000. Housing was at a premium, and beds in tents were rented by the hour.

But Luling had something most boom towns did not have—a rough, tough police force that was not at all sympathetic to crime, vice, and violence. The undesirables moved on through town across the river and settled at a place that became known as "Gander Slue." Gamblers, gunmen, con men, bootleggers, prostitutes, hustlers, and robbers congregated there and flourished twenty-four hours a day. Ranger Frank Hamer put a stop to it for a while, busting up booze and gambling equipment, but it resumed when he left, until the wild surge of drilling slowed and the

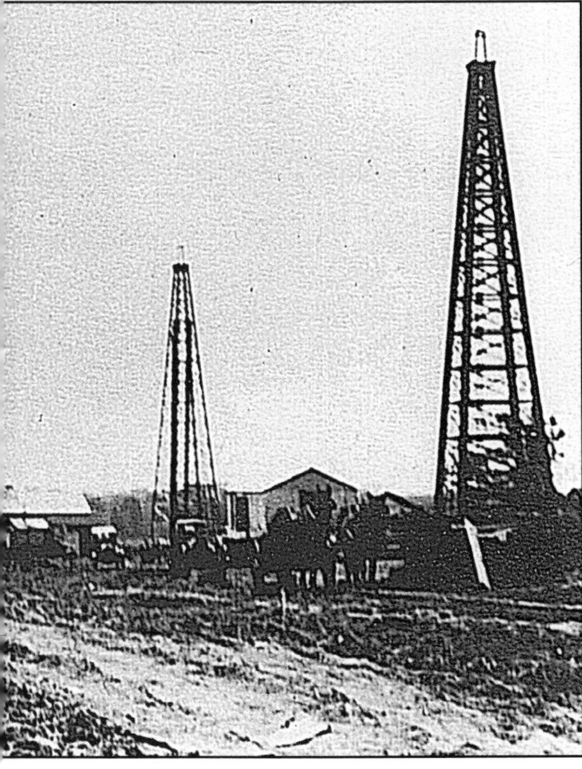

drilling crews moved on. The boom was over, but the town of Luling will never forget the man who made it all possible.

On June 11, 1926, Davis sold his holdings, which covered sixty per cent of the Luling Field and 215 producing wells, to Magnolia Petroleum Company for $12.1 million —$6,050,000 in cash and the rest in payments that would be paid out of future production, giving him a total worth of about $10 million. The transaction was the largest in oil industry history up to that time.

A few days earlier Davis had announced his plans to hold a barbecue, open to the public, to share his riches with residents of Caldwell and Guadalupe counties. In that segregated day, white citizens lunched on 100 acres along the San Marcos River south of town while blacks attended a picnic on the "Negro" school grounds to the north. "At least fifteen thousand guests consumed 12,200 pounds of beef, 5,180 pounds of mutton, 2,000 fryers, 8,700 ice cream sandwiches, 85 gallons of ice cream, 7,000 cakes, 6,500 bottles of near beer, 28,800 bottles of soft

(Left) Luling's effective police force kept undesirables out of the town. They gathered instead at this site known as 'Gander Slue.' (Below) Edgar Davis's United North & South Company had 113 producing wells in Guadalupe and Caldwell counties in 1924. Steel derricks were used after a tornado destroyed the wooden ones in the area.

On June 11, 1926, Edgar Davis sold his holdings, which covered sixty per cent of the Luling Field and 215 producing wells, to Magnolia Petroleum Company for $12.1 million. The transaction was the largest in oil industry history up to that time.

drinks, and unknown quantities of beans, potato salad, pickles and coffee."

Davis's employees were also overjoyed when they learned that the boss was issuing generous bonus checks. The amounts, determined by the length of service, totaled $1.75 million, including a sizable check to Verne Woolsey.

Some of the payments were twenty thousand dollars or more, an amount that made a person very wealthy in those days, when a new car cost about six hundred dollars. A number of the more conservative locals saved their money, but Tubby Weaver said, "Most of those guys had never had more than a couple of dollars in their lives, and they'd spent it all in just a few days. The Ford and Chevrolet dealers sold out and had to bring in more cars with that spurt of money."[33]

Two years later another important chapter in the history of the Luling Field was written.

On March 29, 1928, *The Luling Signal* reported, "Damage in the oil field west of Luling in the hail and windstorm Friday night totaled approximately $200,000. Magnolia Petroleum Co. was the biggest loser. Eighty of their derricks were blown down, the roof of the Rios warehouse blown off and windows in the big power plant broken. Hailstones as big as pullet eggs were reported."

Roughneck Gerald Lynch gave this more personal account of the storm in his book, *Roughnecks, Drillers and Tool Pushers:*

"Our rig was shut down. We had the only steel angle-iron derrick in use in the Bruner Field...We settled in for a long, dull night. Little did we know what was in store.

"We were using three boilers. Possum Parker, the fireman, had cut the fire out of the middle boiler when we came on tour at 6 p.m. The steam had gone down and he and I drained the boiler. Then we removed the manhole cover and prepared to wash the boiler to remove the accumulated scale and dirt.

"Dee [another crew member] was asleep in the boiler shed, a 4-by-12 sheet iron roofed structure. Doc [Milligan, who had hired Lynch,] was reading a book in the tool house. Polecat [Tomlin, another roughneck] was somewhere about. He was a restless soul and roamed around a bit."

"There were wooden derricks all around us, fairly close to us....

"Sometime after 10:30 p.m. the wind got really wild. Possum and I crawled into the firebox in case there was hail.... Then we heard a roaring noise and all hell broke loose. Derricks began to crash down. Doc ran out of the tool house and jumped in his car, which promptly flew into a ditch and turned over. Dee got up to run and a 10-foot length of sheet iron caught him about the hips flat wise. He sailed off into the woods like a big bird tiptoeing along the ground. Polecat was not in sight.

"The storm was over in a few minutes. Our boiler hadn't budged so Possum and I crawled out. We could hear Doc yelling. His car was lying on its side. It was raining very hard, and the car was rapidly filling with water. Dee came out of the woods, wet and scratched up. Polecat also wandered up. He had crouched down beside the motor. He was wet but unhurt. We rescued Doc by setting his car upright so he could open the door. His little Chevrolet wasn't badly damaged....

"Our derrick was still standing and our lights still burning. No other derricks were lit, and it was still raining very hard."

They checked on Penny Spencer's rig nearby. It was down, and Spencer had a broken arm and dislocated shoulder. Two hands were frantically digging in the wreckage of the doghouse where a man had been sleeping off a hangover. They were afraid he was dead. As they fired up the car to take Spencer to the hospital, the man in the doghouse staggered out and asked , "Wha's goin' on?"

The drunk had slept through the whole ordeal. The injured Spencer was so angry he threatened to kill him.

"It was a wild and eerie night," Lynch wrote. "Twenty-two derricks blew down. Two men were killed and many were injured. Damage to rigs was tremendous.... Our steel derrick was the only one left standing in the swath of the storm that came through Bruner Oil Field. I will always believe that the storm made believers out of a few drilling contractors. They had, as usual, resisted changing over from wooden derricks, for all manner of reasons. However, the steel ones came into more general use after that tornado—at least around Luling."[34]

In his book Lynch had kind words for Edgar Davis. "He was a good, decent but very odd man," he said. "We have seen all too few like him in the oil patch."

There were all kind of stories about Davis. He was called the wildflower man, Lynch said, because it was said he liked wildflowers

(Above) An exhibit in Luling's Central Texas Oil Patch Museum tells the story of Davis's discoveries. (Top) Edgar Davis (center) surrounded by a group of friends. They are (from left) David Figart, Hazel Munster, Tommy Caylor, Catherine Davis, Davis, Inez Griffin, Lorraine Crockett, Otis Smith, and Bruce Pipkin. (Opposite top) Oil flows gave workers like Frank Riley a greasy bath.

and sometimes would pluck a flower and say, "Here is my oil well." He said Davis was greatly admired by the men who worked for him. Most of the other roughnecks felt he was opinionated and stubborn and "wasn't playing with a full deck," but nobody dared criticize him lest his workers beat the detractor about the head.

"Unfortunately," Lynch said, "I didn't share in that deal" when the guys on the Davis rigs got the big bonuses "so I just kept on working."

Davis did not forget the citizens of his adopted town. He donated two city parks and two country clubs, one for whites and one for blacks. He established the Luling Foundation to promote scientific agriculture at a cost of $1 million. He made other bequests, always specifying that there should be no bias against race, sex, or religion. Davis also established the $1 million Pilgrim Foundation in Brockton, Mass., his hometown, for charitable purposes.

His interest in artistic pursuits and a belief in reincarnation led him in another direction—to Broadway. In the 1920s he commissioned a friend, J. Frank Davis, to write a play, which he would produce, using the transmigration of souls "to illustrate the theme of mankind's achiev-

Felt hats, not hard hats were the customary attire for drilling crews in Edgar Davis's day. In *Folklore of the Oil Industry*, Mody Boatwright wrote that early workers thought no more of their safety than their employers did. 'Indeed,' he wrote, 'the danger they were exposed to was a source of pride.' These men are (from left) B. H. Chamness, Pat Fuqua, Paul E. Bainbridge, L. H. Sobotik, G. Hardeman, and R. H. Giesky.

Wells like this were still producing in the Luling area three quarters of a century later.

ing a state of well-being and good fortune." The result was "The Ladder," which Davis continued to support in spite of negative reviews and a lack of theatergoers. By February 1928 the show had set two records: It was the longest-running show in New York, and it was the all-time top Broadway money loser.

Davis's artistic endeavors in Texas met with more success. In the late 1920s he financed a series of competitions in art to reflect the beauty and toughness of Texas wildflowers. Although some publications in New York and Chicago scoffed at such regional art as "unimpressive," the contest drew a "pilgrimage" of distinguished artists and attracted national attention. The 1929 contest resulted in 129 finalists, ranging from modernist to traditional, and the judges concluded that Davis had made San Antonio the "art center of the South."

The year 1928 also saw a new field in the Edwards formation—Salt Flat, which was discovered in May. While it was less productive than the Luling discovery field, it increased the geologic knowledge of the formation and pointed the way to another, more important discovery the next summer.

That was the Darst Creek Field, which came in when the Texas Company Dallas Wilson No. 1 found oil on July 10, 1929. United North & South had bought some leases in the area in October 1928 and purchased three more after the discovery. As the new decade opened, both major and minor companies were optimistic about their opportunities, and soon Davis's company had seventeen wells on 180 leased acres producing more than 5000 barrels a day.

Davis began selling the crude to the Louisiana Oil and Refining Company but soon decided to sell that company the entire Darst

The impossible may take a little longer

Corpus Christi oilman Paul Haas recalled going through Edgar Davis's papers searching for royalty owners after his firm had bought properties in the Luling Field.

They found a record of a wildcat Davis had drilled in Matagorda County. It was drilled to total depth and produced salt water.

Davis had ordered casing to be run, a highly unusual practice when there were no oil shows. They pumped the well, which yielded salt water. He ordered the crew to keep pumping. Nothing but salt water. Two, three weeks passed. More salt water.

Then, quite unexpectedly, the well began flowing oil. Lots of oil. Davis had had a discovery on his hands.

"It's not supposed to work that way," Haas said. "No one would do that. But he did. And it worked. You might explain that. I can't."

—Paul Haas

Creek property for $2,250,000. Louisiana Oil and Gas paid $500,000 down and planned to pay the rest out of profits from the oil. However, the company was unable to fulfill its obligations to Davis after the Texas Railroad Commission issued two orders severely limiting the field's allowable production. He took back the property in 1931 and returned the down payment. Darst Creek remained the mainstay of the United North & South Development Company until 1951, the year of Davis's death.

Through organization, drive, honesty, generosity, determination, and faith, Edgar B. Davis became an entrepreneur and highly successful capitalist, accumulating huge wealth in three different businesses. Perhaps his greatest legacy was the discovery of the Edwards Lime. It set off vigorous exploration to find the lucrative shallow production. Luling production helped push Texas ahead of Oklahoma in total oil production.

Davis's discovery well was still producing at the end of the twentieth century but was plugged and abandoned because the casing, in use for more than three-quarters of a century, was beginning to crumble.

Although Vintage Oil Company of Oklahoma owned most of the operations in the area in 2003, the area's importance in the history of the Texas industry had been noted.

John R. Sandidge, geologist for Magnolia Petroleum Company, said the economic impact of discovery of the Edwards formation "has been of major importance throughout the main production area, extending from Caldwell to Webb Counties, a distance of 165 miles." It opened a chapter in the history of the oil business, he said, which was as important to the [South Texas] area as Spindletop was for the Gulf Coast.[35]

Another important chapter was written by two doctors from Nebraska—Pratt and Hewit. They came to Refugio County, bought leases, and in 1920 drilled a discovery well in a rich area that was to make quite a few millionaires.

A happy crowd celebrated the 1920 discovery well on the Rooke Ranch, near Refugio. Dr. Pratt is seventh from the left, Dr. Hewit is sixth from the right. (Opposite) This replica of an early 'filling station' stands in the Central Texas Oil Patch Museum in Luling.

Another important chapter was written by two doctors from Nebraska, who seemed an unlikely pair to follow Killam and Davis to gamble their savings on the Texas Gulf plains. One was head of the School of Medicine at the University of Nebraska at Lincoln. The other was a dentist. Likely neither had ever had a callus on his hands. Yet they came to Refugio County, bought leases, and drilled a discovery well in a rich area that was to make quite a few millionaires.

The doctor was named Pratt and the dentist, William E. Hewit. Their Pratt-Hewit Drilling Company drilled the 1920 discovery well on the Rooke Ranch near Refugio.

They were followed in turn by George E. "Wingy" Smith, a wildcatter whose discovery of the Greta Field marked a high point in his career, and by Hugh Roy Cullen, a cotton broker turned oilman whose negotiating skills would help him open one of South Texas's richest oil fields—the Tom O'Connor.

"Refugio became a boom town," George Hawn, grandson of Dr. Hewit, said. "Basically you had the start of the gas and oil industry there. It started out gas. Then it went to oil. You had people like Pat Rutherford,

Herman Heep, Hugh Roy Cullen with his Quintana Oil Company, Joe Wolf, Arnold O. Morgan, and his brother Bill Morgan."

George's father, Richard H. Hawn, a geology graduate of the University of Nebraska, married Hewit's daughter in 1917.

"Mother and Dad moved down from Burkburnett in North Texas. He worked in the oil field there as a roughneck. They came to Refugio County about 1920, moved out on the Rooke Ranch, and took a lease on the ranch in 1921. Dr. Pratt and Dr. Hewit were the two who raised the money to buy the lease and drill the first well.

"Dad was a geologist, and they developed the Rooke Field, in which ultimately seven hundred wells were drilled," he said. "This was the very first producing oil field in Refugio. If the hole had been dry, the doctors would have returned to their medical practices.

"My dad, who was secretary and geologist for the Refugio Oil and Gas Company, took a 5600-acre lease on the Lambert Ranch for $1.50 an acre," Hawn said. "This became part of the Tom O'Connor Field in the 1930s."

The two doctors who started it all were to become bitter enemies—although not over money. Rather they both fell for the same young damsel.

"After Granddad and Dr. Pratt had a falling out around 1927 over their lady friend, Dr. Pratt threw Dr. Hewit off the board," Hawn said. "Dr. Hewit sued to get back on the board of Pratt-Hewit Oil Corporation."

James R. Daugherty of Beeville was his attorney. The suit was tried in Houston, and his grandfather lost.

"Granddad was bemoaning the fact that he had been thrown out of the company and had lost his suit and didn't have a partner any more, and Judge Daugherty said, 'How about me?'"

So they formed Hewit and Daugherty, which went on to drill 250 wells in the Tom O'Connor Field in the 1930s and 1940s. They were good wells—many were still producing a half century later.

"The wells were multipays," Hawn said. "Hugh Roy Cullen had 900 wells in Quintana in Tom O'Connor Fields. Hewit and Daugherty had 250. I believe Hewit and Daugherty's interest was about 8.5 per cent. So that would have been around 3,000 wells in the O'Connor Field."

The field is one of the largest in the country, covering much of the area between Refugio and Victoria on land owned by pioneer Texas families, including the Lamberts, O'Connors, Welders, Maude Williams, and others.

Hewit and Daugherty "had three steam rigs. Big old boilers. After they were finished with them in the mid forties, they sold them, to Bill Clements [later governor of Texas] and Paul Turnbull and Frank P. Zoch, Jr. They formed Southeastern Drilling Company—SEDCO," Hawn said. "They did a great job putting together a super drilling company offshore and onshore."

Hawn's father wrote in a letter that they were trucking oil to Mirando City to be refined. A little railroad and later a pipeline ran from

'Unworthy' field proves its worth

Somerset Field south of San Antonio was the "poor boy" field discovered in 1913 by water well drillers. After the first World War nearly a hundred derricks dotted the landscape. The shallow sands weren't deemed worthy of serious development, but the field was a training school for many future oilmen.

Major oil companies passed on the production, predicting that the sands would soon be depleted. Geologists of the day said the field was too young geologically.

The same view had been offered for the Mirando sands discovered by O. W. Killam in 1921.

But two small refineries, Simp's Pioneer and Grayburg Refinery, were processing nearly 16,000 barrels from the field daily by 1923. After 40 years, the field had produced some 14 million barrels.

Nancy Heard, *San Antonio Express* Historical Section, November 7, 1965

Refugio County to Harbor Island, exporting the oil until 1973-74, when the industry turned around and oil began coming in from Saudi Arabia.

"Dad partnered with John L. Sullivan in the thirties. The Mexican government took all their property," Hawn said. "They were drilling for sulphur in Campeche. Dad loved Mexico. They didn't get thrown out, but they didn't have enough money to stay."

In the late 1930s his father organized another company called "Poor Boy Oil Company." Partners were Herman Heep, R. H. Hawn, Pat Rutherford, and one other, all "Poor Boys." Oil was ten to fifteen cents a barrel, and they couldn't sell it at that price.

"As a spinoff in 1940, Dad bought some lumber from a timber company near Victoria, Mexico, not too far from Monterrey. At the time Bob Vaunter and Dad formed the Hawn Sash and Door Company and brought timber from near Victoria. They were making ammunition boxes for .50 caliber weapons," he said.

Some accounts have listed George as a native of Houston. Actually he was a native of Houston, but he didn't stay there very long. It happened this way back in 1930, shortly before the family moved to Beeville, where Hewit and Daughtery was headquartered: His mother was informed that she had a tumor. She went to a hospital in Houston to have the tumor removed.

Flowing the pit, transferring mud from the working mud pit to the reserve pit

Craters and piles of debris, like this at the site of the No. 1 Ayers, marked blowouts.

"That tumor was me," Hawn laughed.

Dr. Hewit's contribution to the South Texas oil industry may be obscure, but his name is preserved in a prime section of Corpus Christi developed by the Hawns—Hewit Place is one of the city's most prestigious addresses.[36]

"Wingy" Smith entered the Refugio scene in 1933, when times were hard. It was tough to find money to finance a wildcat well, but Smith usually managed to come through, no matter what happened.

He had lost one arm in a rig accident. Years earlier he had received his nickname when a new crewman saw him throwing his remaining arm around as he gave orders.

"Who is that one-winged so-and-so?" the man asked.

From then on "Wingy" was Smith's name.

Wingy didn't seem to have a lot going for him. Geologists from major oil companies told him there was no oil at the spot near Refugio where he was drilling. He managed to beg and borrow to get enough money for drill pipe. His partner, John M. O'Brien, furnished the tent, and Smith's wife did the cooking. The crew was content to work for food and for pay if and when the well came in.

On May 22, 1933, the No. 1 O'Brien came in at 4100 feet, for an initial flow of 530 barrels per day through one-fourth-inch choke. They had discovered the Greta, opening the Greta Field and one of the largest producing areas in Texas. Wingy was prosperous.

Ten years later, Wingy was back to drilling on a shoestring. It was so bad he hid his automobile to keep it from being repossessed. His rig was old and rickety. It was steam driven with an underpowered generator, definitely a poor-boy operation. He was about to work his way out when a gas pocket blew out and the old rig disappeared in the crater.

The resulting lawsuits put him out of business, ending Wingy's wildcat career.[37]

The Refugio and Greta fields had been discovered, but it remained for a former cotton broker, Hugh Roy Cullen, to make the biggest discovery of the area—Tom O'Connor.

A contemporary of Cullen's once said of him, "When they bury him, he'll raise up in the casket and say, "Dig it a few feet deeper, boys."

Drilling a little deeper is precisely what made Cullen one of the wealthiest men in the world. He was a cotton broker and a very good salesman when former oil operator Jim Cheek employed him to buy oil leases, Cullen's introduction to the oil business. Cullen was successful in obtaining leases because he promised the landowners that leases would be developed. He spent three years in the business in West Texas before deciding in 1920 that prospects were better in East Texas.

Many wondered how a cotton salesman could suddenly become a highly successful wildcatter. Some thought he was lucky beyond all reason. Others thought he must be clairvoyant or psychic to find oil where others failed. Many attributed his nose for finding oil to luck.

His reputation for finding great quantities of oil in abandoned fields—to the chagrin of geologists—was not based on luck. Rather it was based on twelve years experience in the oil fields, a good grasp of surface geology, common sense, and knowledge he gathered from an intense study of every geology book he could find. Also he utilized every geophysical instrument as soon as it was available.

He was the first oilman in Texas to realize the possibilities of the gravity meter, which gave a clear picture of the varying density of sub-surface rock at various levels, and he employed the seismograph and the torsion balance, which measured minute differences in the gravitational pull of the earth at different places. He also adapted the electromagnetic surveys used by Professor Schlumberger of France and the geochemical methods of testing surface samples of soil and rock for traces of oil.

In 1920 he started Pierce Junction Oil Company and drilled on the flank of a small gas well on land Gulf Oil Corporation had abandoned. Cullen's well was a 2500 barrel-a-day producer in Pierce Junction Field, which was located near the site where Houston's Shamrock Hilton Hotel would be built. With the money he made, he drilled in Damon's Mound, a losing venture.

With a new partner, rancher James Marion West, he began to drill in East Blue Ridge Field, another former producing field. By drilling

Geologists followed the train tracks

In the early days, the railroads were the best geologists, old-time geologist and oil operator W. Carlton "Tubby" Weaver said.

Geological Society historian Ray Govett agreed, noting that in the coastal area any elevation is preferred to areas where the roadbed would likely be flooded, so the railroad builders chose ridges for their tracks in order to avoid raising the roadway. These ridges were often indicative of underground structures of interest to geologists.

Weaver pointed out that roads in many areas were nonexistent, so it was convenient to move equipment by rail. Many oil discoveries were to be found near a rail right-of-way. And, Govett said, trains were available to transport oil to market.

An example is the Refugio Field, which was discovered because of its geological feature, a hill that Copanes, a Carancahua Indian, chose as the location for a village. The Spaniards chose the site for mission Nuestra Señora del Refugio there in 1749. The town of Refugio grew up around the mission. And the oil field was the first in what is now Railroad Commission District 2.

—Tubby Weaver and Ray Govett

deeper, their South Texas Petroleum Company brought prolific flows of oil from dozens of gushers.

When he suggested drilling in the badly depleted Humble Field, his partner complained that the place was drilled out. A billion barrels of oil had been removed from the field. It was dried up, with more holes in it than Swiss cheese. Slush pits and broken down derricks littered the landscape.

Cullen had studied old well reports for troubles and was able to cope with a strata of "heaving shale," or "Jackson shale," which crumbled and clogged the "screens"—holes in the pipe designed to allow oil to flow in. The gumbo pressed in and froze the pipe, and the wells were abandoned. He cleared the screen by pumping salt water instead of mud into the well. He ordered the driller to work the pipe up and down like a pile driver. The pipe dropped into the oil sand. The well blew in, producing more than 5000 barrels per day.

A biography by Ed Kilman and Theon Wright noted that Cullen, who had the equivalent of a fifth-grade education, received a Doctorate of Science from the University of Pittsburgh for the "originality of thought, daring and vision in the development of methods of drilling deep wells" that enabled him to drill the Humble Field wells through the "heaving shale that had defied oil well experts for three decades."

It was pointed out that he couldn't rightly be called a wildcatter because his most spectacular discoveries by 1930 were within established fields—Blue Ridge, Humble, Pierce Junction. He found another huge reservoir in Rabb's Ridge Field in Fort Bend County.

After the partners' success in the Humble Field as South Texas Petroleum Company, West wanted to sell Rabb's Ridge for $20 million. Cullen reluctantly agreed and dissolved the partnership. His next adventure, with his new Quintana Oil Company, would be his greatest achievement.

Most of the phenomenal success Hugh Roy Cullen achieved was in existing fields, but in 1934 he was certain he knew the location of one of the largest oil pools in history in a field yet to be discovered. Others had the same idea, but there was one obstacle they had not overcome—Thomas O'Connor of Refugio County, who did not take kindly to oilmen disturbing his cattle with machinery and messy drilling operations on his ranch, one of the largest in the area. Like Cullen's, O'Connor's ancestors had fought in the Texas Revolution. They were colonists who received generous land grants when they came from Ireland to Texas with Empresario James Power. The O'Connors acquired other grants and created a highly profitable ranching operation.

Heavy oil production in the Greta, O'Connor-McFadden, and Refugio fields made Tom O'Connor's land look mighty inviting, but O'Connor wasn't interested in talking leases. That challenge was right down Cullen's alley. An expert lease hound, he didn't approach O'Connor

History linked Cullen to Quintana

The story of the Quintana name involves Hugh Roy Cullen's family history. While he was fishing on the Brazos River near Freeport, he asked a guide about some ruins on the shore. The guide told him the ruins had been Quintana, a port in the early days of Texas settlement. It was the leading port in Stephen F. Austin's colony.

Cullen, whose family was deeply involved in Texas history, liked the sound of it.

His grandfather, Ezekiel Cullen, had come to Texas in time to participate in the Siege of Bexar in December 1835, when the Texians captured the Alamo from the Mexicans. Later, as a member of the Congress of the Republic of Texas, he sponsored the Cullen Act, which started land endowments for public schools and universities, laying the groundwork for the Texas public education system.

He didn't like the site Sam Houston had chosen as the new capital of Texas. He said it would be better to legislate from tents in the high and healthy section of the country "than to inhale this poisonous atmosphere; to drink polluted water; to be subjected to the deprivation and want of comfort incident to life in Houston." The capital was moved from Houston to Waterloo, later renamed Austin.

Ezekiel Cullen was a lawyer, jurist and purser in the United States Navy. He died in 1882, a year after his grandson's birth.

New Handbook of Texas, 2: 439.

directly as others had done. He called and made an appointment to meet in O'Connor's office with the rancher's attorney present. O'Connor listened and said he would be in contact with them. Cullen's plan wasn't working until he contacted Chad Nelms, an oilman who had worked with Cullen and was a friend of O'Connor's. Through this contact Cullen won the right to survey two 7500-acre blocks for one dollar an acre.

The first well was being drilled when Wallace Pratt, the famous Humble geologist who can almost be called the patron saint of American geology, showed up at the Quintana office. He had been refused permission to make a gravity-meter survey on the ranch. Cullen showed him survey maps and told him he had a commitment to drill two wells. He offered Humble a 50-50 interest in the project if it split the cost. Pratt turned without a word and started for the door. He turned around, looked into Cullen's eyes, and walked out. Ten minutes later the telephone rang in Cullen's office. It was Pratt.

"All right, Roy," he said. "We'll take half." It was a good decision.

Chad Nelms, who had doubts about the well, asked to sell his share for twenty-five thousand dollars, a decision that proved terribly painful. Twenty minutes later the well came in, opening one of the great fields in Texas. The date was June 26, 1934.

Portrait of Hugh Roy Cullen from the Archives of the University of Houston. Cullen gave large sums of money to establish the university to educate the children of working families.

The potential of Quintana Petroleum Corporation No. 1 Tom O'Connor was not astounding. It was producing 545 barrels per day on three-eighths-inch choke at 5580 feet. However, other wells in the field would be much more productive. The field would produce more than half a billion barrels of oil.Refugio County's total oil production from 1928 through 2001 was 1,301,070,204 barrels, much of it from the Tom O'Connor Field.

O'Connor, with his wealth escalating at a rapid rate, was much more kindly disposed towards Cullen. He granted him a total of eleven more leases. Gushers were opened up along a five-mile strip. As luck would have it, another company drilled just across the fence from the discovery well. It was dry.

Cullen, already a multimillionaire, was still playing the game and adding to his bountiful fortune, but fate was about to deal him a cruel hand in the tragic death of his son, Roy Gustave. During his cotton-buying days, Roy had met Lillie Cranz of Schulenberg and married her after a four-year courtship. The couple had four daughters and the one son, his father's pride and joy. The younger Cullen had studied engineering at Rensselaer Polytechnic Institute in Troy, New York, and at Houston's Rice Institute. He had worked his way up in the business from roughneck to chief of field operations. He directed drilling at the second well in the Tom O'Connor Field and in 1935 was vice president in charge of Quintana Field operations.

In 1936 he went to Edinburg, where a leased rig was having trouble. The elder Cullen advised him to be careful because the rented

Hugh Roy Cullen and his wife (left) on stage receiving a humanitarian award

rig was not as solid as their company rigs. Roy Gustave found the crew trying to free drill pipe stuck at 7000 feet. As the driller was talking to the elder Cullen on the telephone, he ordered his son down from the rig because it was old. The pipe seemed to be pulling free, but the movement was the rig sinking in the mud. Young Cullen yelled, "Pull her wide open!"

Bracing on the legs of the derrick snapped, and the rig began to collapse. A tangled mass of metal and wood rained down. A hose snapped and threw the driller clear, but the wreckage fell on Roy. He lived two days but never regained consciousness.

Hugh Roy Cullen was in a state of shock. As he recovered, he reflected on his life and decided to give away most of his fortune, much of it in revenue from oil production.

"Giving away money is no particular credit to me," he said. "Most of it came out of the ground, and while I found the oil in the ground, I didn't put it there. I've got a lot more than Lillie and I and our children and grandchildren can use. . . . My wife and I are selfish. We want to see our money spent during our lifetime so we may derive great pleasure from it.

"It's easier for me to give a million dollars now than it was to give five dollars to the Salvation Army thirty years ago," he said at one time.

That gift to charity had bounced.

"I didn't know my bank account was so low," he said. "I had plenty of credit in those days—to buy cotton with—but I didn't have much cash."

While he left his family well provided for, he set up a foundation that gave away some 93 per cent of his fortune.[38] His public philanthropies, which enriched universities, particularly the University of Houston, schools, hospitals, museums, the Houston Medical Center, and many other entities, totaled more than $180 million at the time of his death. One of the buildings at the University of Houston is named for Ezekiel Cullen and another for Roy Gustave.

If Hugh Roy Cullen was lucky, he passed his luck on to the residents of Houston and Texas.

In the meantime, the Welder family of Victoria was becoming involved in production in Victoria and Refugio counties. In 1923 Plymouth Oil Company was organized to develop nearly 15,000 acres obtained from Rob Welder, a son of John J. Welder, on land John J. had bought from rancher Thomas Coleman. On February 10, 1930, Humble made the first discovery in Victoria County on the ranch of Jay Welder, another son of John J. It was a rush of gas from 5600 feet, followed by a flow of oil. The Welder A-1, in San Patricio County near Sinton, turned out apparently dry in May 1934, but Schlumberger used its electric log and found gas sands that had been missed.

Mike Griffith, field superintendent for Plymouth, said, "They picked up a 30- to 40-foot section of gas sand at 5400 feet that the driller had drilled through without knowing it. I think we can attribute the A-lease developments to Schlumberger. [Rob] Welder was so delighted that he presented each member of the drilling crew with a check for five hundred dollars. . . . The union between Welder and Plymouth through the years was cordial and mutually beneficial."

Although wells on Welder A and B leases met with only fair success, the C-1 Welder came in on Easter Sunday, April 24, 1935, with a potential of 336.68 barrels of 32-gravity oil a day. This opened a chain of developments that led to the completion of hundreds of wells flowing from seven different pay levels.

Welders' roots deep in Texas history

The Welders were a noteworthy family, tracing their roots back to Felipe Roque de la Portilla, a Spanish officer who was sent to Mexico in 1788. After Native American tribes and the Mexican Revolution frustrated his efforts to establish a colony near the site of present-day San Marcos, Texas, he was a prominent ranchero near Matamoros.

His daughter Dolores married Irish empresario and Texas statesman James Power, and their daughter married John Welder, whose family came to Texas as members of the colony of Dr. John Charles Beales, an Englishman. John Welder, a successful rancher, was assassinated in 1887, but John J. Welder, his son, continued his father's efforts and by 1900 had amassed vast amounts of South Texas land.

Welder family files, Victoria office of Patrick Hughes Welder, Jr.

(Below) Cattle and petroleum coexisted on Welder ranches. Here cattle graze in front of a recycling plant.

In early November of 1922, this well marked the beginning of gas development near Corpus Christi. A *Caller* story of November 9, 1922, said it was brought in by W. L. Pearson and Saxet Co. on the Dunn Tract. It blew out at a depth between 2300 and 2400 feet, sent mud, water, sand, and stones more than 200 feet in the air. Shale in lumps varying in size from two to eight and nine inches in diameter was thrown out by the force of the explosion, and a quantity of soapstone or blue gumbo was expelled. The pressure continued for an hour, then choked off.

In May 1936 the district offices of Plymouth were moved from San Antonio, and work progressed on building a district office building and a camp for employees just north of Sinton. At its height the camp contained more than twenty homes, a swimming pool, and a community building. Plymouth had 147 producing wells in San Patricio County by the end of 1938, and other companies had come in to establish other fields.

After becoming interested in oil and gas development on his ranch, John J. Welder's youngest son, Patrick Hughes Welder, formed the Portilla Drilling Company. In 1985 there were 504 producing oil wells in San Patricio County. One well, which had been brought in June 4, 1936, and was still producing in 1986, had brought in 88,153,838 barrels in that fifty-year period.[39]

As the pioneers spread the production of oil across the area, natural gas also assumed an increased importance. After encouraging results in White Point Field, the first gas production in Nueces County came in 1922 from the Saxet Field, across Nueces Bay from White Point. To provide gas service to Corpus Christi, which was using an artificially produced gas, John Ginther of San Antonio tried to promote a search for natural gas west of the city. City Commissioner Warren Chapman and Joe Downey helped arrange a lease with John Dunn for the project to W. L. Pearson of the newly formed Saxet Company. The well came in with a roar in early November of 1922, but it blew out and was not brought back as a producer until April 13, 1923. A week later, on April 20, the city of Corpus Christi began using natural gas for heating and power.

By mid-1926, 589 miles of pipelines tied the Southwest Texas area to Houston, which had been importing gas from northern Louisiana. Promoters included the Houston Gulf Gas Company, formed by Odie Seagraves and William L. Moody III, who had options on gas wells in Refugio County, and the Houston Oil Company, "which found gas while looking for oil in the Refugio area and also built a line to Houston."

In 1927 the Saxet Company made a contract with the Houston Gulf Gas Company to supply the City of Houston with 25 million cubic feet of gas daily. In addition to his geological work, Dr. W. Armstrong Price helped promote financing for the project, but he and Pearson, his associate, found that getting backers was a difficult task.

"Mr. Pearson had to go to Chicago because New York bankers wouldn't touch the program. Encouraged by gas production in Saxet Field and in Refugio and Edna, the Moody Seagraves group and the Houston Oil Company agreed to build the pipeline but required a daily production of 25 MCF [thousand cubic feet] by year's end.

"You've never seen busier people than we were in the winter of 1927," Price said, "trying to get production up to meet the contract's quota—5000 cubic feet a day for Corpus Christi and 25,000 for Houston by January 1, 1928.

"The numerous wells drilled by Pearson in 1927 and later by Republic Natural Gas Company and others opened up billions of cubic feet of

Oilmen could have winged it, too

In bad weather, when dirt roads were impassible out around the Cotulla area in the early thirties, messages literally flew between ranches, even without electricity and telephones.

A pioneer oilman recalled that the ladies on the ranches kept pigeons and sent messages by them. When they were in town, they swapped pigeons and then attached messages to them when they needed to communicate.

It would have been a valuable service to an oil rig on the ranch when it badly needed to contact someone in town.

"The birds could fly the thirty-five miles in half an hour," the oilman said. "When they came back to their home loft, it sounded a bell when it landed, alerting the owner."

The Bell System had a competitor it didn't even know about.

—Tubby Weaver

(Below) Dunn No. 2 opened Saxet Field to gas production in 1922. (Bottom) Eight years later the discovery of oil at the Saxet Oil & Gas Company No. 6 Dunn five miles west of Corpus Christi opened oil production in the field and electrified the city.

natural gas of the highest grade in what came to be known as Saxet Field," he said. "This was the first development of gas in the county. . . ."

In another year New Braunfels, Seguin, San Marcos, and Austin were connected to South Texas gas.

These gas discoveries also made the area look promising for significant oil production. "By the end of 1927, activity in the whole region had heated up, and leases, once bought for fifty cents an acre, were now trading for more than ten times that amount. . . . ."[40]

In 1929 a flat tire led to H. M. Grimm's discovery of the Agua Dulce Gas Field. Grimm, who had been a clerk in a dry goods store, started roughnecking in 1908 after reading in a Chicago newspaper about the opening of the Glenpool in Oklahoma.

He came to South Texas in the early 1920s. Working out of Laredo, he drilled wells from Rio Grande City in Starr County through Jim Hogg, Zapata, and Webb counties. He had worn out two pairs of shoes doing surface geology in western Nueces County and was driving on a back road, crossing Pintas Creek near Agua Dulce, when a tire went out.

"The creek was up from heavy rains, and I could see gas bubbling up through the water, kicking the water a foot high at time," he told *Caller-Times* oil reporter George Hogan years later.

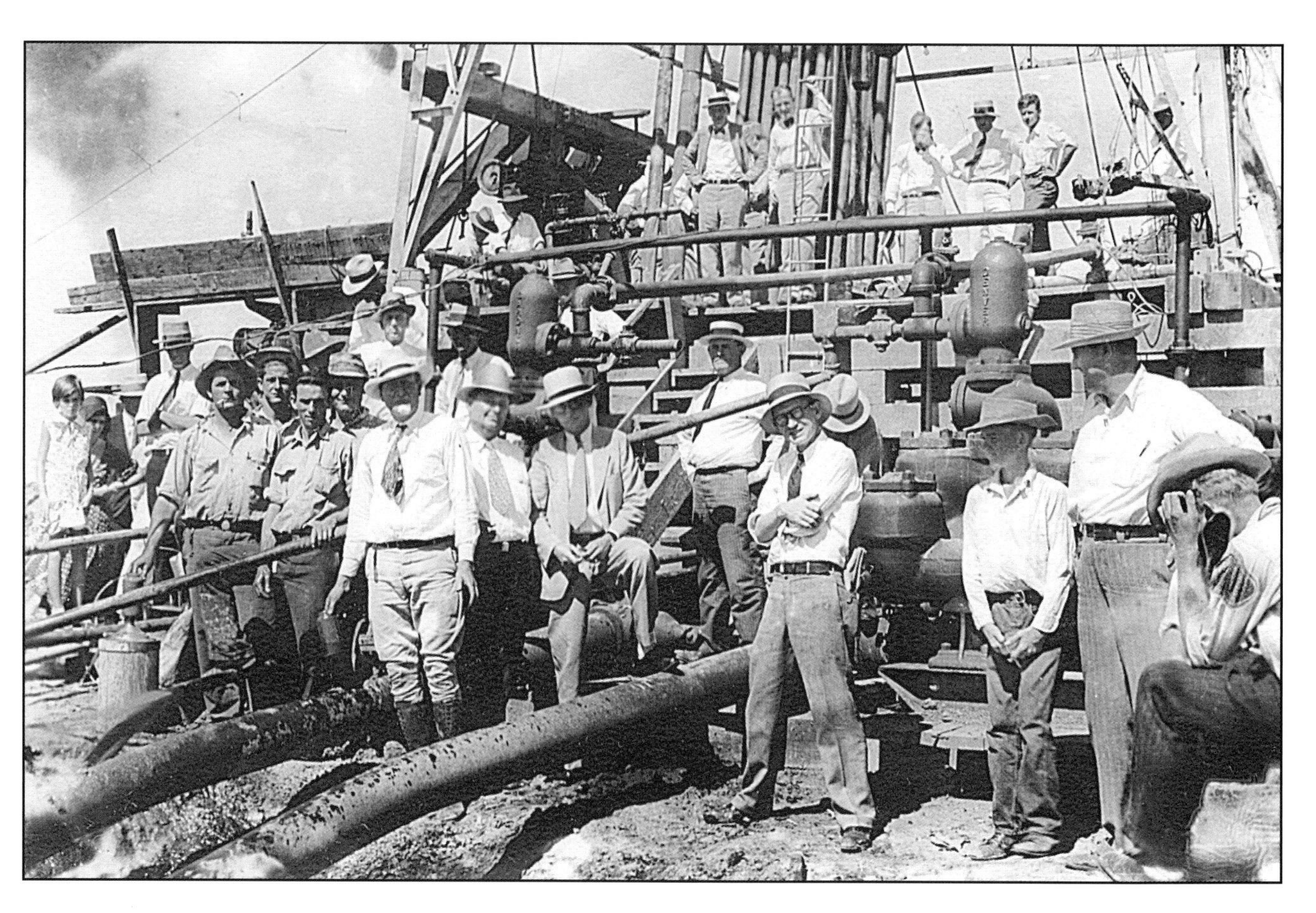

W. H. "Harry" Wallace allowed Grimm to take over a 5000-acre lease for an interest in production. Apparently a self-educated geologist, Grimm seemed to excel as a promoter. He organized a stock company, keeping a quarter share for himself and a quarter for his partner, John Morris, who provided the rig for the discovery. Joe Mireur, Maston Nixon, Bill Marks, and Wallace bought the remaining half. They formed the Agua Dulce Oil and Gas Company and drilled seven wells. Grimm would later be an inventor, owner of a refinery, and a wildcatter.

The Grimm and Morris No. 1 Garrett well was completed in 1928 for an original production of 75 MMcf of wet gas per day from a depth of 2023 feet. Houston Oil Company No. 1 Comstock came in several months later as a significant oil discovery.[41]

It was just ten days short of a year later, on August 16, 1930, when newsboys were prowling the streets with an Extra edition of the *Corpus Christi Times* and screaming, "Oil Is Discovered! Oil Is Discovered!"

Every vehicle in town, loaded with spectators, rushed out the shell road to the Saxet Oil & Gas Company No. 6 on the Dunn Lease, five miles west of town and south of Nueces Bay.

Ever since Spindletop three decades earlier residents had dreamed of having their own gusher—their own oil boom. Spectacular blowouts at White Point and other area locations had drawn national attention, but there had been no oil discoveries. Now, as the well flowed, the crowd cheered. It was like they were on holiday, attending a sporting event. They had dreams that Corpus Christi was about to become a boom town.

But it didn't happen, at least not as they envisioned. Just a month later Dad Joiner opened one of the greatest oil pools in history in East Texas, and the entire Texas oil crowd went there to seek their fortunes. Later the Saxet Field and others to the west would take on major significance for the new Port of Corpus Christi and all of South Texas. But that was yet to happen as the crowd celebrated the excitement of the city's first oil strike.

That strike was located by Dr. W. Armstrong Price and drilled by W. L. Pearson, with his recently organized Saxet Oil & Gas Company. Dr. Price had come to Corpus Christi in 1927 as, he said, the only geologist to set up office in town.

"Others," he said, "were more or less passing through."

That could not be said of Price. He remained in Corpus Christi, engaged in research and writing technical papers up to the time of his death in November 1987, during his 98th year. A native of Richmond, Va., he was a Ph. D. graduate of Johns Hopkins University.

In 1922-23, before coming to Texas, he was vice president and geologist with the Appalachian Oil and Gas Company in West Virginia and aided in discoveries that doubled that state's known reserves of bituminous coal. He was thirty-eight when he came to Corpus Christi via Houston in 1927, "lured by the land and the oil under it."

Discoveries finally brought boom to Corpus Christi in mid-1930s

Oilmen across the country were watching development in Corpus Christi for years, but it was not until 1934-35 that they really took notice.

The city was a late bloomer in the oil-boom business, partly because of the oceans of oil so wastefully produced in the famous East Texas Field in the early thirties, right after W. L. Pearson brought in the Saxet Oil & Gas Company's No. 6 John Dunn.

But by the mid-30s, it was Corpus Christi's time.

First came the Southern Mineral Corporation discovery, opening the new Saxet Field. The SOMICO No. 1 Ocker was completed for 800 barrels a day in 1934. In April 1935 the field was expanded nearly to Clarkwood with the Texan Royalty Company No. 1 Kocurek, a 600-bbl. discovery, followed by the SOMICO No. 10-A Isensee, 735 barrels per day. On November 2, 1935, J. K. Culton brought in his No. 1 H.B. Baldwin, opening the Corpus Christi Field.

This grabbed attention across the state. When it was learned this was in an area divided into many small tracts, the race was on. The town filled with lease hounds, operators, promoters, drilling contractors, oil-field supply men, workers, and others looking for action. Housing was at a premium, and the town was booming. Those who came too late to get in on the Corpus Christi play spread out and in rapid order found great promise in the Clara Driscoll Field.

Supply and equipment houses flourished as companies scrambled for leases. Clarkwood, Agua Dulce, Banquete, Luby, and Flour Bluff fields put Corpus Christi on the map as a major producing area.

Vonda Davis, "Corpus Christi An Oil Center," *Corpus Christi Caller-Times*, October 16, 1966

The White Point and Saxet fields were both plagued by high pressures and blowouts. (Top) A Saxet well blows out. (Middle) The rig collapses. (Right) Fire envelops the wreckage. (Opposite) l to r) Dr. W. Armstrong Price, Michel T. Halbouty, and Corpus Christi Geological Society President Don Boyd in 1967.

Many in his profession in those days had discounted the oil and gas potential of the Texas coast as being "too young geologically" to be a likely area for exploitation. But Dr. Price knew better. Some of his fellow geologists called him the best surface geologist they ever saw.

"Armstrong Price could detect the most subtle part of uplifts that you could imagine," said O. G. McClain. "I marveled at his ability to interpret surface geology from as little evidence as he had to go on. It blows my mind that he could do it."

"The flatlands and the shoreline here were new worlds for me," Price said in a 1976 interview. "I had been primarily a rock geologist. One of the things that interested me most was that the low coast of sand and clay really had not been studied by anybody. It was something fresh and new. It afforded many problems that had to be solved. There was a whole new geology before me. There were very few ways then to find the oil we knew was under the ground."

In Texas he became involved with White Point exploration. Hydrogen sulfide leaking from a spring or water hole on the west side of the White Point peninsula had led to exploration for oil, as it was believed there was a salt dome formation in the field. Between 1913 and 1926 thirty wells had been drilled there for oil. All were gas wells, and extremely high pressures had destroyed several of them.

Dr. Price worked on White Point Field, helping quell oil fires. He found that drilling there failed to produce evidence of a salt dome, but he felt that there should be oil on the south side of Nueces Bay, opposite the White Point gas production. And with the Saxet Oil Company No. 6 Dunn, he finally staked out an oil discovery.

Dr. Price maintained that the nearby Pioneer Oil Company No. 1 Meaney rightly deserved to be called the discovery well because it had tested oil. He claimed that the Pioneer well "initiated modern industrial

Price found honor in defeat

As a young man Armstrong Price was an athlete, playing basketball and baseball at Davidson College in North Carolina and lacross and polo at Johns Hopkins. In a 1976 interview he said, "You might not believe it, but I was tall as a basketball player. I was six feet two inches tall. These seven-foot players of today would have stepped on me."

Since they had no basketball at Johns Hopkins, he played on the lacrosse team. To him it was an honor to have been soundly defeated by the Carlisle Indians and their great Jim Thorpe.

"He was the greatest athlete I ever saw," Dr. Price said. "The Carlisle Indians were perfectly clean and gentlemanly unless you got rough. Then look out."

Later when he was working for the U.S. Geological Survey and teaching geology at the University of West Virginia, he coached the junior varsity basketball team.

"I've never been a fan," he said. "I do not enjoy watching sports. I only liked to participate in them."

Bill Walraven, "At 96, area's first geologist still hard at his studies," *Corpus Christi Caller-Times*, April 11, 1985.

Extensive drilling took place in Saxet Field west of Corpus Christi.

development" in the Corpus Christi area. However, it had blown out before completion and had to be abandoned, so the Dunn well is on record as the first producing oil well in Nueces County.

Price located other discoveries in Baldwin and West Saxet fields in Nueces County and Weslaco and Mercedes fields in Hidalgo County.

In 1936 he headed the Texas Academy of Science. He was active in the South Texas Geological Society, which met in San Antonio. In 1942, with travel curtailed by World War II, he helped organize the Corpus Christi Geological Society and served as its first president.

In the 1940s his interest shifted to marine biology and oceanography. Internationally renowned as a marine geologist, he founded and chaired the department of geological oceanography at Texas A&M University and taught the course for five years.

"When the oil business later got worse and worse, other opportunities came along, and I changed what I was doing," he said. "My avocation became my profession. I had attractive opportunities to leave, but I remained."

He became an activist in dune protection for storm barriers and an expert on the effects on coastal regions of waves, winds, tides, hurricanes, currents, water organisms, vegetation, and much more. Long before it was fashionable, his study of tides and ocean levels caused him to be concerned about global warming.

"I started out studying one thing and was able to branch out laterally into other related areas. I also found true friends, a delightful climate, and a peacefulness here," he said. "I never regretted staying. You become married to your research, and you have no set hours. Very few men can plan a career and follow it through a lifetime. But I think I have."

In 1934 Dr. Price was instrumental in attracting the first industrial plant to Corpus Christi's new ship channel. At that time cotton was the prin-

Many contributed to area prosperity

Although Maston Nixon was a master builder who played a major role in developing the modern city of Corpus Christi and also a representative of the oil industry, his first oil deal was one he always regretted.

Born in Luling in 1896, he picked cotton and drove a wagon for spending money, which he said he "never spent." He moved to Petronilla, southwest of Corpus Christi, where his family had a farm. During a drought he sold his cattle and, at the urging of a banker friend, bought a 10-percent interest in the Dixie Oil and Refining Company in San Antonio. During World War I he was the youngest artillery captain in the Army. Two men came to see him while he was drilling recruits and asked if would sell his interest in Dixie. They offered him $46,500 cash or shares of stock. He chose the cash.

"Thereby I made one of the major mistakes of my life," he said. "The stocks were Humble Oil and Refining Company."[1]

In 1919 Nixon was raising cotton on the Petronilla farm. On Saturday, September 13, he drove his new Dodge car—with just 800 miles on it—to Corpus Christi, where he went swimming with friends from London before spending the night in the Nueces Hotel. He left early Sunday morning on the San Antonio and Aransas Pass train, as he had business Monday morning in Gonzales. By the time of his business appointment, his car, parked in a Water Street garage, was destroyed and his friends were dead, victims of the disastrous 1919 storm that struck the city that Sunday night.

Nixon returned to ranching but found a real talent in promoting. As head of the Robstown Chamber of Commerce, he promoted the sale of more than 100,000 acres of farmland to Midwest farmers. In Corpus Christi he teamed with H. L. Kokernot of San Antonio to build the first "skyscraper" on the Bluff, a twelve-story office building named the Nixon Building.

Then came the Plaza Hotel, with an assist from Robert Driscoll and the Plaza Hotel group in San Antonio. He helped Clara Driscoll in

Two unidentified survivors stand in flooded Peoples Street after the 1919 hurricane.

Maston Nixon's promotional skills led him to the oil business and eventually to the office of chairman of the Chamber of Commerce Industrial Committee, a position in which he was instrumental in attracting industry.

constructing the Driscoll Hotel. The buildings completely altered the Corpus Christi skyline.

Nixon's promotional skills led him to the oil business and eventually to the office of chairman of the Chamber of Commerce Industrial Committee, a position in which he was instrumental in attracting industry. He heard that Eastern industrialists were interested in locating plants near a source of fuel and access to water transport, where they could find, as one investment banker put it, "cheap horsepower at tidewater on the right side of the Panama Canal."

One interested company was American Cyanamid, which was considering building a plant for its Southern Alkali subsidiary. American Cyanamid insisted on a guarantee of a twenty-five-year supply of natural gas. Dr. Armstrong Price helped prepare a report that showed 75 billion cubic feet of gas reserves in the White Point Field and 100 billion cubic feet in Saxet.[2]

A financing problem from a short money supply during the Depression was solved when the Columbia Division of Pittsburgh Plate Glass Company joined American Cyanamid and provided funds to complete the project. The management made a contract with H. K. Ferguson Company to supply key personnel for the construction of the plant at an attractive fixed fee. Mike Gallagher and Ben Garza helped the Ferguson representative find construction laborers from the local Hispanic population.

"They did an excellent job. In fact, Mitchell [the Ferguson supervisor] said they were the best construction workers he had ever supervised," Nixon said.

Southern Alkali, later Columbia Southern, a subsidiary of the two companies, began operations in 1934. Nixon formed a sister company called Southern Minerals Company, also a subsidiary of PPG, to provide natural gas for the plant. The company was reluctant to enter the oil business, since it was interested only in gas but relented at a New York meeting after Nixon offered to put his own money in the venture.

"None of those present at the meeting could foresee that the first well we would drill in the Saxet Field, Ocker No. 1, would come in producing 26-gravity oil from a newly discovered 4,800-foot oil sand," Nixon said.[4]

Not only did it come in, it came in as a gusher, spewing high above the derrick. The successful search for gas reserves led to the development of Southern Minerals as an oil and gas producing company, which bought leases covering several thousand acres and including two gas wells from Gulf Coast Corporation and 1250 acres in Saxet Field from Houston Oil Company. Southern Minerals' success in turn spawned the Southern Pipe Line Corporation, organized to transport crude oil.

The Saxet discovery also strengthened a theory of Nixon's that gas and oil are associated and that oil, heavier than gas, could be found down dip from the gas.

Houston contractor George Echols drilled this test well in the 1930s on the grounds of the Corpus Christi Country Club, located at that time on the west side of town. (Left) Plant and offices of the Southern Alkali Corporation

Ports grew rapidly in importance

Corpus Christi: "This port, though only four years old as a deep water shipping point, has witnessed phenomenal development of trade. The controlling depth is 22 to 28 feet. Exports are principally cotton, sulphur, lead and ores and cottonseed products. As a cotton shipping port it ranks next to Galveston and Houston. Its imports consist chiefly of crude oil and miscellaneous manufactures. Extensive improvements have been made in the past year. Coastwise shipments are cotton, sulphur and miscellaneous manufactures and raw products, while principal domestic receipts are canned goods, bags and bagging, iron and steel products, and miscellaneous manufactures."

Port Aransas: "The port is situated on Mustang and Harbor Islands at the mouth of the channel from Corpus Christi to the Gulf. Termini of several pipelines from West Texas oil fields are located here, and this port is said to lead the world in the shipment of crude oil. It was reported in 1930 that the port would be improved and developed for general commerce through expenditures of $10 million or more in private capital."

The Texas Almanac, 1931, Page 101.

During this period Maston Nixon operated out of an office in the north parlor of Centennial House, the oldest Corpus Christi structure in existence at the time and the only historically significant structure in the city still standing on its original site at the end of the twentieth century. In 1934, the year that Southern Alkali opened as the first industrial plant in Corpus Christi, the city had two refineries, two or three pipelines, and three or four producing fields. The port had only one oil dock, and there were no oil field service industries.

Thanks to Southern Minerals, or SOMICO as it was called, the whole picture had changed by 1937. By 1936 Nueces County had become the leading oil producer in South Texas. By 1937 there were 118 producing fields and a total of 6,090 wells in the area, with a daily production of 189,390 barrels of oil. The port had expanded to three oil docks, Avery Point had a facility with a seven-cluster dock, there were eight refineries with two more under construction, there were two absorption plants to use the natural gas being flared, PPG had built a chlorine plant, and two carbon black plants were being built to use the tail gas from the Southern Minerals absorption plant.

By 1938 there were twenty-nine pipelines, twelve storage terminals at the port, and shipments of petroleum products had risen to a total of 38,389,533 barrels for the year.

By 1951 SOMICO had 292 oil wells and 51 gas wells in the Gulf Coast area. By 1962 the firm operated in Louisiana, New Mexico, and West Texas, as well as the Gulf Coast.

Much of the credit for SOMICO's success must accrue to geologist O. G. McClain, who led its exploration from 1938 through World War II until 1945, when he resigned and became a consultant. Some of his major discoveries were in Richard King Field, Agua Dulce, Pettus, Falls City, and the Swan Lake Field near Port Lavaca. A group of farmers pooled 1,616 acres forty miles southwest of Corpus Christi in Kleberg County, and McClain found the first oil in Stratton Field. This led to oil development on the King Ranch, extending Stratton Field under ranch property.

Maston Nixon served as an area vice president of the Independent Petroleum Association of America, a director of the American Petroleum Institute, president of the Texas Mid-Continent Oil and Gas Association, and a member of other professional groups.

Always known as a hard worker, he was in bed by 8:30 p.m. and accustomed to getting up at 4:15 in the morning. He had the morning paper, the *Corpus Christi Caller*, delivered at 4:30 and read it thoroughly, allowing time to get over any annoyances at what was going on in the world. He would then have one drink of Scotch whiskey and go back to bed at 5:30 for the most restful sleep of the night. He was up again, had breakfast at 7:30, and was at work at 8.[5]

“That’s right,” McClain recalled. “He was always the first man in the field.”[6]

Nixon retired in 1965 and died in 1966.

The industrial plants he enticed to the city because of its deep-water location and bountiful supply of natural gas brought a bonanza of jobs, causing a surge in population. The ripple effect was the birth of the modern city of Corpus Christi.

In addition to Nixon and Dr. W. Armstrong Price, another who deserves a good deal of credit for bringing industry and its resulting prosperity to the area is Guy Warren, whose leadership led to the formation of the Chamber of Commerce Area Development Committee.

Warren was a World War I veteran who set out after his discharge to seek his fortune in the oil industry. Wearing a scruffy civilian coat over his Army uniform, he looked over the milling crowds of workers, promoters, prostitutes, and wagons carrying pipe and oil field equipment. World War I was over. Just mustered out of the Army Air Corps at Call Field near Wichita Falls, he had heard there was a chance to make a million dollars at Burkburnett. The young lieutenant arrived there in January 1919 and was about to receive his baptism in the oil business in Northwest Field.

He had studied mining engineering for three years at the Upper Iowa University in Fayette, Iowa, and had his first taste of Texas when he received flight training in Austin.

After leaving Burkburnett he gained experience in the oil fields of Corsicana and Mexia. He became general superintendent for the National Petroleum Company and later was elevated to vice president. He became an independent operator in 1928 in Michigan, Louisiana, Illinois, and Texas.

It was in Corsicana in 1929 that he found a girl who ended his flying days. Miss Gazzie Suttle told him she would never marry anyone

The oil dock at the Port of Corpus Christi shortly after its completion

This shot illustrates the diversity of activity in Corpus Christi in 1936. The Corpus Christi Refining Corporation is under construction in the background. A producer had recently been completed on refinery property in the adjacent Corpus Christi Field, and storage at the plant indicates a terminal.

"reckless enough to go buzzing around in one of those silly flying machines." He gave up flying. They were married 55 years.

In 1934 Warren, R. S. Bond and D.W. Forbes formed Renwar Oil Company—Warren with the syllables reversed—in Dallas. The company entered the South Texas area in 1935, drilling a number of wells in the Clarkwood townsite, and opened offices in Corpus Christi in 1936. With three geologists, they eventually had production in Nueces, Refugio, San Patricio, Jim Wells, Brooks, and Kleberg counties and other points.

Warren, who remained in the Army Reserve, was called to active duty in World War II. He received a number of medals, including the Bronze Star and the Legion of Merit, helped set up the Allied Military Government in Italy, and later served as military head of that government for the southern third of the country.

Back home, he served as president of the Texas Independent Producers and Royalty Owners Association and director of the Texas Mid-Continent Oil and Gas Association. The Roughneck's Club, a group organized to recognize pioneers in the business, named him Chief Roughneck for 1958. He served as president of the Chamber of Commerce, Rotary Club, and Lower Nueces River Water Supply District board. As Chamber president, he took the lead in interesting other community leaders in a plan of action to solve Corpus Christi's growth problems, a movement that led to the formation of the CofC Area Development Committee.[7]

"Oil has added materially to the tax income of the city, county, and state," Warren said in 1956. Local schools had benefitted, too, since the oil and gas industry paid approximately 52 percent of all the state's school taxes.

At that time an estimated one out of twenty persons in Nueces County was employed in some phase of the oil and gas business, ranging from office employees to oil field workers and service station attendants. Local payrolls of the oil and gas industry were estimated well in excess of $25 million a year.

"Farming has been enhanced by royalties, rentals, and bonuses paid for leases," Warren said, "enabling the farmer to have capital which he can turn into more and better agricultural production."

According to Warren, oilmen had an advantageous position because they could cope with the community's needs and problems without fear of economic repercussion.

"The oilman has little to gain or lose from the economic standpoint if he takes a position of leadership," Warren said. "This atmosphere of freedom isn't enjoyed by all."[8] Guy Warren died in 1984 at age 88.

Another local oilman who left his mark on the Corpus Christi skyline was Sam E. Wilson, Jr. Ask anyone from his era who was the biggest character they remember, and most will include Wilson as one of their candidates.

Many stories are told about him. He started at the bottom—the very bottom—digging ditches.[9] He was to become one of the leading drilling contractors, operators, and property owners in South Texas. He never lost his toughness. It was said two big, muscular roughnecks accosted him on the sidewalk outside Nixon Café on the Bluff in Corpus Christi. They had been fired from one of his rigs and claimed they had received no pay.

"If you don't pay up, we're going to take it out on your hide," one of them threatened.

"I'm not paying you a dime. And if you want my hide, you just come and get it," he replied.

Sam was a bit overweight and seemed no match for the young toughs, but he was surprisingly agile and quick. He knocked the first one

This shows the types of Christmas trees used and the proximity of derricks drilled on small leases in the Corpus Christi Field. In the center front is a Christmas tree with a gauge on top. Relief valves on the side of the Christmas tree are attached to flow lines.

down and beat the other to a pulp before they decided they really didn't want that money.

Then there was the story of the $2 million check. Sam had sold the production in one of the fields he had opened and had the check to prove it. Of course, the banks in Corpus Christi could cash the check, but Sam got the most out of it. At the slightest provocation he would show that $2 million check. Practically everyone in the Plaza Hotel lobby and the Nixon Cafe had seen it.

A couple of his friends decided to cool his enthusiasm. They called his wife, Ada, and told her about the $2 million. Now Ada, who founded the Ada Wilson Crippled Children's Hospital, had a domineering way. She was one person Sam wouldn't challenge. She took away his bragging rights and his check. His friends kept out of his way for a considerable time.[10]

Sam was born in Jackson, Tennessee, in 1898. When World War I started, he was attending Columbia Military Academy. After joining the Army, he was assigned to the newly formed air service and sent to Princeton University for training. He transferred to the Royal Air Force in England, where one of his instructors was Capt. R. A. Brown, who later shot down the famous German ace, Baron Von Richthofen. Wilson spent eleven months in Europe.

When World War I started, Sam E. Wilson, Jr., was attending Columbia Military Academy. After joining the Army, he was assigned to the newly formed air service and sent to Princeton University for training. He transferred to the Royal Air Force in England, where one of his instructors was Capt. R. A. Brown, who later shot down the famous German ace, Baron Von Richthofen.

Released from the service, he went to Burkburnett, which was in the midst of an oil boom. He first dug ditches, then hired out as a roughneck for two years.

In 1921 he married Ada Laverne Rogers of Memphis, Tennessee.

When oil was discovered in El Dorado, Arkansas, leasing was in full swing. Sam declared himself a leasing broker and made enough money to lease a rig. His first effort was a success as a field extender. Other tries were all failures. He formed a company in New Orleans and then came to Texas half a million dollars in debt.

Then his luck changed. He organized Bay-Tex Oil Corporation in 1939 and brought in wells in Agua Dulce, Refugio, and Aransas Pass fields. During World II Sam received a windfall from the government to drill gas wells to supply a carbon black plant north of Aransas Pass with the natural gas it needed for use in the production of synthetic rubber for the war effort.

Sam Wilson was president of Wil-Tex Oil Corporation, the SEW Oil Corporation, and the Sam E. Wilson, Jr., Oil Corporation. He also had extensive real estate holdings. He acquired the Nixon Office Building on the east corner of Leopard Street on the Corpus Christi Bluff and changed its name to the Wilson Building.

H. L. Kokernot, who had financed the 12-story office building in 1927 for Maston Nixon and had also assisted Nixon in building the Plaza Hotel, later the White Plaza, had deeded it to the Baptist Foundation. Many cotton broker offices were located there. However, the Baptists were so upset over the fact that the Nixon Café served beer that Kokernot ended up selling the building to Wilson.

Sam wanted to build an office tower on the west end of the block to serve the dozens of oil company offices that were moving to town. Humble Oil and Refining, with development of the King Ranch leases under way, encouraged him by offering to lease ten floors of the building for a minimum of ten years. This agreement allowed him to borrow the rest of the money for the 17-floor Wilson Tower.

Tubby Weaver claimed credit for extending the height several more stories. "I told Sam he was chicken for allowing Clara Driscoll to build her hotel taller than his building," he said. "Sam thought about it and he added those little buildings on top to top the hotel. He put a big lighted 'W" that revolved on the very top. He stopped it from rotating when he found out what it was costing him. I think it fell off during one of the hurricanes."[11]

Sam had a lot of production on Mustang Island at one time and eventually owned much of the island. Much of that land later became the Mustang Island State Park.

About the same time that Maston Nixon and Dr. W. Armstrong Price were attracting Southern Alkali to the city and Guy Warren was forming Renwar, another event took place in South Texas—in Nueces County, to be exact—that would also contribute in large measure to the prosperity of the area. It was the beginning of the cycling of natural gas. The origins of the process went back to the time of World War I, when the waste of reservoir energy left about 80 percent of the oil in the ground and massive production caused oil prices to drop as low as five cents a barrel in some isolated cases. In the 1930s the waste brought about the Texas statutory proration laws, which decreed that gas produced with oil could not be flared.

According to *The History of Nueces County*, "Objections to proration arose because gasoline is manufactured from the casinghead gas produced with crude oil. Gas wells at Agua Dulce produced condensate, which is also the raw material for gasoline."

The Clymore Production Company constructed a gasoline plant after completing seven wells on the England farm, but forty-five million feet of gas was flared daily, a complete waste. The Texas Railroad Commission, the regulatory body for the petroleum industry, ordered the wells shut in, but the company claimed that they were gas wells and therefore not subject to the proration laws. After a lengthy battle in federal court, the RRC won and the wells were shut in.

However, in 1934 Ed V. Foran had noticed vapor condense on his car's windshield and decided that condensate could be extracted from gas by pressure and temperature changes. He moved from Big Lake in Reagan County to Corpus Christi, where the firm of Parker, Foran, and Knode was formed to construct a plant to cycle gas at Agua Dulce. Plant

Sam and Ada Wilson

equipment was supplied by the National Tank Company, and J. P. Walker, president of that company, agreed to finance the project. After signing the quarter-of-a-million dollar contract at Galveston, Walker laid down the pen and said, "Now, just where is Agua Dulce?"

They purchased a lease on 640 acres of the England farm and drilled two wells, one to produce and the other to inject the gas back into the sand. The plant began operations in 1937, and the huge cycling business was born.[12]

A minor boom followed but slowed as the Depression began to be felt and the great East Texas Field siphoned oil interests out of Southwest Texas for a period. It would be years before pipelines would solve the problem of natural gas marketing. When Atlantic Refining Company opened the Mustang Island Field with a gasser, No. 1 State Tract 436, on April 1, 1941, it seemed an April Fool's joke. Atlantic shut the well in and continued the search for oil.

The Clymore Production Company drilled seven wells on the England farm. This is a Clymore well on the Ollie Purl Lease.

The Railroad Commission, whose order had precipitated the lawsuit over the wells, survived years of struggle, confrontations, and violence and saved the Texas oil industry from itself, at the same time setting standards for regulatory agencies everywhere. Many prominent Texas officials have been associated with the RRC, but none with more brilliance than Ernest O. Thompson, who created the most efficient energy-regulatory agency in the world.

In a wave of reform, the commission had been founded to "deal with inequities that grew up with the railroad industry within the state" after a constitutional amendment was passed in 1890 at the urging of Gov. James Stephen Hogg. Although a law regulating the drilling, casing, plugging, and abandoning of oil and gas wells was passed in 1899 as a result of wasteful practices in the Corsicana Field, it was not until 1917 that the business of the RRC included energy in the regulation of movement of oil and natural gas through pipelines as common carriers.

Then came the fabulous East Texas Field, which was discovered through an unlikely alliance between a 70-year-old small-time wildcatter who was a little more than shady in his financial dealings, and a colorful, 300-pound plus, sombrero-wearing friend.

Columbus Joiner, convinced there was oil in East Texas, scrimped, begged, and borrowed money and futilely drilled wells for three years until he opened his Daisy Bradford 3 on October 30, 1930. He was ably assisted by a rotund, 300-pound friend, Joseph Idlebert Durham, who knew little geology but assumed the name of Dr. A. D. Lloyd, a nationally know geologist, and exhibited charts and maps to entice investors.

Joiner created the biggest oil boom ever. He had discovered the Woodbine Sand, which encompassed an area forty miles long and ten miles wide. James A. Clark, author of Thompson's biography, *Three Stars for the Colonel*, wrote: "It was Oil Creek and Spindletop and El Dorado, Greater Seminole and Burkburnett, Glenpool and Signal Hill, Ranger and Desdemona, all in one great fairway." Joiner, now called "Dad," had a gusher on his hands, but he was over his head in debt and even deeper in legal problems. One lease had been sold eleven times.

Enter Haroldson Lafayette Hunt, Jr., a gambling-hall owner who became an oilman when the Ku Klux Klan threatened to burn his saloon in an anti-vice campaign. Hunt offered to pay the debts and settle the legal claims. After considerable wrangling, Joiner let Hunt have the 5,000-acre largest and richest oil field in Texas for $\$1^1/_3$ million, most of which would be paid from production. As late as the 1940s, the lease was paying Hunt a million dollars a week. He died in 1974, leaving fifteen children from three wives to contest his considerable estate.

The wild boom was a godsend to poverty-stricken dirt farmers crushed by the Depression. Tyler, Henderson, Longview, Kilgore, and Gladewater suddenly became cities. Dozens of wells were completed daily, and the field was producing a million barrels a day.

All over the country the oil business shut down as the glut caused oil prices to plummet. The first order to limit production in East Texas was issued in 1931 and met with a fire storm of criticism, lawsuits, and injunctions. Operators considered their well sites private property— no business of the government—and a federal court promptly nullified the order. It would be April of 1933 before the commission issued an order that would be upheld by the federal courts.

Because their experts did not believe geology indicated a favorable structure there, major oil companies had moved out of the area before the discovery, leaving the field open to hundreds of small operators. Many of these wildcatters were poor farmers, new to the game. They had never heard of rules or the Railroad Commission and had drilled at a breakneck pace for six months. Many were armed with pistols, rifles, and shotguns. Some built barbed wire fences to guard well sites, and one built a concrete bunker.

The Railroad Commission was unable to curb the growing disaster. By June 1, 1932, there were 1,000 wells in the field. By 1933 there were 12,000. Oil prices dropped to as low as ten cents a barrel. The Legislature seemed too confused to act. Through the Railroad Commission Gov. Ross Sterling ordered 1,200 National Guardsmen to shut down the field. The federal court then ruled that the National Guard had no authority to act on orders from the RRC.

Sterling then re-issued the orders from his office personally. His basis for action came from a law Gov. W. P. Hobby had signed in 1919 that gave the state more control over its natural resources, improving on an 1898 law passed to curb waste. Even so, independent operators challenged the governor's right to maintain martial law in the absence of riots and insurrection, and a three-judge federal court agreed with them.

Joiner: Discovery not just a lucky strike

C.M. "Dad" Joiner, finder of the famous East Texas Field in 1930, got his hackles up when writers said he made the discovery because of a wild guess or a stroke of luck.

He wrote he did not choose the location as "a haphazard" guess or a "hunch."

Though his funding was limited, he "had the wholehearted cooperation of the citizens of Overton and the farmers of the immediate territory and several thousand acres were placed under lease."

Only one geologist agreed with his research, he said. All the others failed to agree with his findings. Choosing the spot "followed many months of painstaking and weary work, checking surface formations, the tracing of fault lines and as much other scientific research as was possible under the strained financial circumstances under which I labored at the time."

Two tries failed and the third was sorely endangered, and a paleontologist checking cuttings from the formation disagreed with his findings, which made the raising of funds for the completion difficult, Joiner wrote. But the Woodbine Sand was tapped and "the assembled crowd went wild with enthusiasm and the flood gates of potential wealth were opened and East Texas came into its own."

C. M. "Dad" Joiner, "Discovery of the East Texas Oil Fields," *The Historical Encyclopedia of Texas*, 44.

Thompson was hero of World War I

Ernest O. Thompson had been self-motivated from early childhood in Amarillo, getting up at 4 in the morning to deliver newspapers, then establishing a newsstand and working at various jobs. In high school he was a successful businessman, selling Overland automobiles.

He had a taste of the military life at Virginia Military Institute and learned shorthand and typing at a business school before enrolling at the University of Texas. While attending law school, he worked for Judge Reuben R. Gaines, a former Texas Supreme Court justice, as chauffeur and secretary.

Thompson traveled with the judge and his wife to Europe in May 1914. They were in Austria when Franz Ferdinand, the heir apparent to the Austrian throne, and his wife were assassinated in Sarajevo. They witnessed the funeral procession in Vienna and were almost trapped by World War I. Their car was confiscated in Switzerland, and they barely made it to Paris, where they saw the first French battle casualties arrive. The trip was so stressful that the elderly judge died a month after returning home.

When the United States declared war, the entire law class volunteered for officer's training camp. Their diplomas were delivered to them. Because of his military training at VMI and some service in the National Guard, Thompson became a captain. At the battle of the Argonne he was promoted to major. He was so successful with a revolutionary use of massed machine guns in the final battle of the war, at Meuse-Argonne, that a captured German officer thought the Americans were using a new secret weapon. The action won him another promotion, to lieutenant colonel, the youngest in the American Expeditionary Forces. Back home he opened a law practice and was elected mayor of Amarillo.

James A. Clark, *Three Stars for the Colonel.*

Although it was clearly unconstitutional, Sterling resorted to delaying tactics to keep the troops in the field while the case was appealed to the Supreme Court. They were finally withdrawn in December 1932 after the court ruled against the state.

In the meantime former Gov. Pat Neff, who had been trying to cope with the East Texas uprising, resigned from the commission to become president of Baylor University. Sterling, a former president of Humble Oil and Refining Company, looked for someone capable of handling the volatile situation. He chose Ernest O. Thompson, the mayor of Amarillo.

Sterling chose him because, as Amarillo mayor, Thompson forced both the gas company and telephone company to lower rates. The gas company, which had been paying five cents a thousand cubic feet for gas and selling it to the city for 45 cents, refused to discuss a compromise. Thompson activated an abandoned pipeline to a nearby gas field and brought gas to the city. The move cut electric charges by more than half. He urged citizens to leave their phones off the hook and threatened to cut down telephone poles on city property. Phone charges were slashed. He turned the city dump into a park, streamlined the government, and made Amarillo a model city.

At the time of the appointment, in June 1932, Thompson asked if he could take the position temporarily, until the term expired in January. Sterling told him that the troubles would not be over that soon, that he would have to run statewide for election to the commission, and that "... if an incompetent or a crook should win the election, Texas will be doomed." Although Sterling lost a reelection bid to Miriam "Ma" Ferguson, Thompson was elected.

He was soon impressed with the work of E. O. Buck, a young engineer who had developed a formula for setting allowables that took into consideration both acreage and bottom hole pressure. Buck and other engineers worked out a grid map of the East Texas Field, and Thompson, using his military training, worked out a plan that ended with 323 key wells marked on the huge map.

In November 1932 the Legislature had passed a market-demand law, which gave the commission the authority to space wells, grant or deny drilling permits, and limit the production and above-ground storage of petroleum. In December the commission ordered a two-week shutdown of the East Texas Field. On December 21 Thompson and a group of enforcers closed in all the wells but six. Those six were fenced in, and the operators chased the officers off at gunpoint.

Thompson decided to end the standoff. He, a Texas Ranger captain, and several Rangers approached the barricaded well sites. At each they asked if the armed defenders wanted to shoot it out. None did. They knew of the Rangers. The shutdown was complete. In controlling the East Texas Field, Thompson had succeeded in restoring order and stabilizing the industry.

In addition to the opposition it had drawn, the setting of monthly allowables of how much each well may produce pegged to market demand—known as proration—also created a new problem—"Hot Oil."

There was no law setting a penalty for tapping lines to steal oil. Some operators ran a flow line into their own casing below the surface so the production could not be gauged. Others used "left hand" valves that turned in the opposite direction so an investigator would lock open a well he thought he was closing.

In 1932 the legislature passed a law making such practices a felony. At Thompson's request President Franklin Roosevelt issued an executive order banning shipment of hot oil in interstate commerce.

In the early 1930s there had been a strong movement towards federal control of the industry, a move that won support from some of the major oil companies. In 1935 Thompson, with assistance from Roosevelt and Texas Governor James V. Allred, organized an Interstate Oil Compact Commission for cooperation among the oil-producing states in exchanging ideas and advancing technology. This quieted the drive for federal control.

Thompson also tried to convince Texans that gas was valuable. He had been familiar with natural gas problems in Amarillo, where the first great gas field was discovered in 1918. As mayor there he had worked to get industries to come to the area, where the gas was practically free. None came.

He realized early on that natural gas would have no value unless a demand could be created for it. Until then the commission couldn't stop wasteful production practices—by 1934 gasoline and carbon black plants were releasing half a billion cubic feet of gas a day into the air—and a valuable resource was being frittered away. He worked to get industry to come to Texas and tried to get pipelines to extend their lines to the East, where coal prices were high. He also realized that gas wells could be profitable if the liquids were removed and the stripped gas returned to the reservoir.

Thompson turned down many opportunities to get into the oil business himself. As Clark, his biographer, put it, "His mission was to prevent physical waste. . . .

"His one obligation . . . was to the people of Texas. This great natural resource was theirs; it was valuable and irreplaceable. It was like water in a well; it would play out in time. And if the well were not operated with a thought toward reservoir pressure, it would play out before all of the water had been drawn. Oil was something like that, only more so. . . . And every barrel of oil or every cubic foot of gas wasted was that much lost forever."[13]

Though Thompson led the commission through its turbulent years, all the while preaching conservation, other commissioners also

Gov. Ross Sterling, a former president of Humble Oil and Refining Company, looked for someone capable of handling the volatile situation in the East Texas oilfields. He chose Ernest O. Thompson, the mayor of Amarillo.

Billy Blake

Tall tale gave column its name

Billy Blake of *The Corpus Christi Caller-Times* first named his column "Done in Oil." But he soon discovered that the oil columnist at the Houston Press was already using that name for his column.

Hugh Sullivan of Hiawatha Oil Company claimed he had fallen from a rig over in Louisiana and landed head first in the dope bucket and the soft grease softened his fall.

The bucket on the drilling platform was always handy for greasing equipment.

Blake really didn't believe Sullivan, who had a penchant for practical jokes, but the name "Dope Bucket" seemed like a good heading for his column.

Billy Blake, "The Dope Bucket," *Corpus Christi Caller*, July 1938.

made valuable contributions. In 1930 there were no geologists and only one engineer on the commission's staff.

Olin Culberson, first a chief examiner and later chief of the gas utilities division, had conducted sixteen investigations of natural gas companies as a staff member. Gas interests were not happy, and pressure groups got him fired in 1939. The following year he defeated twenty other candidates to be elected railroad commissioner. He served twenty years.[14]

As the controversy raged in East Texas over the commission and its orders, drilling for oil continued in South Texas. The King Ranch had been involved in some of the earliest South Texas drilling operations, though not for oil. The process began in 1881, when Robert J. Kleberg, like other ranchers of the period, hired equipment to drill for fresh water. Most of the land of his vast empire was useless, for his cattle were dying of thirst.

"Where I have grass, I have no water," he lamented. "Where I have water, I have no grass."[15]

Kleberg became so desperate he contributed to a program of setting off explosives to produce rain. It did rain at nearby Alice but only sprinkled at his ranch.

For years he hired drillers whose equipment failed. In 1898 he found the equipment and the man to run it. T. H. Herring brought a Dempster No. 6 Combined Hydraulic Rotating and Cable Drilling Machine from Beatrice, Nebraska.

On June 6, 1899, the first well was completed at 532 feet. As Kleberg watched the cold, clear water pour from the pipe, tears ran down his face.

The *Corpus Christi Caller* noted, "It matters little whether oil or artesian water is found, as in either event, an industrial impetus will follow that will astonish all and the future of the country will be assured."

Kleberg wrote the manufacturer, complimenting him on his machine. On January 8,1900, Kleberg said he had ten flowing wells, flowing millions of gallons of water a day. Six years later there were sixty-seven wells, some of them flowing with the help of windmills. The speed and precision with which the wells were completed represented a major victory for rotary drilling.

Other ranchers followed suit. The water table was found from 400 to 700 feet. Soon there were hundreds of artesian wells over the area.

"The men wondered why I cried when we finally saw what we had all been praying for," Kleberg said. "But I knew that once a definite source of water was available, I could induce railroad construction, which in turn would lead to the development of South Texas."[16]

Continued on page 120

They couldn't even agree on their smokes

It is said rotary well drillers and cable tool drilling crews competed heatedly in the early days and wanted nothing to do with each other.

They reportedly walked on the opposite side of the street and had their own brand of tobacco. The cable tool men went for the Mail Pouch brand while the rotary men preferred Beech-Nut.

An apocryphal story tells of a time when the cable tool crowd was beginning to be outnumbered by the favored faster drilling opposition. A group came to town, and one of them said, "Look at that. Now the SOBs have their own club," pointing to a sign advertising the Rotary International.

The ancient Chinese used a crude cable device to drill brine wells, as salt was a valuable commodity. In the early 1800s cable was still used to drill brine wells, which were said to have been abandoned in disgust when oil flowed from them. A brine driller hired by Col. Edwin Drake used his rig in the first well specifically drilled in the search for oil.

As late as 1948 one in three wells drilled in the United States was drilled with a cable tool or spudder rig. They were the pioneer tools in the business, but they failed in the early attempts to drill at Spindletop because of shifting quicksand. It was a rotary rig that completed the famous discovery.

Basically the cable rig repeatedly drops a large iron chisel-shaped drill as water is poured in to loosen shale, and a long pipe bailer with a valve on the bottom is lowered to bring up cuttings and clear out the hole. The process is slower than that of the rotary rig, but it is cheaper because it can be operated by two men. Often spudder rigs have been used to set surface casing with a rotary rig moved in to complete the operation. However, the oil discoveries at Luling where the city was drilling water wells were cable driven.

The method is not feasible for use in areas with substantial water flow or in areas with high gas pressures. Cable tools have been effective in breaking through rock and other difficult formations. They were also used in very porous formations where drilling mud was absorbed in a rotary operation and not returned to the surface. Cable tools required more casing to seal off water stratas and were seldom used in wells deeper than 6000 feet.

The need for cable tools to penetrate in very hard rock formations vanished with the development of the fishtail bit, which allowed rotary rigs to drill faster and penetrate rock formations. Labor intensive work of the tool dressers, who pounded the red hot drill with heavy sledgehammers to reshape it when it became dull, was no longer needed, and the specialty—like that of rig builders—became an anachronism.

(Top) Decaying bull wheels symbolized the passing of the standard cable tool rig. (Bottom) An early day rotary assembly with rock bit, roller reamers and stabilizers

"They Still Drill With Cable Tools," *The Humble Way.*

Drilling for water had saved the King Ranch, and drilling for oil—through a relationship with Humble Oil and Refining Company—would, more than forty years later, assure its financial security. As Robert J. Kleberg Jr., head of the King Ranch family, once said, "We operated the ranch for eighty-five years without oil and did well. With oil, we do better."[17]

Humble Oil Company was formed through a merger of the properties of a number of Texas independent oil operators: Ross S. Sterling, later governor; his brother, F. P. Sterling; W. S. Farish and R. L. Balffer from Balffer and Farish, H. C. Wiess from Paraffine and Reliance oil companies; Walter W. Fondren and C. B. Goddard, producers; and lawyers L. A. Carlton and E. E. Towns. It was chartered to operate in Texas in 1917. In 1919 Standard Oil of New Jersey bought half interest in Humble for $12 million. Standard soon had a majority interest but allowed the Texans to continue operating under the Humble name because they were able to cope with the regulatory and legal systems of Texas.

This well, drilled by the Kleberg Oil and Gas Company, came in on February 7, 1920, but yielded only a small showing of oil.

In 1919 Humble leased a few hundred acres on the King Ranch and drilled four shallow dry holes. Because of this, Farish, who was Humble's president, was strongly opposed to any further drilling in the area. The lease was allowed to expire in 1926, the year after the death of Henrietta M. King, matriarch of the ranch family.

Back in 1910 Wiess had drilled a dry hole at Piedras Pintas in Duval County with his short-lived Cactus Oil Company. On February 7, 1920, a Kleberg Oil and Gas Company well drilled south of Kingsville yielded only a small showing of oil.

Humble was told the surface geology did not indicate any uplifts, there were no oil seepages or indications in water wells or any indications of "any favorable area for drilling wells for oil or gas," and "there was no justification for a test well to be drilled."

Despite all the gloomy forecasts, one person believed there were many fields of oil and gas under the vast ranch rangeland. That person was Wallace Everett Pratt, who was the prototype of what every geologist would aspire to be.

Pratt was among the first to convince the oil industry that geology was essential to its success. His statement: "Oil is first found in the minds of men" is the most quoted ever as applied to exploration.

"The conviction of our best minds that little or no oil remained to be found has continuously handicapped the search for oil," he said. "Unless men can be convinced that there is more oil to be discovered, they will not drill for oil."

Pratt was a scientist with the heart of a poet. He actually composed poems as a sign-off in letters to his friends, yet he was an astute businessman and innovator. He was an active conservationist and environmentalist who discovered oceans of oil.

Born in 1885 in Phillipsburg, Kansas, he worked as a hotel night clerk and on the Kansas Geological Survey while attending the University of Kansas. He earned bachelor's and master's degrees in geology and another degree in engineering.

He worked five years for the Bureau of Science, worked in the Philippines from 1910 to 1915, and joined Producers Oil Company, a subsidiary of Texaco, in 1916 in Houston. He joined Humble Oil and Refining Company as its chief— and only—geologist in 1918, when it was a small, year-old Houston company. Early on he worked to change the oil industry view of geologists and helped pave the way for business to accept a scientific and geological approach to oil exploration. He assembled a staff of ten geologists and became a major factor in making Humble a major oil company.

He accomplished this by urging the company to lease huge tracts of untested land where it could drill without competitive leasing and drilling. He personally negotiated many of these leases. In some, including the King Ranch, early drilling was not immediately successful, but he had faith, which turned out to be justified, that equipment yet to be devised would prove him right—that there was tremendous production deeper than the earlier shallow tests indicated.

Pratt, who was responsible for major discoveries in East, Central, and South Texas and other areas, said he did not find all that oil himself. He gave credit to the men he hired. He hired Humble's first geophysicist, pioneered use of geophysical equipment and a wide variety of other scientific ventures, and created a geological research laboratory. When he joined Humble, the company had 32 million barrels of reserves. When he left in 1937 to become a member of the executive committee of Standard Oil Company of New Jersey, Humble's parent company, Humble had 1.9 billion barrels in proven reserves.

As an environmentalist, Pratt advocated wider spacing in well sites, prevention of water pollution by tankers and refineries, and prohibition of gas flaring. He personally purchased ranchland in the Guadalupe Mountain area of West Texas and later donated the land to the National Park Service, with the provision that the government also furnish land that ultimately became the Guadalupe Mountains National Park.[18]

Pratt retired from the company boardroom in 1945 and moved to his home in a West Texas canyon, where he was a consultant, adviser, writer, and philosopher. He maintained correspondence with geologists he had worked with and younger ones whose work interested him.

"It was never easy to answer his letter," said Amos Salvador, author of Pratt's Memorial in the American Association of Petroleum Geologists Bulletin. "He would always ask what do you think about this situation or technique that someone wrote about in a certain article. Then you would have to go and read the damn article before you could write an intelligent response."[19]

The statement of Humble geologist Wallace Pratt: 'Oil is first found in the minds of men' is the most quoted ever as applied to exploration.

By the time of his death on Christmas Day, 1981, at age 96, Pratt had won every prestigious award offered for his contributions to science and the art of exploration, to his profession, the industry, education, his country, and his fellow men.

In 1921 Pratt persuaded Humble to lease a big acreage that experts had rejected west of Mexia, and 175 out of 180 wells there were successful. He predicted huge reserves that Humble later developed in Alaska. For years while he waited, he formed a close friendship with Robert J. Kleberg, Jr. Because of Humble's reluctant leadership, there was small chance of doing business with the ranch.

At the time Humble was involved in buying and shipping oil in Southwest Texas as well as marketing its own products. When the Port of Corpus Christi was dredged in 1927, fill from the channel formed an island that became known as Harbor Island. Humble established a shipping terminal there with storage tanks, pumping stations, and dock facilities and from 1929 to 1945 operated a refinery at Ingleside.

Oil from the Refugio and other regional fields was moved through pipelines to the terminal, where Humble's supertankers of that day loaded quickly and sailed on their way to the East Coast. The first tanker was loaded in October 1928.

Henrietta King's will specified that the ranch would not be partitioned among her heirs until ten years after her death. In 1934 Kleberg saw that inheritance taxes and debts left by Mrs. King's death would total more than $3 million. The Great Depression was on, and cattle prices had hit rock bottom. To make matters worse, South Texas was gripped by a severe drought. It was time to talk serious business with Pratt, by then Humble's director in charge of exploration.

The geologist had to use all of his persuasive powers to convince the Humble board of directors to lease the entire acreage of the King Ranch, but he succeeded in negotiating a twenty-year agreement, which was signed on September 26, 1933. In it Humble received exclusive exploration and drilling rights on 1,133,156.31 acres, which included all the land left to heirs of Henrietta King plus some other tracts. The ranch received an annual bonus of thirteen cents an acre for drilling rights, the usual royalty of one barrel out of every eight of all oil extracted from the property, and, at 5-percent interest, a cash loan of $3,223,645, which was used to clear the ranch's debts.

Humble management played it safe in 1938, drilling around the ranch perimeters opposite established fields. They completed the first oil well on the ranch as an extension of the Colorado Field in Jim Hogg County. The next was drilled in the Luby Field in Nueces County, then in the lucrative Stratton Field west of Kingsville, and then, in 1941, in the Seeligson Field. Others were in the Willamar Field in Willacy County and in the part of the Tijerina-Canales-Blucher Field on ranch property in 1945.[20]

The first wildcat drilled on the King Ranch was the No. 1 King Ranch Borregos. Drilled to 8549 feet, it produced oil at the rate of 40 barrels a day but in January 1946 was completed as a gas well producing 18 million cubic feet of gas a day.

It was 1945 before the first wildcat was drilled on the ranch. It was the No. 1 King Ranch Borregos, the discovery well of the Borregos Field, which came in on December 13. Drilled to 8549 feet, it produced oil at the rate of 40 barrels a day but in January 1946 was completed as a gas well producing 18 million cubic feet of gas a day, with thirteen barrels of condensate per million feet of gas.

In 1959 Humble marked another first on the ranch with its King Ranch-Alazan No. 11, the company's first quadruple pay oil well. In 1960 Humble built what was at the time the largest natural gasoline plant in the country. Located west of the ranch headquarters, it separated natural gas from liquefiable hydrocarbons from the ranch as well as from other fields in the area. A 30-inch pipeline had been installed in 1959 to run from the gas plant to Humble's Clear Lake plant, connecting to Port Arthur's oil and petrochemical plant area. A network of gathering pipelines from eight Southwest Texas counties was tied into the trunk line.[21]

In 1927 Humble started a pioneering seismic program with a group known as Party Six. Seismic technology, similar to that which measures earthquake intensity, basically measures the intensity of a sound or energy wave from the surface of the land down thousands of feet and then back.

The Party Six group, led by W. R. "Foots" Feather, was organized with the advent of reflection technology. A few months after joining Humble, Feather participated in the industry's first discovery of a major oil field through geophysical work. It was the Sugar Land Field, "found by tracing the borders of a deep salt dome with refraction surveys using 400 pounds of dynamite per shot. Previous surveys had used 200 pounds."

As one tanker churns through the channel, two others take on cargo at the Humble docks at Harbor Island. The tank farm in the background held Texas and Southwest Texas grades of crude.

The first Party Six ranch crew, shown in front of the trailers soon after moving onto the ranch, included (left to right, front row) L. A. Doyle, camp cook L. W. Hancock, a Mr. Butler, L. J. Epperson, W. R. 'Foots' Feather, D. R. Spell, and Bobbie Allen. (back row) J. E. Gregg, S. B. Holland, C. W. Ingraham, H. L. Voelker, G. F. McReynolds, S. A. Teasley, V. J. Perez, Jack Starr, P. B. Powell, W. E. Phelps, R. C. Goodwin, Eddie Miller, and C. A. Durham. They were the first seismograph crew to work on the King Ranch.

For the next forty years Feather would lead seismic crews in many U.S. states, in Europe, and in Asia. The group was the oldest continuously operating reflection seismograph crew in the world.

In 1938 Party Six began a comprehensive geologic mapping program on the ranch—a project that would last fifteen years. It was a formidable task, for there were no roads in the vast area, which included grassy plains, black farmland, hills, thick mesquite thickets, swamps, salt-water inlets, fresh water lakes, groves of live oak trees, and desert-like sand dunes.

Feather said he was cautioned to "bend over backwards" to cooperate with the owners of the ranch. It was during the first week that he learned not to set up a shoot near a water well, because of concerns that the shooting might ruin the well or frighten cattle away.

"We had just gotten onto the ranch," he said, "and everything was going smoothly until we got about two hundred feet from a well. Up drives Mr. Robert Kleberg, Jr., himself, and that was the end of that. We bundled everything up and headed out the gate, I really thought, right then and there, we had had it."[22]

The crew learned to appreciate the rules and philosophy of ranching. They learned to stay out of pastures where cattle-breeding operations were under way and to avoid damaging timber. Humble installed smooth, rounded-pipe guardrails around the wellheads instead of angle iron and bumper-type gates so that cattle would not be injured. Crewmen also had to be careful not to disturb the wildlife. Conditions improved with construction of paved roads. Even so, some locations could be seventy miles apart on the ranch. Party Six lived out

A herd of Santa Gertrudis cattle appears unfazed by the rig behind them. The Humble crews were careful to cooperate with King Ranch owners and follow ranch rules.

of trailers for fifteen years. The crew often spent weeks in the trailers without leaving camp because some locations were too remote to permit travel to town. Brush had to be cleared, and there was constant danger of rattlesnakes.

The geologists' job was more difficult because they had no records of production or dry holes in the area as points of reference. As the years passed, technology in the search rapidly improved. With new theories and new equipment, the work was a continuing education.

Jack Starr, who took over from Feather, said, "After reflection surveys replaced refraction as the standard tool for geophysical exploration in the early '30s, little technology change occurred prior to World War II. But change came rapidly after the war, and Party Six soon was using a lot of new equipment."

Jack Clements, later a Party chief, said the group was among the early users of RTUs—remote telemetry units—when they were marketed in the late 1970s.

(Above) The crew sets off a round of simultaneous pattern shots on the King Ranch in 1954. They got off as many as six shots a day. (Opposite) The Party Six caravan makes one of its final journeys along a ranch road.

"Prior to that time all of our instrumentation was on one truck. The geophone stations were connected to the truck by heavy cable. And laying this heavy cable across the countryside could become quite a chore, especially when crossing rivers and highways. This was eliminated by remote telemetry."

Starr was a bit nostalgic when Party Six pulled out of the ranch. "Those trailers had been home to us for many years, and it was the saddest day of my life when we watched them pulling out for the last time."[23]

While it might seem strange that a man could be nostalgic about life in trailer homes, a look at life in the early oil fields and oil field towns makes it understandable. Early drilling crews worked in the most primitive of conditions, exposed to the heat of the South Texas sun, chilled by north winds, and often with little protection from monsoon-type rains.

If they were lucky, they lived in tents that didn't leak. Sometimes they lived in their cars or slept on the floor of the rig doghouse. It was worse if their families chose to follow them to the well sites far from niceties of bathrooms, grocery stores, and other comforts.

A look at a 1931 Texas highway map gives a clue to the problem. The road south from Riviera was dirt and sand with no improvements. U. S. Highway 96—later U. S. 281—was paved from Alice to McAllen. And west of that, all the way to Laredo—nothing. Drilling crews followed trails or tracks in the wilderness or made their own in vehicles carrying spare tires, shovels, block and tackle, and other equipment to get out of bog holes or loose sand.

The north boundary was the road from Laredo to Hebbronville. That road was unpaved in 1921. When O. W. Killam founded the town of Mirando City, it took all day to drive the fifty-five miles to Laredo through the sand. One old veteran of the brush country suggested that many locations were selected near the Texas-Mexican Railroad tracks for access rather than geology. It seemed easier to reach a site by railroad track rather than chopping through the brush.

Conditions were worse than most in Freer after Cezeaux, Trussell, and Putman drilled a well on the William P. Norton farm two miles south of Freer near a water well dug in 1878 for the U.S. Army cavalry. The discovery, in 1928, started an immediate oil boom in Government Wells Field.

The boom was slowed by discovery of the huge East Texas Field, but it became a full-fledged boom in 1932. The Great Depression was in full bloom, and workers came by the thousands, many with their families. There was no housing for half those who came. Some rented cots by the hour in makeshift hotels. They slept in barns, tents, cars, shacks, cardboard lean-tos, and even in abandoned streetcars. Some had to make do with a mesquite tree for a roof. One man was tickled to show his wife to a shack with a dirt floor, even though they had to share it with bats.[24]

James Mosier, who spent thirty years, off and on, working in the oil fields, gave a vivid description of his childhood in Freer and other aspects of a roughneck's life.

He said that when he was a kid, his daddy was firing boilers in Freer. The nearest paved road was in Alice. They had salt water piped

'Those trailers had been home to us for many years, and it was the saddest day of my life when we watched them pulling out for the last time.'

—Jack Starr

(Bottom) Driller Lawrence Mosier, James Mosier's uncle, is at the left in the crew of an Al Buchanan rig. (Below) A standard rig, near Sinton in San Patricio County

into all the houses. The nearest fresh water well was two miles out of town, "so you went there to get drinking water.

"And fleas were so bad. I don't mean in spots. They were all over town. They just covered your face as soon as you stepped out the door. Walk in the grass and they climbed up your legs. Hundreds of them," he said.

The streets of Freer looked perfectly level, "but there were chug holes knee deep," he said. "You hit it and dust would fly and cover you with powder. The joints there had slot machines, and it was wide open around there. They had a story in *Life* magazine and there was a picture of Freer, Texas, on the cover. . . . The sidewalks were boardwalks. Some were planks in the mud. The beer joints stayed open, I think, all night. And it seemed like there was a killing nearly every night. It was a tough place. Really was. But the people at Freer were used to hard conditions, and they could take care of themselves. I remember two guys killed a full grown mountain lion with a knife."

"Tough" reminded him of rig builders. "They were hard, tobacco-chewing, tough people. They worked hard. My stepfather was the oldest rig builder that J. Ray McDermot had on his payroll. His name was John Runnels. McDermot went into a partnership with Glasscock.

"My stepdad had two brothers killed in falls building derricks. They didn't use safety lines, and they walked across those little narrow girders like they weren't even there," he said. "They climbed around like acrobats and hoisted steel. A lot of them looked for fights.

"When they came out with power rigs and jackknife rigs, they cut out rig building. You never see a standard rig any more. It used to take days to move out. The new rigs can be gone over night," he said. "They were scary when you get up there working derrick. There's not much holding you up on them. When he latches on the first stand of pipe and that thing drops about a foot and twists and cuts a dido. . . . You get used to it. I was used to the old standard derricks. You have a safety belt, but it don't do a lot of good. If you fall from the railing, you fall about fifteen feet and it whaps you against the derrick.

"Sometimes people are sort of stupid about the oil fields.," he said. "I've seen them sell bottom hole information to people when the well was already dry. Any roughneck on the floor could have told them it was dry if they had just gone over and asked them. They came out in a big Cadillac. The company was still coring like they were looking for pay dirt. And it was already dry.

"I hired out when I was almost seventeen years old. I went to the railroad first, but I was laid off much of the time; and my uncle broke me in to roughnecking in Atascosa County."

He had four brothers. All worked in the oil fields, three as drillers and one as a drilling superintendent.

"In those days you started right in," he said. "You didn't stand around and learn. You learned on the job. My uncle would kick me in the seat on the way home and tell me what I was doing wrong. If I complained about the rough treatment I was getting, he'd say, 'Well, I don't want to show any partiality.'

"I did all the dirty work. If there was a job he didn't want the hands doing, he'd make me do it—not being partial. I was a boll weevil, so I was breaking in and they were teaching me. So I was the goat," he said.

One of the toughest jobs of all was running water lines to supply steam boilers. "Sometimes you had three weeks work just running pipeline, connecting pipe with 36-inch wrenches through the cactus and mesquite. Man, that was tough work. It was hard—and hot.

"I've worked through some hard times. I was so tired I couldn't even get to sleep when I got home," he said. "It was hard work. The roughneck has to be a little of everything. He's a plumber. He's an electrician, carpenter, painter, or whatever. You did all your own work on the rig. Of course, a lot of people were electrocuted or shocked on the job."

He said unions were unpopular in South Texas. "They didn't know what a union was. They would say, 'I don't want anybody telling me what to do.'"

But, he said, the oil companies and the drilling contractors could treat them any way they wanted. "You were at their mercy. If a driller didn't like you, he could fire you, whether you were doing good work or not. You had no recourse. If he had a relative he wanted to feed, he could put him in your place. They didn't care if you got hurt or wore a hard hat or anything else.... Lawyers would have had a heyday if they had been around in them days."

Cementing wells was a backbreaking job for roughnecks. There would be acres of cement sacks stacked ten feet high that had to be hand-

'The roughneck has to be a little of everything. He's a plumber. He's an electrician, carpenter, painter, or whatever.'
—James Mosier

The streets of Freer looked perfectly level, James Mosier said, but there were chug holes knee deep. Mosier lived in Freer as a child while his father was firing boilers.

(Below) A cement hopper (Bottom) Preparing to cement (Opposite top) Halliburton cementing or mixing mud to control a well. Cement sacks and pipe racks are stacked on a platform alongside a wooden catwalk. (Opposite bottom) Cement sacks for a surface string

hauled to the cutting table and dumped into the hopper. Halliburton could take that cement faster than you could put it in the hopper, Mosier said.

"There was no safety protection at all. You breathed that dust that stopped up your nose," he said. "It got in your eyes and you were covered with the dust. You stand there until you nearly fall over. Your eyes are burning, and you think the driller is about to take mercy on you and tell someone else to take your place. This was after you had worked sixteen hours running long string. It was a horrible job. These roughnecks don't know how easy they have it today."

The Halliburton man was expected to bring a file to sharpen the sack cutter and a pound of coffee. The coffee was important, because if he didn't bring it, he didn't get any help.

"Those big steel joints to hook onto the wellhead so they could pump the cement were real heavy," he said. "The most wonderful thing was when Halliburton got those trucks to drive up there and you could sit down and watch them load cement. You don't get much sitting time out there in the oil field."

His uncle's rig was working in West Texas in extreme cold. The men furnished their own cars and carried water cans. They drove forty miles to the job. There were five men to a crew, and they took turns driving. His uncle hired drunks and floaters.

"They smoked cigarettes in the car so we rolled the windows down. One of them said, 'We might as well knock the rest of these windows out if you want them all down.'

"My uncle told him if he didn't like it, he could get out and walk. When we got to the rig, the guy rolled up his clothes and rode with the other crew back to town. My uncle was the driller. He told me, 'All right, get on that elevator.' The elevator takes you up to the monkey board, and I jumped on it as it went by. That's when I became a derrick man, also known as the attic man because he's on top of the rig," he said.

He decided to come back to South Texas when the monkey board had three inches of ice on it and "the guys down below scattered like quail when icicles rained down on them."

He worked for a number of contractors in Atascosa County. "I was working for Joe Beard, and the boss got sick. The pusher asked me, 'What kind of driller are you?'

"I told him I'm not a driller. I'm just a roughneck."

"He said, 'You're my driller.'

"I said, 'Well, it's your equipment.'"

He was running the rig and figured anybody could do that if they had been out there as long as he had. "It was a matter of responsibility. You had to hire your own crews, keep up with payrolls and all that stuff."

Once a company man came to the rig "all drunked up" and ordered him to fire one of his roughnecks. "I told him the man was a good hand.

Ed Mosier is second from the left in this photo of a rig crew.

I hired 'em and if they don't suit me, I'll fire 'em. If you want us to leave, we're gone." He expected to be fired, but wasn't.

"I tried to avoid the drunks. If you didn't, you'd always be short handed.

"When I was a kid, the roughnecks followed one driller. They had a little two-wheel trailer with a box on it. You put all your earthly possessions in there, and you'd be liable to go plumb across Texas to the next location. That was my life when I was a little kid. When the rig was shut down, pop was on dry watch. He stood guard to make sure nobody stole anything. He'd wash the boilers down while he was on day watch.

"Around the rig we had so many blister bugs they would stack up. You'd wash 'em off the floor, and the beetles, grasshoppers, crickets—you name it—washed down in the cellar. The stench would get so bad we would have to go down there and clean out all that mess.

"We drilled a little ol' well there at Cross, Texas, for Massingale and made a well and flared it off. The crickets were three feet deep under that flare. And the blue quail. If you had a shotgun, you could have filled a truck with them. They were having a ball feeding on those crickets."

Mosier managed to accumulate enough time with the railroads to retire. He had started as a fireman in an old steam engine. "In 1989 they had a buyout, and I took it," he said. "Better than a gold watch. I'm so thankful I can drive by a rig now without asking if they're shorthanded. I worked in the oil fields thirty years, off and on. And look. I still have all my fingers. I consider myself fortunate."[25]

A great many old roughnecks can't make that brag.

The conditions that Mosier described finally eased at Freer, but still the honky tonks and gambling halls flourished. Rangers made raids, but many had been appointed by Governor Miriam "Ma" Ferguson and some were of questionable character. Even so, in 1934 a dozen Rangers swooped down and put a stop to the illegal activities.

It wasn't long before the shady characters returned to the Free State of Duval. Oilman Bob Smith told A. E. Bennett, former Texas Ranger and private ranger for a number of oil companies, that gambling and liquor were using up workers' paychecks, causing a hardship for their families.

Bennett accepted the job of chief of the five-man Freer police force. He went to each of the gambling houses and gave the owners ten days to close down. He was ignored. He issued another warning after five days. Again he was ignored. This time he and his men went to each place and picked up all the money, then gave it to local churches. Then he commandeered several large heavy oil-field trucks, looped heavy cables around the hot spots, and proceeded to turn them into heaps of firewood.

The situation improved shortly. Standard Oil Company of Texas built a camp called Alta Mesa on land leased for drilling on the W. W. Jones Ranch about thirty-five miles southwest of Falfurrias. The houses, some two and others three bedroom, were neat and sturdily built.

Even after the boom was over in Freer, residents of the town did not get paved streets until 1938 and fresh water until 1962.

Some oil companies, however, had begun building housing for workers as early as 1920. Killam built a boom-proof town for families,

(Below) Two boilers on this rig indicate that it was probably a shallow well. (Bottom) Freer shows signs of the boom. Residents did not get paved streets until 1938 and fresh water until 1962.

(Top) A Humble camp had dormitory-like buildings and individual homes. Screen porches offered a slight respite from the South Texas heat. (Below) The gateway to the Plymouth Oil Company camp near Sinton

with churches, a school, and housing for 100 workers at his Misko Refinery. Magnolia Petroleum Company built a camp soon after buying out a large segment of Killam's production. Other companies followed suit. Eventually neat rows of bungalows appeared across the countryside near oil facilities, usually near a small town where families could find provisions, schools, and churches.

Drilling reached a frenzy in the fifties and sixties, and the influx of workers was overpowering. Humble provided houses for employees at the Kelsey Bass Camp in Jim Hogg County, Stratton Camp in Kingsville, and Borregos. Plymouth Oil Company and the Superior Oil Company built camps at Sinton, and other companies built camps in remote areas where there were no conveniences.

The Kelsey Bass camp was more luxurious than most. It boasted a swimming pool, tennis courts, and a grocery store. Rent and utilities were very cheap, allowing workers to save money, but residents were always aware they were in wild company. Cattle sometimes lusted for the green lawns and flower beds. One mother

reported that the snake barrier in the fence was damaged and her teenage son was bitten by a rattlesnake. Fortunately, thick corduroy and quick first aid saved his life.

Enjoying life in town for shopping or attending church involved a ninety-mile round trip, and many of the roadways were not paved. There was a lack of privacy in the camps, but friendships flourished and there was a sense of family among residents.

In 1954 Dick White, as chief geologist, lived in Plymouth Oil Company's large camp north of Sinton. "The kids went to the Sinton schools," he said. "We had five or six kids at the time. We had a four-

(Above) Rattlesnakes and other varmints sometimes found their way into the rural-area camps. (Left) Sometimes residents could take advantage of such amenities as tennis courts and swimming pools.

As towns grew, highways spread out, production in some fields dwindled, and travel became easier, the day of the company communities ended.

bedroom house, gas and water was supplied. We were charged forty dollars a month, which was pretty good for those days."

As the company's production was not near any town, it was cheaper to build a camp close to the production. The company might support the local schools or in some cases build a school so it would be convenient for their employees.

"The company owned it and ran it, but it was sort of loose. If you had a complaint or wanted to move to a larger house or a smaller house, they were agreeable," White said. "If you were next to people you couldn't stand, you could ask for another house when it became available. You were allowed to keep certain pets. One of my daughters had a horse. She couldn't ride it on the campground, but there was ranchland all around. There was a place for a garden if you wanted to do that.

"It was to the company's advantage to keep lives of employees as pleasant as possible. Superior Oil Company had a camp nearby, and it had more amenities than we did. There was one other camp in the area. They were important for the community."[26]

The wife of one pumper gauger who lived in Alta Mesa found two reasons for satisfaction with the housing.

"Your children got to be children longer and weren't pressured to grow up so fast," said Tommie Lee Edwards, whose husband moved his family to the camp in 1961. "We moved out there because we had a big family. I knew we could save enough to send the kids to college."[27]

As towns grew, highways spread out, production in some fields dwindled, and travel became easier, the day of the company communities ended, but in their day they were a vital part of South Texas oil history.

In 1938, six years after the big boom struck Freer, the same dog bit Alice. H.H. Howell is credited with being the man who opened the Alice Field in 1938 and created a boom that made Alice a bustling oil community, but the man behind the discovery was actually an obscure young petroleum engineer named Jack Pratt, a graduate of the University of Texas. He was no stranger to the oil patch—he had worked as a roughneck at Luling and other Texas oil fields and helped bring in the discovery at Seguin.

In 1936 he was visiting his mother, Mrs. Gene Wilson, on her farm six miles south of Alice. The countryside was baked by a persistent drought, and he noticed a crevice had opened, running the full width of the farm and beyond. He believed that the crack marked a geological fault, indicating a potential oil-bearing area. After he had a San Antonio firm run a magnetometer test that confirmed the possibility of oil, he began leasing a block of 6,800 acres involving thirty-one landowners.

Pratt spent a year and a half cajoling, urging, and paying as much as five dollars an acre. Others demanded a one-eighth royalty. In 1964 he recalled that he had milked cows and performed other chores to win over reluctant owners.

In 1938 the leases were all obtained, and Pratt signed a contract with H. H. Howell. The well was drilled, and it showed oil but not in commercial quantities. Pratt was broke, Howell was broke, and the leases expired.

Pratt laboriously got the leases renewed. Joe Browning of Corpus Christi came to the rescue with some financing.

(Opposite) Plymouth Field in the 1930s (Below) Oil well supply companies like this one helped Alice become the 'Hub' of the South Texas oil industry.

W. Carlton 'Tubby' Weaver

'I stopped [by a wooden derrick where a well was being drilled] and managed to get some oil on me. I never got it off.' –W. Carlton 'Tubby' Weaver

As the story goes, the well had to be drilled quickly before the money was gone. In his haste, Howell drilled a hole that angled off at about thirty degrees. It was later determined that had he drilled a straight hole, it would have been a dry hole.

Howell wanted to quit, but Pratt urged him to keep drilling. According to a report by the *Alice Daily Echo*, Howell did something unusual for that day. He exploded three charges at different depths at the same time to save money. The well, drilled to 5300 feet, flowed at the rate of 385 barrels.

Life in Alice changed dramatically. Cafés became hotbeds of financial transactions as speculators poured in. Soon trucking firms, oil-field supply houses, and construction crews came. The population tripled in short order, and Alice was well on its way to becoming the "Hub" of the South Texas oil industry. Howell became quite wealthy. He later moved to California, then to Florida, and eventually retired in Seguin.

Pratt shared his profits with Joe Browning and lived in Alice five years. He was instrumental in forming both the Chamber of Commerce and Rotary Club there. More than 100 wells were completed in the new Alice Field. He later worked with Homco in Corpus Christi and A-1 Bit Tool Company. During World War II he was one of fifty-eight petroleum engineers who built the largest refinery in the world at Abadan, Iran. He was later in the real estate business in Kerrville.[28]

Another South Texas oilman who, like Pratt and many others, played an important role in the Allied war effort, was William Carlton Weaver—better known as Tubby. His part in the production of carbon black, necessary for wartime production of synthetic rubber, was just one of many positions he held in a long and varied career.

When Tubby was a kid near Luling, Texas, he rode his horse past a wooden derrick where a well was being drilled near his home in the Salt Flat Field.

"I stopped there and as I watched, I managed to get some oil on me. I never got it off," the 94-year-old oilman laughed in a 2003 interview.

He was practically born a businessman. When he was six, he and his grandfather formed a partnership raising hogs and cattle. They split the profits. In addition, he milked cows and sold the milk to local restaurants.

When the eccentric Edgar B. Davis opened the Luling-Branyon Field in August 1922, Luling's population grew from 1,500 to more than 10,000 overnight.

"My grandfather and I knew all those original people in the oil business and took them into our home," Tubby, who was thirteen at the time, recalled. "We made pallets for them on the porches. They were glad just to have cover over their heads."

Among the many oil workers he met was geologist Vernon Woolsey, who was responsible for the discovery of the field and who gave Tubby his first lesson in practical geology.

"He pointed out to me the fault evidence in the bank of the San Marcos River," Tubby said. The reservoir proved to be narrow, but it extended for ten miles.

Another of the men who came to Luling was Earl P. Halliburton. "They used to take the pipe and wrap this linen stuff around it. When the pipe was set, stuff would fall in on top of it and seal it," he said. "Then Halliburton and his assistants came along in the Luling and Mexia Fields in 1922. The wells at Luling had about a hundred feet of surface pipe, and Halliburton cemented the long strings. The Luling Field was about 2200 feet deep, and they'd cement the bottom of it. And that's where Halliburton got started—in Luling."

When Tubby was ten, the family spent the summer in his grandmother's two-story home near the water on North Beach in Corpus Christi. It was September 1919, the year of the great hurricane.

"I had to get back to school Monday so we left for home on Friday," he said. "The storm swept everything away Sunday. The house was gone. My grandmother would never go to Corpus Christi again. All of her friends were gone."

Coating and wrapping a pipeline

When he was fifteen, he worked summers helping stake locations for wells, pipelines, and storage tanks. One summer he made money driving prospective leaseholders to Laredo and Refugio, where oil was coming in. This was no easy task, for there were few roads in South Texas and none of them paved. Caliche was about the best the motorist could hope for. It was much easier to ride the San Antonio and Aransas Pass (SAAP) Railroad.

In 1927 he spent the summer working for Smith-Clark Oil Company, a Corpus Christi brokerage company that bought leases for Humble and Gulf oil companies. That fall he entered the University of Texas to major in geology. In 1930 he withdrew from school to help his uncle, who had a rig working in the huge East Texas Field.

Tubby Weaver at age 23

"I was the geologist for this company we formed," he said. "I cored 85 feet of oil sand." He estimated the production would have been 45,000 barrels a day.

"Well I was rich. Just a kid and I had scored 85 feet of fine sand. I said, 'Hell, this is easy.' A sad thing. As rich as I was, oil went to ten cents a barrel. The Depression was on. Then a little boy crawled up on the separator and fell off, and it killed him. They sued us for $50,000. Oil wasn't worth anything. We hadn't paid for the well. So we lost that."

With no job and nothing to do, he returned to the University and finished his bachelor's degree. He managed a fraternity house and made all the members pay what they owed. "I had to whip a couple of them," he said. He graduated in 1932.

It was at the University he picked up his nickname but not because he was fat. It was to distinguish him from his football-playing roommate, who was "Tuffy."

Tubby, his uncle, and Rosie Roark from Arkansas founded Brown Drilling Company and persuaded Roark to hock his wife's diamonds. They used the money to buy a new drilling rig. "Worst mistake I ever made," he said.

"I was a pretty good all round roughneck, and I was a geologist of some sort when I got out of school," he recalled.

On his first job a geologist told him he would find the location by turning right at a red flag on the road south of Tilden in McMullen County and it would be "clear as a bell." Unfortunately, the geologist had never been to the location, which was two or three miles through thick mesquite brambles. Desperate to get a road for the trucks hauling his rig—he had a quarter interest—he went to Tilden to find laborers. A man in the street said there were none and threatened to fight Tubby after Tubby refused to drink with him.

"I got out of the car and was taking a pretty good whipping in the middle of the street," Tubby said. "You tie onto one of them ol' cowhands who were used to throwing a full grown bull, a city boy's not as good as he thinks he is."

The only thing that saved him was a man who came out of a store and recognized Tubby as his cousin. "He's one of us," the man said.

Tubby learned that the man in the street who had no love of strangers in McMullen County was the county judge. After Tubby was identified as "one of us," the judge took two truckloads of migrant laborers and personally directed them in clearing the way to the well site in record time. He also insisted that Tubby bed down in the jury room of the courthouse instead of the rude tent at the well.

Brown Drilling Company continued to drill in Duval County. It was not an easy life. To feed the crews, Tubby picked up 100-pound sacks of beans for three dollars each and 100 pounds of sugar for the same price, along with bags of cornmeal, potatoes, rice and dried fruit—items that did not require refrigeration. He bought Brer Rabbit molasses by the case.

"I worked those men in twelve-hour shifts. . . . They worked for me, and when they didn't have a job, they stayed with me and I fed 'em," Tubby said. Both crews stayed at the rig, so he was feeding twelve men and providing a tent for living quarters.

"We ate beans, and I killed so many doves I wouldn't eat a dove for twenty years. Bankers I knew came down to the rigs to get something to eat," he said. "They ate the birds. We ate a lot of javelina hogs. The big ones smelled, but the little ones you couldn't beat. I'd get some corn and lay it down along the fence row and wait. Shells cost a nickel, and if I couldn't get those birds lined up, I wouldn't shoot. Now those sand

Traveling around South Texas was no easy task, because even if roads existed, they were primitive.

hill cranes were delicious. You pull out the breast, it was the best. That's against the law now."

"Then the law [National Recovery Act] came along in 1933 and said that you could work a man but six hours and you paid him five dollars."

He caught the men throwing junk into the hole to prolong the project and fired them all. They filed an NRA complaint.

"You don't realize that during a depression there isn't any money," he said. "An official said there was nothing I could do about it. My trial was set for Friday, and on Wednesday the Supreme Court declared the law unconstitutional. Timing is everything in the oil business. That was one time that timing was right for me."

Shortly after this a driller experienced with gasoline rigs but unfamiliar with Tubby's steam-driven rig ended Tubby's drilling career. When the pipe stuck in the hole, the driller jammed the clutch instead of easing up on it.

"He couldn't pull the core barrel out," Tubby said. "I saw him getting madder and madder. He had the motor running wide open. I tried to get there. I got as far as the V on the derrick, and he put that clutch in and that whole derrick came down. When you see all that steel falling, it's scary—steel, the traveling block, and all that. None of it hit me. The roughneck didn't get hit. When it all quieted down, we took off running like hell."

About this time his uncle drove up, and Tubby yelled, "I'll give you my goddam interest in this well. Every damn bit of it. I'm leaving."

He went to Mississippi to see his girl, Adele Howie, whom he had met at the University of Texas, but he lost there, too. She was marrying someone else. However, there was a redeeming factor. His Mississippi

Scouting had its moments

Corpus Christi Caller-Times Oil Editor Billy Blake reported that a number of oil scouts had been taking flying lessons. Several of the "dope grabbers" had already soloed.

"Don't be surprised, you drillers," he wrote, "if some day an airplane zooms over your rig and its occupant hangs his head out of the cockpit and shouts, 'How deep are you?'"

*****

One company devised a new way to inform oil scouts that no information was being given out and they were not welcome at the well site.

"A friend of ours was startled to see a remarkable replica of a man hanging from a tree near a gate leading to a Jim Wells County wildcat," Blake wrote in 1938.

"Closer inspection of the dummy revealed a sign that read, 'This one got the dope, but he didn't get away.'"

"The Dope Bucket," April 28, 1938

A collapsed rig is an intimidating sight.

sweetheart was soon divorced. She and Tubby were married in 1946 and had many happy years together until her death in 1999.

There had been other rig accidents in which men were killed or hurt because there were so few safety provisions on small rigs. "Most injuries are a result of carelessness," Tubby said. "I had some bad injuries on a drilling rig. Fused my spine and all that."

In 1934 he went to work in San Antonio as a geologist with Tidewater Oil Company under chief geologist A. I. Leverson. He was geologist, landman, and oil scout.

By this time South Texas had become a hotbed of drilling activity. All the companies had scouts gathering information about various prospects. They were expected to give reports on all well activity in their district. As Tubby knew them all and was well liked, they elected him Bull Scout. This meant he rode herd on some sixty-five scouts, presided at scout checks, and made assignments as their leader.

Even though he had a wealth of achievements, an association with many of the most influential men in the industry, and many discoveries to his credit, he regarded being Bull Scout as one of the highlights of his career because of the pure joy of it.

He had no trouble keeping the troops in line. No one dared question his preference to personally scout the Laredo area.

A new member couldn't see why Tubby should reserve this cushy assignment for himself. An older member cautioned him not to question Tubby. "There is one assignment you would not want him to give you," he was told. "There is no road, but there are some thirty-five gates along the way, all locked."

That was the assignment faced by any scout who did not perform to Tubby's satisfaction.

Tubby himself knew how to negotiate such hazards. He carried keys to most of the locks. He also had tools to lower the fence, then reinstall it, or he could cut the chain with bolt cutters and replace the lock with his own. He also carried a reliable compass.

"I got stuck in the sand many times between Alice and San Diego," he said. "I had a block and tackle to hook onto a tree and pull the car out."

When a company decided to withhold information about a well, it was a challenge for the scouts to find ways to learn what was happening. Tubby, an accomplished horseman, once donned full cowboy regalia, including chaps, hat, and a bag of Bull Durham, to saunter around a rig, surreptitiously gathering small core samples and hiding them in pockets under his shirt. He was once accused of taking a log that showed a company had been giving scouts false drilling reports, then returning it through a hotel room transom. This was never proved.

In 1936 Tidewater drilled a Duval County well that was not a successful venture. The company gave Tubby, chief geologist and

Quest sometimes turned tragic

Not all South Texas landowners were interested in oil. In fact, some were downright hostile to promises of riches.

It was in 1936, not long after the Charamousca Field was discovered in northwest Duval County.

Money was being made and competition for leases was stiff.

"I learned that if those people there didn't want you to come on their property or if they didn't want to talk to you, you didn't go in," Tubby Weaver, who was geologist with Tidewater Oil Company at the time, recalled.

"I knew those people. I lived among them and could talk to them. I spoke their language. They were really poor, but a lot of them really didn't care about oil. Their people had lived on the land for many years, and they didn't want rich Americans taking it away from them.

"Tom Banks, chief geologist for SRC, a Tom Slick company, and a man from Austin went down there and talked to a man and offered to buy half of the royalty on his land. He said he didn't want to sell anything. They kept hassling him, and he told them to leave. He said not to come back—he didn't want to talk to them any more.

"I said, 'Tom, don't go down there again. That man is mean, and he doesn't want you there.'

"Tom and this fellow went to George West and had some coffee and said, 'Let's go back and offer that guy some more money.'

"Well, they went back and drove up to the house, walked up on the porch, and knocked on the door. The man pulled out his shotgun and killed Tom Banks. And he killed the man with him, and then killed himself.

"We know all this because the man's wife was standing right there and heard all the conversation."

—Tubby Weaver

South Texas oil scouts and landmen in Corpus Christi in 1936. Tubby Weaver is seventh from left, second row.

landman, its share in the project. This allowed Brown Drilling Company, still owned by his uncle, to drill a very productive well that opened the Charamousca Field. The uncle had retained Tubby as a partner in spite of his proclamation that he was leaving

In 1937 Getty Oil Company acquired Tidewater, and Tubby, who had just brought in dozens of wells in Duval and Webb counties, moved to San Francisco to work for Bishop Oil Company, developing gas sources for a cement plant that supplied cement for the big bridges and dams of the West. He returned to Texas in 1939 and worked as geologist with Midstates and Lockhart Oil Company until 1941.

Early in World War II the Japanese cut off the supply of rubber from Southeast Asia, creating a critical need for synthetic rubber. Carbon black was an essential ingredient in creating synthetic rubber, and natural gas was the key to producing carbon black. United Carbon hired Tubby as chief geologist to facilitate drilling for natural gas and arranging transmission lines to a carbon black plant erected near Rockport.

"I became a combination pipeliner, landman, geologist, and engineer," Tubby said. He was able to help local drillers in their search to find gas and others to market gas that had been wastefully flared in area fields.

After the war he worked for Tennessee Gas Transmission Company until he was offered a lucrative position in Houston. Because his wife wanted no part of Houston, he became an independent and a permanent Corpus Christi resident.

"I was a pretty big operator," he said. "I drilled as many wells as anyone in South Texas, and I brought in a lot of those [oilmen] that went on up."

Houston Mayor Oscar Holcombe was one of his partners. Tom May of San Francisco was a partner for thirty-five years, and Fred Sharp a partner for twenty-five. Weaver and Sharp drilled many wells. There were many others, including Bob Kirkwood and Arnold O. Morgan.

Tubby was associated with many South Texas oil and gas field discoveries. Some of them were Charamousca, 1935, Duval County; Munson, 1938, McMullen County; Adami, 1939, Webb County; Bonnie View, 1944, Refugio County; Helen Gohlke, 1951, Victoria and DeWitt counties; Biel, 1956, Webb County; La Huerta, 1957, Duval County; Que Será 1958, Webb County; Ocones (Yegua), 1957, Duval County.

William Carlton Weaver was a charter member of the South Texas Geological Society in 1936 and of the Corpus Christi Geological Society formed when travel was restricted during World War II. His peers bestowed a number of honors on him for his contributions to the profession.

He sold off his properties when his wife became ill but continued to be active in geological affairs until his death on June 30, 2004.

"It's a sad thing when you live as long as I have," he said in 2003. "You outlive your whole family. I've always liked young people. It's a good thing I did because all my contemporaries are all gone. It's been my good fortune to spend most of my professional life in South Texas, one of the most important oil- and gas-producing areas of the United States. I've found oil people are good people. They aren't the trash some people accuse them of being."[29]

What's in a name? Part 1

Que Será Oil Field was a fond memory for Tubby Weaver. Another field name had been chosen, but one already bore that name. His daughter was playing her record player, and "Que Será" was the song.

"Daddy," she said, "why not name it Que Será?"

Que Será it became.

—Tubby Weaver

*****

Ira Brinkerhoff, chairman of the South Texas Geological Society nomenclature committee, announced in 1938 that the name of the De Soto oil pool discovered by Arkansas Fuel Company would be changed.

The company had no particular preference for a name, so they had left it up to the fee holder, Mr. Cuellar. He had been as surprised as anybody when it was named De Soto.

Perhaps Brinkerhoff was new to South Texas. It had sounded like Cuellar said "De Soto." What he said was "Tesoro," "treasure" in Spanish. A treasure the discovery was for Mr. Cuellar. And the name was changed to Tesoro.

"The Dope Bucket," July 17, 1938

Adele Weaver (second from right) accompanied her husband, Tubby, (second from left) on this visit to the field.

Among the "good people" in Tubby's description was Tom Graham, who also received that description from his brother-in-law. Graham was married to Virginia Deaver, daughter of a Texas Ranger and former El Paso sheriff who was a certified veteran of the Indian Wars, and her brother, Eugene Deaver, worked as a roughneck for more than thirty-five years, most of them as a derrick man.

"I started with Tom Graham, back with the old wooden derricks," Deaver said. He remembered Alice as a mud town with bad whiskey, Freer as a rough boomtown.

"We lived in tents. Ate lots of jackrabbits," he said. "They were tough but good in stew. Ate armadillos. They were good, too, but a little greasy. Lots of beans and potatoes."

"Tom Graham was a good man. He treated everybody honestly. You know, he was the only one who paid his bill at the water department."

Tom Graham (center) accompanied Fleet Admiral Chester Nimitz in a parade after the Allied victory in World War II. Judge L. M. Fischer is at right.

Deaver had learned the business by sitting on the step and watching the drilling crews work until he saw how every job on the rig was done.

"I hired out," he said. "They said, 'How long have you been working?' I said, 'This is my first day.' I could do any job on the rig. I was powerful when I was young—I picked up a four hundred-pound bale of cotton and walked off with it.

"Driller came out there one night drunk. I took over and drilled that hole 262 feet. Driller woke up and said, 'I don't believe it. I don't see how you did it.'"

That was his first experience as a driller.

Deaver said that Tom Graham changed after his son Jake was killed in France during World War II. "After his boy got killed, he never took much interest in life."

Tom Graham

Another of Graham's sons, Jack, became a successful Corpus Christi realtor. The reason that Jack Graham didn't follow the footsteps of his father could be that he had no taste for the business.

"When I was a baby, they warmed my bottles on the old boilers," Jack said. "They cored, then would lay the core out in this trough. They'd put this doodlebug light on it. Doodlebug light was purple, and the oil in the core would show up green. Then Dad made me smell it. And then he made me taste it. I ate so much core before I was fourteen, I thought I was going to turn into one."

An anecdote told by one of Graham's partners, Harry Mosser, could explain Graham's dislike for the taste of oil. The two of them tasted a core. The taste was tantalizingly close to oil. They tasted again and again. Suddenly Mosser felt ill but was too embarrassed to admit it. He hoped he wouldn't be seen. He ran to hide behind a bush to throw up so nobody would spot his weakness. Unfortunately, Graham was already behind the bush. He was throwing up, too.

Jack was fourteen when his father died, in 1948. "He drilled a lot of wells and he drilled a lot of dry holes," Jack said. "They went back in later and made good wells out of a lot of them."

Tom Graham was born in Gonzales County in 1892. In 1900 his father, a Methodist circuit rider minister, died and the family moved to Alice, where Tom graduated from high school. He worked in banks for ten years and then went to Mexico, where he became interested in the oil business.

Some reports say he was involved in acquiring leases for Humble Oil Company. He was also reported to have had a business relationship with Pancho Villa, whom he considered a friend. He told many anecdotes about the Mexican folk hero.

"I guess it was when he was working for Humble that he went down to Mexico," Jack said. "I have two half brothers and two half sisters and they lived in Mexico. He could speak Spanish fluently."

'Dad closed more deals on the golf course than he did in his office. I can remember Jimmy Demaret, Byron Nelson, and all those old golfers who used to come to the house after tournaments. They'd have a big party at the house.'

—Jack Graham

Graham returned to South Texas and the oil business in 1925 as superintendent of the land department of the Houston Oil Company. He moved to Corpus Christi in 1931.

Jack Graham told a story that illustrates his father's generosity. Gus Tsesmelis, owner of the Alice Café, was a very close friend of Graham.

"He was always getting Daddy leases," Jack said. "One of his greatest ambitions was to own a Cadillac. One day we walked in there, and Dad pitched him a set of keys to a brand new Cadillac parked outside."

Graham drilled a hundred wells for La Gloria Corporation in the Willamar Field in Willacy County.

"In the Valley he was impressed by a roustabout who was carrying two sacks of cement up on the drilling rig floor. That was Jack Modesett. That's where he got his start—with Dad," Jack said. "Jack Modesett and my brother Burdett bought out the company and Jack Modesett Sr., and it became Mokeen Oil Company with Joe Kennedy—President John F. Kennedy's father. Modesett was the Mo in Mokeen. He was killed in a car wreck on the way to Port Aransas.

In 1941 Graham was president of the Rontex Oil Company, the Kiran Oil Company, the Tom Graham Oil Company, and the Jim Wells Pipeline Company and chairman of the East Alice Recycling Company. He also served as president of the Corpus Christi Country Club and the Chamber of Commerce. He was an avid golfer.

"Dad closed more deals on the golf course than he did in his office," Jack said. "I can remember Jimmy Demaret, Byron Nelson, and all those old golfers who used to come to the house after tournaments. They'd have a big party at the house."

Jack told of a young newcomer who asked Tom and Guy Warren how much they were playing for. Graham replied, "Twenty-five a hole." Upon losing, the fellow was worried to death because he didn't have any money.

"He said, 'Mr. Graham, how much do I owe you?' He answered, 'Seventy-five cents. Twenty-five cents a hole.' My dad said any time you gamble on golf, it's not a game any more."

Graham was primarily a drilling contractor, but he did some wildcatting and owned production from several hundred wells. The blowout near Clarkwood, one of the largest in the area, was Graham's rig.

Blowouts were just one of the hazards the rig workers faced. Drinking was a bigger problem in those days than anyone will admit, Deaver said. It was the cause of many accidents and injuries. He himself quit drinking after he was thrown in jail.

He remembered several narrow escapes. One time he grabbed a bare electric wire and was severely shocked but held on and didn't fall from the derrick.

"The Good Lord was taking care of me," he said.

(Opposite) Tom Graham on horseback (Below) Location rigging up to drill a relief well for the blowout at left

One morning an explosion blew the doors and windows out of the doghouse when he lit the boilers, but he was unscathed. On still another occasion an angry driller caught him between the tongs, trying to kill him. Graham fired the man.

"I've said all my life I'd rather have Jesus Christ sitting beside me than all the money in the world," he said. "He's been with me ever since."

Deaver, who was 95 in 2003, outlived all the guys he worked with. Asked during an interview that year if he got any exercise, he replied, "Yeah, I play dominoes."[30]

While Tubby, Tom Graham, and Deaver all had deep Texas roots, other adventurous souls continued to be drawn to area oil fields. One of these was John Crutchfield, who came to South Texas for just a nickel an hour.

Crutchfield majored in petroleum engineering at the University of Oklahoma because jobs were scarce and people in the oil business had jobs. He learned that he could make five cents an hour more in Texas than at home, so he got a job with Humble Oil and Refining Company as a roustabout in the Greta Field near Refugio, then on a drilling rig near Freer and Cotulla.

Like roughneck James Mosier, who had lived there as a child, Crutchfield had vivid memories of Freer.

"In the late thirties and early forties there were two hundred drilling rigs operating in Duval County," he said in a 2003 interview. "Freer wasn't like some of the boom towns back in the 20s, but it was about all the boom I could handle."

(Above) Eugene Deaver, brother-in-law of Tom Graham, at his Corpus Christi home in 2003 (Right) Oil and farming interests coexisted on the plains of South Texas.

Most of the roughnecks at Freer chose to live in Alice, where there was housing, and drive the thirty miles over a caliche road with cavernous potholes and clouds of choking dust.

"Roughnecks were well paid, making $1.25 an hour, and they had the best credit in town," Crutchfield said. "We moved to Alice and immediately began charging our groceries at the corner store, and they'd never seen us before.

"In the late thirties," he said, "seismograph was developed, mainly to locate salt domes. We just drilled more wells in those days. If you drilled ten wildcats and got three producers, you were considered to be very successful. Now you have to make seven out of ten to be termed successful."

He said that Corpus Christi had a population of 59,000 in 1940, with "a lot of major oil companies in town."

After five years with Humble, he established a consulting firm to evaluate oil and gas reserves. James O. Lewis, one of the first petroleum-engineering consultants in the industry, helped him get started. Lewis was considered the dean of the profession, Crutchfield said.

"Some of it was reservoir engineering and [using the] best practices in producing," he said.

"We had three major independents here," he said—La Gloria Corporation, Southern Minerals Corporation, and Renwar Oil Corporation.

He worked as a consultant for the independents, most of them, he said, very successful.

'Freer wasn't like some of the boom towns back in the twenties, but it was about all the boom I could handle.'
—John Crutchfield

John Crutchfield

'There was a lot of interaction between different professions—geologists, engineers, landmen, scouts.'
—John Crutchfield

During World War II he was a member of the Military Petroleum Advisory Board, which helped determine the maximum production from fields without damaging the reservoirs.

"During World War II the Donald Carr Field was producing 40,000 barrels a day. It was very rich in toluene, which was essential in the manufacture of high octane gasoline," he said. "The Tom O'Connor Field was increased to 80,000 barrels a day.

"When I came to Corpus Christi first," he said, "oil was 90 cents a barrel, and even after the war it only got up to a dollar and a half. The best price that I ever sold oil for was $3.65. We operated profitably when oil was selling for $3 a barrel. I think we did almost as good as people today with $28-a-barrel oil.

"I remember when Sun Oil Company increased the price of oil from 90 cents to a dollar and a half a barrel, I'm telling you, there was a real ovation up on the Hill," he said, referring to the office buildings above the bluff in Corpus Christi.

He remembered when 90 percent of the people on "the Hill" were in the oil business. "We met at the Nixon Café, and there was a lot of interaction between different professions—geologists, engineers, landmen, scouts. It was a great place to be.

"We had our share of wildcatters here. They tended to come and go. The independents were here to stay unless they were bought out by one of the majors."

Mythology was what made the search for oil interesting.

"It appealed to a lot of people. Because of mythology, people got in the oil business who shouldn't have been there in the first place. And part of my consulting business was rescuing some of those people from the bad decisions they had made," he said.[31]

In April 1946 Horton Pruett joined Crutchfield in the firm of Crutchfield and Pruett. In 1952 they hired James O. Lewis's son, Donald D. Lewis, who was working for Standard Oil Company of Indiana.

Looking back on his days in the industry, Crutchfield said "of course" he would do it again.

"If the sands were good, it was fun," he said, "kind of like a poker game."[32]

If Crutchfield's achievements were compared to a poker game, he probably held his best hand in the early 1960s—at an oil field called Prado, where luck, skill, and determination would come together to lay the groundwork for one of South Texas's most successful business endeavors.

Luck, according to A. O. Morgan, was a necessary ingredient. He probably believed in luck more than most oilmen.

"My theory has always been that I would rather have luck than brains because I can hire brains," he said in an interview in 1978, three years before his death.

"When I discovered the Arnold David Field on the Chapman Ranch," he said of the well that was to make him a fortune, "I staked the location in Section 113 because 13 had always been my lucky number, and that's how we discovered the field. That's luck. No science at all."

He conceded it wasn't wise to enter risky projects without partners, as partners spread the risk. He blamed government policies for making the business difficult for the independent operator, and he was correct in his prediction that "the independent is short-lived. He won't be here long unless something is done. And yet the independent has been the backbone of the oil business. Independents have discovered more oil and gas than the majors ever thought of. That's because they like to get out and take chances."

Even in 1978 it was very difficult for a wildcatter to strike it rich, he said.[33]

Born in Belton, Texas, Morgan lost his father when he was two. He studied mathematics and comptometer in night school and became a warehouseman for Coast Drilling Company, which was owned by his uncle William F. Morgan. William F. Morgan was the president of United Gas Company and a partner of William L. Moody III, O. R. Seagraves, and T. P. Morgan in numerous leases near Refugio.

Replacing brake bands on drawworks in the summer of 1937

The firm was co-owner of pipeline facilities that gathered crude in the Refugio area and transported it to Harbor Island for loading aboard ships. The pipeline was sold to Atlantic Refining Company in 1930. The next year Seagraves and the Morgans merged their properties to form Republic Natural Gas.

A. O. worked for Republic as a warehouseman, in the land department, in the geological department, and as a scout. In 1933 he worked for John L. Sullivan doing well work, taking cores, and running casing and completions. Ted Scibienski also learned well work with Sullivan's Texon Oil Company.

Morgan also worked for Clymore Oil Company as landman and engineer and bought and sold crude and distillate. He became an independent as Morgan Minerals in 1935 and discovered the Clarkwood Field and the West Sinton Field. William F. Morgan owned Mills Bennett Oil Company, which had huge successes in the Laredo area. William Morgan then sold out and bought a huge ranch.

Sullivan was a jewel of a joker

When oldtimers were asked to name the biggest characters they had known in the old days, most all of them said John L. Sullivan and his son-in-law, Bevo Garnett.

Both were big and loud and widely known. Sullivan made a lot of money and lost a lot on horses. He and Bevo had about twenty rigs working all over South Texas.

Bevo had a reputation as a practical joker, though his jokes could not always rightly be called funny.

One was at the expense of Jimmy Dellinger, of Brown and Root.

A group of local oilmen had gathered at Taylor Brothers Jewelry Store, looking for Christmas presents for their wives. It was cold and they were all wearing heavy coats.

When Dellinger got home, he reached his hand in his pocket and pulled out diamonds, rubies, watches, rings and all sorts of jewelry—a fortune.

Dellinger told friends, "If I had been stopped coming out of that store, I could never have explained all this stuff."

Bevo probably got his laughs from a distance.

—Tubby Weaver

(Right) Cementing surface casing in November 1935. A. O. Morgan became an independent as Morgan Minerals that year. (Opposite) Humble oil-field workers show their support for the war effort in 1942.

A. O. drilled many wells in South Texas. He sold his holdings to Houston Natural Gas Company in 1963 and later headed Morgan Enterprises and Tejas Minerals. After the Japanese attack on Pearl Harbor on December 7, 1941, he served in the Marine Corps in 1942 and 1943.[34]

The U.S. entry into World War II brought difficult times for the entire nation, including the oil industry. The loss of foreign markets for crude oil shipped through the Port of Corpus Christi caused a drop in activity, and, except for extensions of existing fields, there were no significant discoveries in the region during the war.

Even so, on June 21, 1942, a *Caller* story reported that several important deep tests under way in the Lower Rio Grande Valley oil area provided "the most interesting news in several weeks."

"In southeastern Cameron County, Pure Oil Co. No. 1 Skelton Award Vacancy, about seven miles southwest of Port Isabel in Santa Isabel Grant, will be another deep test drilled in that vicinity by Pure Oil. Fain Drilling Co. moved in a big rig from the Willamar Field in Willacy County to drill the test.

"This will be the first test to be drilled in the Lower Gulf Coast blackout area, and the big derrick will be completely covered by boards to prevent the derrick lights from being seen out in the Gulf. The Lower Gulf Coast has recently been included in blackout or dimout regulation so all lights on the rigs will be covered."

The blackout was a wartime necessity, as German submarines used lights on the U. S. mainland to silhouette their targets in the Atlantic and the Gulf of Mexico. In the beginning they found easy pickings, sinking

The U.S. entry into World War II brought difficult times for the entire nation, including the oil industry. The loss of foreign markets for crude oil shipped through the Port of Corpus Christi caused a drop in activity, and, except for extensions of existing fields, there were no significant discoveries in the region during the war.

(Above) Americans on shore watched the war news and worried as German submarines inflicted heavy losses on Allied shipping in the Atlantic and Gulf of Mexico during 1942 and 1943. (Opposite) Tankers like this one, their Merchant Marine sailors, and the military crews who manned their guns contributed greatly to the Allied victory in World War II.

freighters and tankers delivering fuel to the East Coast. According to United States Merchant Marine statistics, twenty-six tankers were sunk in the Gulf of Mexico during the war and seventy-five off the East Coast, with the known dead totaling 633. Many others died but were never listed as casualties. Quite likely the majority of them sailed from Texas ports.

These men were among America's first casualties of World War II, and they died as bravely as any fighting men who perished on Pacific or European beaches or in battles at sea. These heroic seamen sailed with little or no protection until U. S. forces were able to develop defenses.

As the war progressed, statistics show the progress of the Allied cause. Of the tankers lost in the Gulf of Mexico, twenty-four were sunk in 1942 and the other two in 1943. Most of the ships were sunk with torpedoes, but several went down under additional fire from the submarines' deck cannon.

The U-boat commander who attacked the tanker Olney off the East Coast on January 25, 1942, was apparently serving an apprenticeship. He fired seven torpedoes and missed. He damaged the vessel with shell fire, but she escaped and the crew survived. On May 4, 1942, the tanker Munger T. Ball was sunk in the Gulf of Mexico by torpedoes and its crew decimated by machine-gun fire. Thirty men died.

In the *Caller's* oil column, The Dope Bucket, writer Early Deane, Jr., told of the loss of the Venore, a 550-foot ore carrier torpedoed off the Carolina Coast on January 23, 1942. It often carried oil from Port Aransas to the East Coast. Twenty-two crewmen were rescued. Twenty-two others were listed as missing. Postwar statistics said that seventeen of them died. The heaviest loss of life on a Gulf tanker occurred when the Gulfstate was torpedoed on April 4, 1943. Thirty crewmen and six members of the armed guard died. The last tanker lost in the Gulf was the Touchet, torpedoed on December 3, 1943. Ten members of the armed guard were lost, but all crewmen survived.[35]

On January 2, 1944, *Caller-Times* Oil Editor Nancy Heard wrote, "Labor is hard to get. Materials are scarce, operating costs and taxes are up and other troubles confront the industry, yet, in most cases, the outlook is good.... Oilmen realize the serious need for added petroleum and intend to see that it is produced."

Heard, who had become oil editor by late 1943 and would later take the same position with the *San Antonio Express-News,* predicted a great increase in wildcatting and extension of existing fields. By 1944 more steel was becoming available, but labor continued to be a problem.

"Drilling wells is a job that requires not only strong men, but trained ones," she wrote. "Use of inexperienced or young men has been practiced as much as possible, although this sometimes results in expensive fishing jobs, not to mention injuries to the men themselves."

One drilling superintendent told Chester Wheless of his problems during the war.

"Workers were scarce, and it was hard to fill out a crew every day," Wheless said. "Some of the fellows spent their days off in the beer parlor. He would try to get people, his own employees, to stop drinking and come back to the rig. He was told, 'I haven't used up my money yet. So I don't think I want to work today.' The pay was about a dollar and a half an hour."

The industry was hit harder than most in key labor losses because "not only did a large number of oil-field workers enlist immediately after

Future enemies were Port customers

"One of the things that made Corpus Christi into a prime shipping point was the demand for Miranda crude which was a good lubricating base. Strangely, one of the main customers were the Japanese who were storing it up for the Japanese Navy. The other reason for the Port of Corpus Christi prospering during the 1930s was all the metal junk, scrap they later shot back at us during the war."

—Wilford Stapp

Independents unite to protect their interests

A love-hate relationship has always existed between the major oil companies and independent operators.

The majors could be described as "integrated" companies that controlled lease acquisition, exploration, exploitation, production, shipping, and marketing, often with strong domestic and foreign connections—all under the same umbrella, with an inexhaustible supply of available cash.

A number of oil companies spawned by Spindletop would become majors, including Humble Oil and Refining, Gulf Oil Corporation, the Texas Company, and Sun Oil Company. Standard of New Jersey, assisted by its 50-percent of Humble, grew to a giant that was broken up to sire twenty other majors.

In the beginning the independent was a wildcatter, often uneducated but with earthy experience gained as a roughneck. Financing was a problem. Often the wildcatter achieved it by giving shares to landowners, crew members, and any backers who would gamble on the venture. If the well was productive, the gamble paid off handsomely. More often the wildcatter would have to start all over again with the same process.

World War II drastically changed the business. Before war's end the majors, becoming aware of the vast Mideast oil reserves, had organized the Anglo-American Petroleum Agreement, which advocated "restraint on the use of domestic reserves and larger drafts on foreign oil supplies." They lobbied for congressional support for the agreement on the grounds of "future national security."

Through the efforts of independents "augmented by the intervention of [Texas Railroad] Commissioner Olin Culberson," the treaty was blocked, but the episode convinced Texas independents of the need for a forum to represent their interests.

In March 1946 thirty-nine independent producers, led by Glenn McCarthy and Jack Porter of Houston, met in Austin and formed the Texas Independent Producers and Royalty Owners Association (TIPRO). Another of the leaders was R. L. "Bob" Foree of Dallas, and among the thirty-nine were Walter Henshaw and L. A. Douglas of San Antonio and Guy Warren and Maston Nixon of Corpus Christi.

Information from Lawrence Goodwyn, *Texas Oil, American Dreams*, (Austin: Texas State Historical Association, 1996).

Pearl Harbor, but many did not take advantage of draft deferment after it was offered to the industry."

Heard noted that some of the drilling costs were financed by companies whose main business was not the search for oil—they were concerns with surplus money or those with large war contracts, she wrote.[36]

Heard was correct in her prediction of increased drilling. During 1943 Sinclair-Prairie completed the first oil pipeline out of Southwest Texas, from Corpus Christi to the Houston area, and "by the end of 1943, expanded pipelines and tanker outlets had improved markets for Southwest Texas crude oil. . . . The next year brought the highest level of activity since 1937, with production at an all-time high and drilling up 22 percent over 1943."[37]

One important result of the war was the growth of the petrochemical industry. The first petrochemical installation in the region was the Celanese Corporation of America plant constructed at Bishop in Nueces County to process butanes and propanes recovered by gas-recycling plants at La Gloria in Jim Wells County.

On August 27, 1941, United Carbon Company began operation of a plant at Aransas Pass to produce carbon black, vital in the production of synthetic rubber.

Geologist Tubby Weaver, who was in charge of finding natural gas for the carbon black plant, ran across one woman who did not support the war effort.

She threatened to shoot him if he tried to put a pipeline across her land. Construction equipment was lined up at her fence as she displayed her .30 caliber rifle to emphasize her point.

"I told her she was holding up the war effort," Tubby said. "That pipeline was to carry aviation gasoline for the war effort.

"She said, 'I don't care who it's for.'

"I told her, ' If you don't sign the papers in the morning, I'm going to have the newspaper put a three-inch headline about how you are stopping the war effort.'"

She angrily signed the papers the next morning.[38]

A wartime project that provided an outlet for the growing gas production came to fruition in October 1944, when the Tennessee Gas Transmission Company opened a gas line with a capacity of 200 MMcfd [million cubic feet a day] from the Agua Dulce-Stratton fields of

Nueces County to West Virginia, immediately increasing the value of gas reserves in the area.

Even more important, both to the war effort and to the future of the South Texas petroleum industry, were two other pipelines—the Big Inch and the Little Big Inch. As early as 1940 Secretary of the Interior Harold Ickes realized that German submarines would be a deadly threat to tankers in American waters. In 1941, at his urging, oil industry executives began to plan two pipelines—one, twenty-four inches in diameter, called the Big Inch, would transport crude oil; the other, twenty inches in diameter and called the Little Big Inch, would transport refined products. The lines were the largest single government project in history.

On August 3, 1942, work began on the Big Inch. "A ditch four feet deep, three feet wide and 1,254 miles long was to be dug from Longview across the Mississippi River to Southern Illinois and then east to Phoenixville, Pennsylvania, with twenty-inch lines from there to New York City and Philadelphia. Crude oil was delivered to the end of the first leg, Norris City, Illinois, on February 13, 1943. By August 14, 1943, the Big Inch had been completed. In January 1943 approval was given for the first half of the Little Big Inch; approval for the entire line was given on April 2. This line, beginning in the refinery complex between Houston and Port Arthur and ending in Linden, New Jersey, was completed on March 2, 1944."

Both lines were financed by the government through the Reconstruction Finance Corporation. Together they carried more than 350 million barrels of crude oil and refined products to the East Coast before the end of the war.[39]

Many years later Texas geologist/petroleum engineer O.G. McClain recalled not the early slowdown but the dedicated efforts of the later wartime days.

"You know," he said, "Winston Churchill made the most pertinent comment near the end of World War II when he said, 'The allies floated to victory on an ocean of oil,' and the United States provided most of that oil.... Drilling activity, number of wells drilled, fields drilled, discoveries made just skyrocketed, and that's where our oil came from we floated to victory on."[40]

Drilling activity in the South Texas area played an important role in providing the petroleum products that were vital to the Allied victory in World War II.

New Horizons

It was a great time to be in the oil business. The war was over. The boys were coming home. Price controls were being removed. America was gearing up for peacetime prosperity, which would require enormous quantities of fuel. And South Texas was the place to find it.

Steel mills were hard pressed to keep up with the demand for drill pipe and casing. Huge new wartime pipelines to the East improved marketing prospects for Texas products. For the first time homes and manufacturing plants could use natural gas, making it a more valuable commodity.

Both major and independent firms created South Texas district offices, and cities and towns flourished with a new prosperity, both on and offshore.

Taxes from the new wealth were a boon to the schools and government of Texas, which began to rely on petroleum as a permanent support.

But warning clouds appeared as competition from foreign imports began to threaten domestic production. Discovery sands were located much deeper and at greater costs. The number of independent operators decreased rapidly, particularly during the 1980s, when a devastating slump in the world and domestic markets caused a drastic cutback in the number of rigs active in the country, resulting in widespread unemployment.

Better geophysical equipment and computers improved seismic interpretations and reduced the gamble in drilling operations as the emphasis began to change because deeper horizons in South Texas yielded not oil but natural gas.

Going in the hole. The kelly is down to begin making a connection.

This image of sheep grazing in front of a drilling rig illustrates the coexistence of oil and agriculture on the plains of South Texas.

Even after the end of World War II, flares–and fires–lit up the South Texas skies.

Pipelines brought age of natural gas

Just as the Spindletop gusher at Beaumont ushered in a new oil age and transformed America forever, shortly after World War II a group that would become Tennessee Gas Transmission Company (TGT), later Tenneco, brought another major change to American life—and to the oilfields of South Texas—with the introduction of the age of natural gas.

O. G. McClain, geologist and consulting engineer, said, "There was an event in October of 1945—mark it well—when TGT completed and put into operation the first long distance [natural gas] pipeline. It went from Stratton, Agua Dulce, to West Virginia, 1,265 odd miles, more or less. . . . This was an event that should be recognized as a seminal event, because for the first time natural gas could be transported long distances and serve a multitude of uses in distant places, far from its origin."

In explanation he quoted a friend who said, "There are two factors in business: one is the put and one is the take."

"We could put the gas there," McClain said, "but if you didn't have anyone to take it and use it, that would choke off the put. It's just that simple."

McClain pointed out that hundreds of thousands of miles of gas pipeline had been laid by 2000 and almost all factories and homes in the United States had natural gas available as a fuel. These developments led to new technologies and whole new professions—for geologists, engineers, business people, and many others. In addition, electricity is largely generated by natural gas, a fact that further elevated the significance of the pioneer pipeline.[1]

Furthermore, opening vast new markets through the new gas lines rejuvenated the South Texas petroleum industry, eventually raising the value of a product once thought to be almost worthless. It would take years to overcome pricing and governmental control problems, but natural gas would make the area attractive to new industries and become a new industry itself.

Opening vast new markets through the new gas lines rejuvenated the South Texas petroleum industry, eventually raising the value of a product once thought to be almost worthless.

Ironically, government controls had an advantageous effect. In 1954 the U. S. Supreme Court ruled that the Federal Power Commission had jurisdiction over the production of gas sold in interstate commerce. The FPC had set interstate prices since 1939, and its policy of keeping the price low led to the growth of intrastate gas sales and the further expansion of the Texas petrochemical industry.[2]

Although the road was not always smooth, the emergence of the natural gas and chemical industries, the addition of offshore drilling, and the development of new technologies contributed to a period of great prosperity for the petroleum industry in the area and elsewhere. Major oil companies became international giants, smaller companies achieved unprecedented growth, and South Texas independents flourished.

If there was one man who brought natural gas to America, it was Gardiner Symonds. Symonds had been vice president of the Chicago Corporation before moving to Corpus Christi in 1938. He was "farmed out to form Tennessee Gas to build gas transmission lines in 1940 to move gas from gas reserves held by the Chicago Corporation."

The project had started in 1940 with an incorporated group interested in a pipeline from Louisiana to Tennessee. However, neither financing nor a gas reserve was available. The Chicago Corporation took over a 90-percent interest, got a $44 million Reconstruction Finance Corporation loan from the government, and built the line to West Virginia. *Business Week* magazine reported that Chicago Corporation feared that ownership would result in federal control and sold out.

"Chicago Corp. made $3.8 million from the sale, but lost a vice president, Gardiner Symonds. . . . Symonds chose to remain as TGT president in 1943 when Chicago sold out," the publication said.

But competition for Symonds and TGT soon came from an unexpected source. During World War II the Big Inch and the Little Big Inch had served the country well. After the war the government placed the two pipelines up for bids. Symonds entered a bid but lost out to Herman and George Brown, founders of the worldwide Brown & Root construction firm. They won with a bid of $143,127,000 and formed Texas Eastern Gas Corporation.

"The transaction represented one of the few instances in which taxpayers received full return on money invested in a wartime emergency project."

Texas Eastern later adapted the pipelines to carry gas and built compressor stations to keep pressures constant. Both Texas Eastern and TGT began aggressive drilling programs to bolster their supplies.

In the spring of 1946, TGT hired William "Bill" Vrana, recently discharged from the Army and living in Houston, as a draftsman, with the provision that he would be transferred to a new geology department the company was forming. The transfer came through on July 3, and he began his job as the first geologist hired by the company.

Gardiner Symonds, who played a major role in bringing natural gas to the nation, was on the cover of *BUSINESS WEEK* magazine on January 28, 1956.

By the following year Tenneco had thirty-five exploration geologists in twelve district offices. From June 1946 to December 1953, Vrana worked as geologist and senior geologist in the Houston office. Starting in 1954, he was district geologist and exploration manager in Corpus Christi for six years.

"It was an exhilarating experience to work for a new company in its formation stage," Vrana said. "Everyone in the company felt close to each other."

Completely unassuming, Symonds drank coffee and talked with his employees as equals. "He would show up at the rig before the well was to come in, roll up in a blanket, and would soon be snoring with the rest of us," Vrana said.

Vrana told of one close call he had on a rig.

"The drilling line broke, and the traveling block fell on the floor," he said. "The driller and I were standing there drinking coffee. The driller jumped out, and he and the tool pusher ran into each other and fell onto the floor. There I was right in front of them. I couldn't move. So I just stood there with my cup of coffee, and everything got quiet all of a sudden after all the noise. Nobody got hurt. We were in the right spot."[3]

Bill Vrana

'It was an exhilarating experience to work for a new company in its formation stage. Everyone in the company felt close to each other.'

—Bill Vrana

As Symonds and others ushered in the age of natural gas, many companies still focused their efforts on the search for oil. In the late 1940s Republic Natural Gas Company—the company formed in 1931 by O. R. Seagraves and the Morgan brothers—was one of the rare original companies that was primarily looking for natural gas, Louis A. Beecherl, Jr., retired chairman and CEO of Texas Oil and Gas, recalled. Beecherl was district engineer of the Republic office in Corpus Christi after World War II service in the Navy and graduation from the University of Texas.

In the early forties Republic had the old Saxet Field that supplied the City of Corpus Christi with gas, Beecherl said, and supplied United Gas with most of the gas taken out of the area.

"At that time United Gas was the biggest marketer in all of the Gulf Coast area, Houston, and East Texas," he said.

Beecherl worked for Republic as district engineer until 1954, when he, Dick Andrus, and Herman Wetegrove went into the business as consultants. He stayed with the gas business because Tennessee Gas, Transcontinental, and Texas Eastern all had the pipelines leading from the Coast to Eastern markets and the price was certain to go up. He and Arnold Bruner started Tex-Star Oil and Gas Corporation in Dallas.

"I took over Tex-Star in 1957," he said.

In 1965 he changed the company name to Texas Oil and Gas Corporation because there was a Tex-Star owned by Tom Slick in San Antonio. The company did well. By 1965, the first year Texas Oil and Gas was listed on the New York Stock Exchange, revenues had grown to $10 million annually. Things improved further in 1973, when the price of natural gas shot up from thirty cents to two dollars a thousand cubic feet.

Beecherl retired in 1977 after twenty years as chief executive officer of Texas Oil and Gas. Gov. Bill Clements appointed him chairman of the Water Development Board in 1979, a position he held until 1985. He remained an honorary member of the board of Texas Oil and Gas and was responsible for its continued growth and the eventual sale of the company to U. S. Steel in a stock trade valued around $3.8 billion.[4]

Another South Texas geologist who started his career with Republic was M. H. Major. Major, a recent geology graduate, was disappointed in 1942 when the Army Air Corps rejected him because of his bad eyesight, but it might have been the luckiest thing that ever happened to him. After a brief period as a surveyor with a Houston seismic company, he got a job as a Republic geologist.

"I was delighted to have the job, but it was years before I realized the incredible good fortune of my situation—the timing in the development history of South Texas oil and the nature of the oil business and the geology of the area," he wrote in a memoir in 1992. "Oil could be found at almost any depth, from a few hundred feet to the deepest any rig could drill. Ownership was broken into reasonably small tracts, so dealing could be done, although there were still large ranch tracts available."

Christmas tree on a Republic Natural Gas Company well

The most obvious advantage for him was that there was no competition from other young geologists, for most of them were caught up in World War II. Another was the association with Lewis Kelsey, Republic's chief geologist, who hired and trained him.

During his job interview Kelsey explained that Major would be his assistant and they would be hunting for oil.

"Kelsey didn't use fancy words like exploration," Major said. "We would maintain maps on the whole area, keep up with field development, and recommend lease purchases to the land department. Also he said we would be receiving many submittals from brokers and promoters and we would have to make evaluations and recommendations on these."

The acquisitions might lead to the drilling of wells, Kelsey told him, and then their job would be to supervise the coring, logging, and testing of these holes, which would lead to the discovery of oil and gas.

"I don't remember discussing pay, benefits, or anything except the kind of work we would be doing," Major said.

In early August he received notice that he got the job.

"Republic's office was in a galvanized iron production building five miles west of Corpus Christi on their Dunn lease, near the discovery well of the large Saxet Field that had catapulted Corpus Christi from a small town to a city," he said.

He served his apprenticeship spotting wells and contouring maps. A problem was keeping the maps from smearing from sweat produced

This galvanized iron building five miles west of Corpus Christi was the office for Republic Natural Gas Company. M. H. Major said it was difficult to keep maps from smearing in the hot, crowded offices, which had no air-conditioning.

in a metal building with no air-conditioning, where the geologists, scout-landman, and secretary were crowded into two tiny rooms. The production department crew begrudged them even that.

His first assignment, he recalled, was working up a series of contoured maps on the Flour Bluff Field.

"I finished the job in three weeks and learned more in that time than I had learned in college," he said.

He met hundreds of drillers, promoters, geologists, landmen, operators, and others in the oil patch. They freely traded logs and well information, and they constantly swapped stories about where wildcats were planned and who had run an oil or gas test.

"I would run into people who gathered at well sites at a critical time for evaluating the hole," he said. "I became acquainted with more and more people in the business and learned a lot about production and geology from these contacts, and living two or three days, or maybe a week together, sometimes twenty-four hours a day, we became friends and acquaintances for a lifetime."

In his early days with Republic, Major was closely supervised, but soon, he said, he was expected to know what to do.

"We had engineers to help with these decisions," he said, "but Kelsey didn't want to yield much authority to them. Basically the geologist had the authority over the drilling operations until it was decided to run pipe into the hole and attempt to complete it as a producer. At that point

the roles of advisor and boss changed, the engineers taking charge and the geologists reverting to the role of advisors."

Major explained the basis of subsurface geology as "the construction of structural maps of formations at or near the depth of rocks in which oil or gas is expected to be found. After the early 1930s most wells were electrically surveyed, or logged."

Collections of the logs, which recorded the depth of various formations, were studied, and "a vast amount of new information constantly poured into the offices of the many companies and individuals active in the area."

As a well was drilled, cores or rock samples were taken at promising depths, and if oil indications were found, fluid tests were made to evaluate the productive capacity of the hole. In 1942 the company's main effort was to complete development of the Rachal lease on the west side of East White Point Field.

Major remembered Hugo Allen, who represented H. R. Smith, a substantial independent drilling contractor of Alice. "He introduced himself as they waited for core samples and said, 'Let's go out to the bunk house and talk about girls.'

"He might talk about girls between cores and tests at a well, but everyone knew he was about the best man in South Texas in putting deals together," Major said. "He wasn't a geologist, but he knew geological concepts. He kept Mr. Smith's four to six rigs busy drilling prospects throughout the area."

After about a year in Corpus Christi, Major took a job as a geologist in the Dominican Republic for Standard Oil of New Jersey, but about a year later he returned to Corpus Christi and Republic, where he found some important changes. The geological offices were now in "beautiful bayfront space" on the tenth floor of the Nixon Building, but for Major the most important change concerned personnel. Kelsey had hired as a secretary Ann Campbell, a "beautiful and intelligent business administration graduate of the University of Texas," whom Major had noticed as she studied in the geology library during their undergraduate days. A few months later, on September 10, 1944, they were married.

Major gave a vivid description of another person Kelsey had hired, a scout.

"He was Charlie Bolton, a Seabee, a veteran of the South Pacific war, discharged because of a chronic skin ailment contracted in the tropics. Before the war he had driven a Jewel Tea truck in San Antonio, then went to Freer to work as a roughneck in that area's oil boom of the late thirties. His early discharge was a stroke of luck. . . . Charlie was rough-cut, artificially hearty, and spoke atrocious English that he embellished with difficult words that were usually meaningless to both him and his listener, or he would surprise you by using some important sounding word in a context that proved he hadn't the foggiest notion of its meaning.

"He would tackle any assignment with the enthusiasm often characteristic of ignorance. His persistence paid dividends again and again as he made deals with recalcitrant landowners that most people couldn't even talk with."

Major also recalled the legendary brothers, H. H. and S. H. Howell. S. H. once tried to stop an enraged lady friend from leaving a restaurant. He "followed her outside, trying to explain things and mollify her . . . ," Major said. "She jumped in her car, ran into him, knocked him down, and ran over him. Fortunately, none of the wheels hit him, but he still suffered severe abrasions. 'You'd be surprised how low one of them Cadillacs is built,' he told me later. Their sister, believe it or not, was a tool pusher on one of H. H.'s rigs.

"In the field," Major said, "I was accustomed to the deference accorded my profession, or maybe I just imagined it. The first time I drove up to S. H.' s rig, his big yellow dog walked over, raised his leg, and wet my leg. The tall, blond, sleepy-faced Victor Mature lookalike, S. H., just stood there by the bunkhouse door and watched, his blue eyes twinkling.

"'Don't blame the dog,' he said. 'All he ever hears around here is 'Piss on the geologists.'"

M. H. Major

'[M. H. Major's] knowledge of South Texas geology was so complete that he could look at an electric log without seeing the heading and identify the location of the well within a few miles.'
—Jess P. Roach

After leaving Republic, Major became district geologist for Tennessee Gas Transmission in 1948 in Houston. He returned to Corpus Christi as an independent the following year.

Geologist Jess P. Roach, who worked for Major, said, "His knowledge of South Texas geology was so complete that he could look at an electric log without seeing the heading and identify the location of the well within a few miles."

Major was associated with Frank Zoch, H. R. Smith, and Hugo Allen in many projects. Zoch put drilling deals together, and Smith and Allen were in the well-drilling business. The first well drilled and based on Major's interpretation of the area was a discovery for the Midway-Holst Field in San Patricio County. A few weeks later a discovery well based on Major's interpretation was drilled for the Atlee Field in Duval County.

In late 1956 he became associated with George Taggert, Jr., and Southwestern Oil and Refining Company as a consultant. Southwestern, which was capitalized by the Simard family from Canada, hired Bill Burch after drilling fifteen excellent wells on a deal submitted by Bill Carl and Jeff Carr. Major worked with Zoch and Paul Turnbull, drilling six producers in the Clarkwood area, and was a partner with John Collier, working closely with Southwestern manager Charlie Steen.

Major and Collier were active in exploration near the community of Violet and Suntide Refinery in Nueces County, at Rancho Viejo and Escondido fields in Jim Hogg County, and in the South McCook and State Park fields in Hidalgo County.

Major lamented the passing of the annual Seguin picnic, which landmen, scouts, and geologists from San Antonio to Laredo and Corpus Christi to McAllen held from 1935 to 1955. There were games—baseball, golf, and riding the bucking burro. All families brought food. It was an event eagerly anticipated until an accident ended it. Frank Morrison and Jack Pitcairn, geologists and partners, wrecked their car returning home, and Pitcairn was killed. No more picnics were held.

Geology students at the University of Texas were so close they were almost like a fraternity, for they attended classes together, went on field trips together, attended the same summer field geology course, and formed lifelong friendships and associations.

"Bob Begeman, Wayne Wood, Jim Freeman, Louis Sebring, Benton Stone, Clay Cooke, Bus Bloomer, Jack Kern, J. B. Means, and many other classmates stayed in contact through the years as we took up active roles in the oil business," Major said. "Al Frericks became chief geologist for Forest Oil. We ate lunch together weekly until his retirement. Less than a year later he died while fishing in Laguna Madre."

Last but not least, except in size, was Ed Rowley, Jr., who, Major said, "was exactly five feet tall and perfectly proportioned.

The Seven Sisters area of Duval County

"I think Ed already had a business degree from Chicago," he said. "His dad had bought some leases adjacent to East White Point Field in San Patricio County. When these properties began producing significant quantities of oil, Mr. Rowley thought somebody in the family should learn a little about geology. [Ed] was a little older than the rest of us, and much richer. He had an air-conditioned Chrysler, only the second air-conditioned car I had ever seen."

In 1992 Major wrote this epilogue to his memoir as sort of a remembrance to friends he had known or worked with:

"E. A. Durham, my tennis friend who climbed the Matterhorn, died of leukemia.

"Laid off by Sinclair and unable to find a job in the oil business, Charlie Whited returned to West Virginia to a job checking foundations for the highway department. He died in 1986.

"Lewis Kelsey . . . kept most of the working interest and borrowed heavily to finance a five-well development. Before any of the reserves were marketed, a carelessly drilled Seaboard well blew out. It burned for three months, effectively exhausting the reservoir and bankrupting Kelsey. Not long after this, he suffered a massive stroke that left him paralyzed and speechless at the age of 46. He died in his mother's boarding house.

"Hugo Allen died of Alzheimers disease.

"Cigarettes killed Charles Bolton, ebullient Republic scout and landman, but not before he had realized an ambition to own a Mercedes-Benz automobile."

Hank Harkins' wife and mother died in a fire at their Alice home (originally built by H. R. Smith). "Hank finally died of a debilitating disease similar to Alzheimers.

"Paul Turnbull died an early death, a victim of heart failure.

"George Taggert, who had never wanted to go to Europe, contracted pneumonia and died during a Scandinavian cruise.

"Jacque Simard, head of exploration for Southwestern, died in Canada, still a relatively young man.

"Frank Zoch, 64 years old and suffering from intestinal cancer, parked his car in the parking lot of a funeral parlor and shot himself.

"Jess and Maxine Roach visited Ann and me on March 17 and 18, 1992 . . . . They looked wonderful!

"Charles Steen is actively engaged in the oil and gas business as an independent.

"Bill Harwood is still hunting oil and gas, as well as big game in Africa.

"Gay Barr and her husband Jack Wilkes are retired and living happily adjacent to one of the fairways of their golf club on the outskirts of Austin.

The No. 1 Shaeffer in DeWitt County, the discovery well of the Cottonwood Field, brought in by John, Bill, and George Hawn. Turnbull and Zoch was the drilling company, and the rig was an old Hewit and Daugherty steam rig.

James C. Freeman

James C. Freeman had a number of discoveries to his credit as an independent geologist, with his greatest successes on the O'Connor ranches in Refugio County.

"On July 10, 1975 Margaret Collier found John dead in his car parked by the W. B. Ray High School, of heat prostration or heart failure. He was 58 years old."

In May 1977 Willard Holland Major suffered a stroke that left him partially paralyzed, greatly restricting his geological activities. He died April 2, 1996.[5]

Major's University of Texas classmate Jim Freeman—James C. Freeman—had a number of discoveries to his credit as an independent geologist, with his greatest successes on the O'Connor ranches in Refugio County.

He was hired by Magnolia Petroleum Company's senior geologist, Fred Wilcox, in San Antonio in 1947 and trained by W. W. "Doc" Hammond. He joined McCarrick, Gouger & Mitchell after a year and a half and became an independent in 1951.

"The oil business was booming after a long slow period of the years of the war," he said. "All of the major oil companies and many smaller companies and independent producers were very active in this area. There were over six hundred geologists working from offices in San Antonio, Laredo, Beeville, McAllen, and Corpus Christi."

Most of the majors had offices in Corpus Christi by the time Freeman moved to the city in 1952. He had chalked up a number of discoveries before, but between 1965 and 1975 he went through a slump, which was broken when he was invited by Thomas O'Connor to drill on his ranch. He had drilled an earlier well there in 1955, opening the Greta West Field.

"In January 1975 we [He was in partnership with Lou Flournoy] drilled on the M. E. O'Connor property," he said. "In 1959 I had tried to get Mary Ellen O'Connor to give me a lease on this prospect."

Those years of waiting ended with great rewards. The field produced more than a million and a half barrels of oil and would probably produce for years to come.

"I have thanked them many times," he said. "This field has contributed over $20 million to the M. E. O'Connor Estate."[6]

In the meantime Republic Natural Gas Company had fallen victim to hard times. As a result of the low prices and severely curtailed production of the early 1960s, it had been taken over by Mobil Oil Corporation in 1961.

It took much longer for Tenneco to fall, but the huge conglomerate eventually fell victim to its own excesses.

Bill Vrana said that Gardiner Symonds had a brilliant mind for business and felt shareholders could be protected only by diversifying the interests. This he did, and by 1970 Tenneco assets exceeded $4 billion.

Symonds's influence continued after his retirement in 1968 and his death in 1971 until the gas interests were the smallest part of the corporation. The company became involved in a variety of businesses, including pipelines, refineries, oil production, shipbuilding, insurance, agriculture and land management, chemicals, packaging, automotive, plastics, construction, and farm and construction equipment. Agricultural holdings included citrus orchards, 144,000 acres of irrigated farmlands, and large ranches in New Mexico and Arizona. Other enterprises included a military and space electronics company and the manufacture of Walker auto exhaust systems and Monroe mufflers. Tenneco bought the shipbuilding facility at Newport News, Virginia, and built two nuclear aircraft carriers—the John F. Kennedy in 1969, followed by the Nimitz—plus four nuclear submarines, five 580-foot fully automated, amphibious cargo ships, and two nuclear-powered frigates. The company also held contracts for overhaul of carriers and other naval vessels too large for other shipyards to handle.

It was one of the largest conglomerates in the country, but its growth became painful in the 1980s when the oil business was in the doldrums, interest rates were high, and diversification became a curse.

The petroleum holdings were among the first to go. The empire began to crack when a drastic drop in farm prices caused the J. I. Case Company to fill showrooms with farm equipment that could not be sold at a time when its debts were mounting. Other subsidiaries were sold off in 1988 to pay off debts. And—with all its billions— the giant Symonds had founded was bleeding to death.

Chevron bought Tenneco's offshore holdings for $2.6 billion. T. Boone Pickens's Mesa Limited Partnership bought the Mid-Continent division oil and gas subsidiary for $715 million in cash. Vinal Oil and Chemical Company bought surface and mineral acre access and other facilities in and around Corpus Christi.

In 1991 Tenneco sold its natural gas liquids business to Enron Corporation for $632 million. In 1995 Tenneco moved its headquarters to Connecticut, and the following year the natural gas marketing and transportation division was sold to El Paso Energy Company.

Tenneco's troubles saddened Vrana. "Global economies developed into a highly competitive spirit with megamergers among large corporations. Almost overnight companies found themselves in debt over their heads resulting in a slide backwards. Perhaps Tenneco became one of the victims," he said in 2002.

"In my opinion it was one of the finest companies to work for . . . . The company had one of the best thrift plans and retirement plans available for their employees. Forty-two years after I left Tenneco to pursue business as an independent consulting geologist, I am still enjoying some of the benefits I earned from the company. They provided the basis for most of my experience utilized as a petroleum-exploration geologist."[7]

A Christmas tree of the pre-World War II period

Radcliffe Killam

During World War II Radcliffe Killam served as a PT Boat officer in the Pacific Theater.

While some men, like Vrana and Major, were hired into oil field jobs, others were literally born into the industry. Among these were the Hawn brothers, grandsons of Dr. Hewit, who made important discoveries in Refugio Field, and Radcliffe Killam, son of Laredo oil-field pioneer O. W. Killam.

"The Hawn Brothers partnership began in 1948," said George Hawn in a 2003 interview. "My dad died January 27, 1947. Johnnie and Bill [George's brothers] had just come back from the war. We got together and formed Hawn Brothers.

"I was at the University of Houston and came down to DeWitt County when we were drilling in this field to spend my weekends down there. We had a geologist. We had a landman. We had a couple of engineers. We put deals together just like my dad and granddad had done. Same thing."

About 1973 when the Hawns had to file papers for the windfall profit tax that was in effect, they reported that they had been involved in drilling five hundred shallow-to-medium wells.

Frank Scanio, who ran the Rooke Ranch for more than thirty years before moving to Corpus Christi in 1988, was George Hawn's partner in a number of operations. Scanio was married to Marian Rooke, granddaughter of Roberta Driscoll Rooke, daughter of Jeremiah Driscoll. Jeremiah was the son of Daniel O'Driscoll, Irish veteran of the Battle of San Jacinto.

"Brothers John and Bill died, and I sort of fizzled out," George Hawn said in 2003, but he pointed out the value of the industry to the area.

"Oil saved many ranchers," he said. "I know that 4,000 acres on the Rooke Ranch leased for a dollar an acre. The Daugherties were close to losing their land. They can get as much as five hundred dollars an acre for a lease today."[8]

Like Hawn, Radcliffe Killam followed his father's example by going into the oil business. Although O. W. started out with partners, his organization soon became a family affair. After 1925–1926 his interests remained in his hands or, later, in the hands of Radcliffe and Radcliffe's son David.

For thirty-six years Radcliffe had as partner another family member, his brother-in-law John G. Hurd, husband of Louise Killam. They operated as Killam and Hurd from 1946 until 1982, when they divided the company, leaving Radcliffe in Laredo and Hurd in San Antonio.

As Radcliffe told it, his father was the aggressive founder, who pioneered, took risks, discovered and produced oil and gas, and made the company viable and independent through good times and bad. On the other hand, he said, he was not so prone to take great risks and directed company affairs conservatively.

"The wells in Mirando Valley are going to about 11,000 feet. They are gas wells and are very high pressure and high volume." Radcliffe said in 2002. "My knowledge or expertise stops at 6500 to 7000 feet."

Radcliffe Killam was a great influence on the life of his partner, John G. Hurd. Both attended Harvard University, and Radcliffe introduced his sister Louise to John. Both men had experience in the oil industry, and both became officers in the U. S. Navy in World War II. Radciffe served in PT boats in the Pacific and Hurd on a subchaser in the Atlantic.

Hurd was born July 2, 1914, in San Francisco. His natural father, a German by the name of Von Hochstetter, was a Brazilian coffee planter. Shortly after his marriage, he returned to Brazil and was killed in an uprising. When John was six, his mother married Dr. Hurd, who adopted the boy and became the only father he knew.

John attended Harvard from 1930 to 1937. After graduation he worked for a large San Francisco law firm. One of its clients was Standard Oil of California, which hired him in its land division. Because he was in Naval ROTC in college, he was called to active duty before Pearl Harbor.

He was discharged a lieutenant commander. Radcliffe Killam had just been discharged, too. He called John, they talked about how they were both in the oil business, and the partnership of Killam and Hurd was born.

"My original capital to the venture was six thousand dollars," Hurd said. "His was ten thousand. That wasn't much, but we got free office space with his father, Mr. Killam. We got free assistance from the secretaries and filing and bookkeeping—we got that free. He also paid us $125 a month to help him out when he wanted something done.

"So we pretty well spent our capital on drilling wells. Actually, we only drilled two wells before we struck. I don't think those two wells together cost us over five thousand dollars. Mr. Killam had drilling rigs ... he gave us a real break on that. But he would not loan us any money.

"The wells came in at 2200 feet. This was in the Hughes Field. As I recall, we drilled sixty producing wells and probably half a dozen dry holes. These were pumper wells that started producing two hundred barrels a day, then over their life of more than twenty years averaged forty to fifty barrels a day."

It was premium oil refiners used to blend with their stock and always brought top price. The highest was during the Suez crisis in 1978, when it went up to thirty dollars to thirty-five dollars a barrel. Then it went back down.

Killam and Hurd worked the Mirando area until the 1960s, when they had capital to look elsewhere. First they looked in the Gulf Coast area. They brought in some production in Jackson County and in East Texas and Wyoming.

A Hawn Christmas tree

Hurd's dad had close call

At the beginning of World War I, Dr. Hurd volunteered to serve with the Russian army. He was in the front lines when Czar Nicholas II made him a member of his household, and he became the Czar's personal physician. The United States Root Commission realized that the Bolshevik uprising was about to start, and Senator Root told the doctor he had better get on the train. He did, came back, and went to France with the American Army.

******

John Hurd had wanted to attend West Coast universities that had rowing teams. The elder Hurd forbade it but allowed him to go to Harvard, not knowing it was the hotbed of rowing competition. John rowed until he developed bunions on his posterior and had to pull out. He tried to compete in lightweight football, but his hands were too small. He tried boxing and found he had a glass jaw. The coach recommended fencing. He became a champion fencer and competed in the 1936 Olympics in Berlin.

Hurd Interview at UT

Louise Killam Hurd died in 1955, and John Hurd married again in 1957.

Politically Hurd was a Republican in an area of South Texas controlled by Democratic patrons. After he agreed to help Republicans in Webb County, he was named County Chairman.

"I later found out I was chairman of ten people," he said.

Next he was State Republican chairman for Richard Nixon. This led to his appointment as ambassador to South Africa in 1970, just as the Apartheid and Afrikaner problems were increasing. He managed to mollify both sides for the time he was there, 1970 to 1975. The State Department wanted him to go to Greece, but the government was overthrown. Then he was offered the post in Portugal.

"They had my agreement all written out; it was about to go off," he said, "and I'll be a son-of-a-gun if the Portuguese didn't have a revolution."

After six months he asked to be relieved. Probably facetiously, Secretary of State Henry Kissinger called and said, "John, you are so successful. Why don't you just stay there and start a revolution?"

In 1977 Hurd opened a Killam and Hurd office in San Antonio. Killam chose to stay in Laredo. Hurd found he spent too much time driving to San Antonio from Laredo—"Until you get your planes, the sun gets in your eyes both ways."

In 1982 Killam and Hurd "split right down the middle." In later years the partnership was limited to a Beechcraft airplane.

Hurd employed two geophysical consultants and three consultants in geology, and Hurd Enterprises became a multimillion dollar company, with four ranches totaling some 55,000 acres.[9]

While Killam was born into the South Texas industry and Hurd married into it, Lee Durst was more or less adopted into it. Orphaned at age ten, he was raised by an older sister who was married to G. Ray Boyd, an oil worker who was later a drilling superintendent for several drilling companies in Laredo before forming his own company. When Durst was fifteen, he started roughnecking for his brother-in-law.

"Ray Boyd had several partnerships, including Bob Kirkwood at one time," he said. "Kirkwood and Boyd were partners on a well near Mirando City. They thought they were on Easy Street. They got themselves an oil well. They debated for two weeks over whether to sell it or not. Meanwhile, it went to salt water, and they went their separate ways.

"First time I ever saw Kirkwood, I was about twelve," he said. "He wore a little greasy hat and had a little cigarette clipped between his teeth all the time. In those days water was an extreme problem. They had to haul water in an oil truck and an open-top tank. You have to have water for drilling. Kirkwood was most unhappy. He'd take that little greasy hat, throw it on the floor, and jump up and down on it to encourage his hands to get more water.

"Kirkwood was a very interesting fellow," he said. "He was very dynamic and very determined. He was a hard worker and a good organizer. He developed very fast from those days. He was a good operator and well thought of. Stories about him were that at one time he was a card dealer for the house in Nuevo Laredo.

"He drilled several wells for the Henshaw Brothers out of San Antonio. They had trouble with a pump. The safety valve kept popping off," he said. "The driller went back and took the safety relief valve out and put a bull plug in. The bit balled up and pressure went way up high and exploded the pump—blew it up!

"Well, Walter Henshaw was very mad. He was an ex-boxer. He went down and started chewing the crew out. He said, 'I can whip any one of you.'

"One great big old tough boy stood up and said, 'You're going to have to prove that.'

(Opposite top) From left: Radcliffe Killam, O. G. McClain, O. W. Killam, unknown, and C. C. Miller at a Laredo reception the Killams held June 8, 1951, for geologists on an annual field trip (Below) Sometimes it was necessary to use oil tanker trucks like these to haul water needed for drilling.

Rig builders erecting a rig

"'I said I could whip anybody who worked for me,' Walter said. 'You don't work for me any more.'"

Durst served in the Navy in World War II. When he came home in 1945, his brother-in-law made him an offer he couldn't refuse: "He would send me to get an engineering degree at A&M, and he would take me in as a partner."

He became a partner in 1949. "We merged with another brother-in-law, A. J. 'Dutch' Kinstler, and became Boyd, Durst and Kinstler, Inc."

The first significant discovery they made was on the Lou Ella Wade Ranch three hundred feet from the south end of the Lake Corpus Christi Dam. The discovery, opening the Lou Ella Wade Field, put the company in solid financial condition.

"We opened the test tool on that well at 2 a.m. Sunday and flowed pipeline oil in ten minutes, which was a very good test," he recalled. "Naturally the open acreage prompted us to have a representative in the mayor's office at 8 o'clock the next morning. The mayor was already fielding calls from as far away as St. Louis, wanting to buy that acreage. There was quite a scramble for that Corpus Christi [city] property there, and I suppose we just weren't politicians enough to get it. The lease was made to A. W. Gregg of Houston. That was our first experience and our main stake for several years and upgraded our business considerably."

It was, indeed, a good lease, for they drilled eighteen wells with only three dry holes, but it would have been much better had they acquired the lease owned by the City of Corpus Christi, where Gorman Drilling Company drilled thirty wells without a dry hole.

Durst wasn't a good enough politician to get the lease because, as it developed, the mayor, Leslie Wasserman, was indicted June 2, 1952, for accepting bribes for granting oil and gas leases on city land. Gregg was indicted on two counts of bribery, and a Corpus Christi businessman was also indicted in the case. The scandal brought down the Wasserman administration, and a number of lawsuits were filed against Gregg.

Durst's company found oil and gas all over South Texas. They operated four rigs and moved from Laredo to Alice to Victoria, where the banks were favorable to them. Boyd retired in 1970, and the controlling interest was sold to Maynard Oil Company, with Durst as manager with a 15-percent interest. In eight years the company had eight drilling rigs. Later the company had drilled more than three thousand wells in South Texas.

"Lucian Flournoy and I were pretty young in the drilling business at the same time," Durst said. "We'd sit outside machine shops while our equipment was being repaired and compare work sheets and see how many wells we were drilling and how much success we were making."

He conceded that "Flournoy and Kirkwood outdistanced me because they passed work on to other people rather than trying to do it all themselves. That was the mistake I made.

"In the early days we did our own rig building—had to pour bearings ourselves and put grease fittings on and keep them lubricated. We had antiquated equipment and worked to keep it up to date," he said. "We had a small standard derrick, which the crew tore down and built themselves. In 1945 Boyd bought a 127-foot jacknife derrick."

Another job is torn down. Lee Durst said that when Roy Boyd bought a jackknife rig in 1945, setting up the derrick went from three weeks to one day.

Setting up the derrick went from three weeks to one day.

"I spent many a 24-hour day without going to bed. We were operating two rigs, and I tried to push tools on both of them. They'd be a hundred miles apart, and you'd just go from one to the other to keep them running. When my daughter was about two and a half years old, I came home from a three-day trip in the field and needed a shave and a bath real bad. It was about 9 in the morning. The front screen was latched. I heard little bare feet come padding out of the back room, and my little daughter looked me up and down and turned around and said, 'Hey, Mamma. There's a man at the door.'

"That sort of woke me up that I was neglecting my family.

"Later on in 1963 Corpus Christi geologist Bill Easley and Joe Thomas of Sinton were involved in the discovery of the North Midway Field, which was five miles out of Taft," he said. "Incidentally, some of those wells we drilled in the sixties and early seventies are still producing."

His toughest job?

"One time I waded a flooded creek neck-deep to get to a rig that had been hit by a hurricane near Floresville. The rig had blown over, and the crew was caught out of the hole. They had 5000 feet of drill pipe stacked in the derrick when it blew over. Some of the hardest days of my life were spent dragging that crooked pipe out and stacking it. It was made up in tribbles, and we extended tongs from the back end of winch trucks to break those joints, unscrew that pipe, and run them through a

A crew with a testing tool in 1936

straightener and get a new derrick put there and redrill the well, which we did. Other fishing jobs were hard times, but that was the worst."

He retired in 1983 and became a consultant. In 1995 he turned that business over to his son and tried to play tennis six times a week.[10]

Family connections also brought Jake Hamon into the oil industry, but the circumstances were not happy. When Jake was about twenty years old, his father, a railroad right-of-way buyer, was fatally shot in a barroom altercation. Young Hamon inherited some oil and gas interests and went to Ranger, where there was scrubby production around 1927-29. He and a rig builder he hired got some 10-acre leases, drilled, and made some wells.

But, said Jon Spradley, "Jake had a penchant for taking partners in, so he worked up some of these deals and went to Dallas," where he met Edwin Cox, the father of Edwin L. Cox. The elder Cox was "a smart hombre and had a lot of money." He and Jake hit one deal and then formed a partnership. Hamon was to find the prospects and Cox to put up the money.

By 1963 Hamon was a big independent with seven districts. In that year Spradley became Hamon's district geologist in Corpus Christi.

"Hamon was a fine gentleman," Spradley said. "I was going to stay four or five years and then go independent, but he gave his district geologists an override on everything he did. That five years went into fifteen.

"He loved wildcat wells, and he hated seismic," Spradley said. "He said, 'You spend more on seismic than you do getting down to the point where you're going to log your well. Seismic is like taking dope. The more you get, the more you want. And you don't ever know when you've had enough.'"

Spradley said that Hamon was a subsurface wildcat driller. "We cored and tested in almost every well we ever drilled at any depth. That way you don't run pipe accidentally on a bad hole. We drilled some nice prospects. We found some good Wilcox production down southwest of Hebbronville and the Thompsonville Field, and we found some production down below Mirando City in Aviators Field."

Aviators Field was named for aviators who would come from Laredo to eat Mexican food at Mirando City.

Abel and Buck, out of Laredo, drilled shallow wells in that area, Spradley said. "That was the cleanest crude oil in South Texas—Mirando Crude. It drew a premium price. It didn't have any impurities in it at all. No sulphur. It was like machine oil."

Spradley also had family connections to the industry, but, even so, his entry was delayed. When he was a "little bitty kid," his father had put in a United Gas plant at Agua Dulce, and later his stepfather was one of the pioneer Baroid mud men in South Texas.

Spradley had three years of college in business administration, but during his Army service a friend who was a geology major from USC interested him in the formations and such they saw while on maneuvers in Germany. This stirred an interest in the oil field, and upon his return to Corpus Christi in 1955, he worked with Carl Birch at J&L Supply.

"With J&L," he said, "I was a grunt. I was a truck driver, and I just delivered stuff out to the rigs."

One day on a delivery he went into a trailer house at a drilling rig where a blowout had occurred. When he returned to town, he said, "Carl, there's a guy sitting there in clean khakis in front of a fan reading a magazine. What job does he have?

"He said, 'Well, it's probably not the tool pusher if he's got clean clothes, so he must be the company geologist.'

"I said, 'That's what I want to be,'" Spradley said.

From September 1955 through August 1960 he attended Louisiana State University, earning both bachelor's and master's degrees. He returned to Corpus Christi and was hired by Pan American Petroleum, which had started as Dixie Oil Company, then became Pan American Production Company and then Pan American Petroleum. Later it became AMOCO, a subsidiary of Standard Oil of Indiana.

"I think AMOCO is now BP," Spradley said in 2004.

No sleeping on this man's time

Jon Spradley related a story Jake Hamon had told him about his only trip to Freer.

When Hamon arrived in Corpus Christi, he thought it strange that the train backed into the station. It was December, cold and raining, and he'd had a bit to drink on the trip. He bought another bottle and paid a taxi driver seventy-five dollars to drive him to Freer.

He was told where to find the Cox and Hamon wells.

"There was a shanty with a fire going and a guy sleeping on a cot," Spradley said. "He asked the guy what he was doing. He said he was tired. Jake told him, 'You ain't going to sleep on my time. Take this fifty dollars and drag out of here. You're fired.'

The guy took the fifty, got in a pickup truck, and drove off. It occurred to Jake that the guy might have stolen his truck.

"Another crew drove up to the cabin. It was his men. He told them he had fired that sunovabitch.

"'Was he driving a green truck?' they asked. Hamon replied that he was.

"'That was the Humble pumper,' they said.

"Hamon got back in the taxi and took the next train out.... So far as I know," Spradley said, "that's the only trip he ever made to Corpus Christi."

—Jon Spradley

Family members shared the flare

Jon Spradley did not tell this story. His wife did.

"My husband drilled a well in 1980 near Portland, she said. "It blew out and created a big fire. In fact it took them nine months to put it out.

"I joked with him every day. I'd say, 'There goes my string of pearls.' And the next day, 'There goes my fur coat.'

"For the first time Southwest Airlines chartered a plane. About 120 people were going to Lubbock for a wedding. I was given the job of riding herd on the crowd, seeing that they all got on and off the plane.

"On the way back to Corpus Christi I finished my chores and went into the cockpit, and the captain invited me to sit down. We chatted and talked nearly all the way home.

"Then he said, 'See that light down there. You can see it for miles. It is a perfect beacon for flying into Corpus. It gets us practically to the airport.'

"I looked and saw the lights of Portland and the big flare near it. 'I'm glad you like it,' I said. 'That's my husband's oil well.'"

****

Her husband added another chapter to the story.

"We had a hunting lease down at Big Wells. We bought these two old golf carts from the country club here. Before I could take them to the hunting lease on a flat bed, my boys were riding around the neighborhood on that golf cart and a cop stopped them.

"Jason was about ten and Breck was about eight.

"The officer said, 'Boys,you can't be riding on the street without a license. I might ought to give you a ticket.'

"I think he was pulling their leg.

"Jason said, 'Please don't give us a ticket. That big well that's on fire, that belongs to my daddy, and he doesn't have any money to pay for a ticket.'"

At Pan American he trained under Bob Treadwell and had two "great office mates," Don Boyd and John Hyndman. "They did good subsurface geology," he said.

"We had some good scouts," he said. "Jack Skinner and a bunch of guys, all the scouts or the landmen, started out as gravity men at the Canadian border and came down to the Gulf of Mexico across the United States, running gravity. When they got to the Gulf, they had run out of business, so they had to become scouts or landmen.

"If you knew the right scouts, you could get the dope," he said. "Humble used to have the best set of scout dope here in Corpus Christi. They had good scouts—Mill Bostead, Buck Rogers, Jim Hooten, and all that bunch. They were really good with the independents. They'd let us have that dope. If I ever had a log that we'd run on one of our wells and they called me, I'd give it to them, because they were always good to me.

"Another thing about them—If you wanted to know about an area, just pick a county, and you could get the survey names, the sections, etc. You could go over there, and they had card files on every company that shot seismic on it. You couldn't go wrong using their dope. It was a good back-scratching operation."

While Spradley was with Hamon, Joe Dawson developed the Harvey Field and the South Hardin Field. Spradley opened his own office in 1979. The next year a White Point well he drilled provided him with a massive headache and an unusual opportunity.

"We drilled that well and it was in for damn near a year, doing real good," he said. "Then it blew out after it had been producing. Pipe failure. It burned for nine months and twenty-seven days. We had to drill three relief wells before we finally got in to start sucking gas out of there. Then we turned the big relief well on. We didn't know if it would work or not. We just knew it was close. They have certain

(Right) O. G. McClain said, 'I roughnecked on this well. In fact, it was on Mills Bennett No. 5 Jesus Lopez that I nearly got a finger cut off, which marked the end of my roughnecking days.'

tools they can run. We turned it on, and it worked. And it's still going.

"It was 11,000 to 12,000 feet. The day before we turned it on, we were $7 million in the hole because of the relief wells and everything that went with it. But we had a skimming operation out there. We were putting thirty or forty thousand barrels of water a day into the bay. We had the RRC and the Texas Water Quality Board monitoring the water that went down the ditch to White Point.

"We were skimming condensate off of the water. At East White Point there is a facility that Koch Refining had where they cleaned the ships out and pumped the old oil and sludge into pits. There are two big Permian Oil Company tanks about seven hundred yards from our well, down the road half a mile.

"We leased it from them for $250 a month, and we were skimming all that condensate off the top of that salt water, and we had these pits. We sold the condensate to Permian Oil. We produced about a million and a half dollars worth of oil off of those lakes out there. I don't know how much gas we burned. I hope I never find out.

"That well was my biggest hit because it has produced so far the net revenue $150 million to all the partners. We got our money back. That was the equivalent of a pretty good field because it's probably done about close to 40 Bcf, which is big for this area. That would be my biggest discovery."[11]

While Jon Spradley spent five years making the change from truck driver to geologist, it took only one day to change O. G. McClain's life. It happened when a rotary bushing caught his little finger and

Geologist Joseph Dawson was a D-Day hero

His reputation as a national hero all but obscured the fact that Joseph T. Dawson was a Corpus Christi geologist who worked for Renwar Oil Corporation and Humble Oil and Refining Company before becoming an independent oil operator.

Dawson's wartime exploits are legendary. He participated in landings in North Africa and Sicily and, as a captain, led members of the First Infantry Division through heavy fire, the first to break free from bloody Omaha Beach on June 6, 1944. He rescued a platoon of his men near Paris. His company was decimated but held a ridge overlooking Aachen thirty-nine days against large German forces. Promoted to major before he was discharged, he was awarded the Distinguished Service Cross, the Silver Star, Purple Heart, and many other awards.

His civilian life was filled with public service as city councilman and chairman of the Civil Service Commission, the Arts Commission, the Planning Commission, the Red Cross, and the Corpus Christi Heart Association. He served as president of the Reserve Officers Association of Corpus Christi and was active as vice chairman of the University of Corpus Christi in obtaining Ward Island, which became the location of Texas A&M University-Corpus Christi. He died in 1998 at age 84.

In 1997 the Corpus Christi Independent School District named an elementary school for Joseph T. Dawson, a true hero and a dedicated geologist.

(Top) Sculptor Buddy Tatum's statue of Dawson stands in front of the school.

almost cut it off. "That's the last day I ever worked on a drilling rig," he said.

He already had a bad back and weak knees from roustabout work, and the finger provided the incentive to move on to the job he was trained for—geology—his main interest in life since he was a child.

Born in La Vernia in Wilson County, southeast of San Antonio, in 1911, he had his first contact with the oil business through wildcat crews drilling wells in the area.

"My father had a boyhood friend who was a successful lease dealer," he said. "He served as a model, so to speak, and I would like to study geology. As a boy I would come home with a pocket full of rocks and wondered how things were put together. Even in high school I remember going downtown and visiting with geologists, even though I was just a kid—Oloff Dover, Gentry Kidd, W. A. Maley, Ike Maley, Cliff Owens, Toosey Miller, and others of that genre who lived in San Antonio. I just wanted to find out what they thought and what kind of people they were. It just builds and builds, and there was no question that I was going to the University of Texas and study geology."

At the university in the fall of 1928, he was in the class with Tubby Weaver, who would become his lifelong friend. McClain moved to the University of Oklahoma and took a degree in geological engineering. There were few jobs for graduates at that time, but drilling contractor George Echols of Houston took him on as a "toolie."

"A toolie has the same duties as the tool pusher, but he has no authority....The other thing is the toolie makes straight roughneck wages. He's cheaper than a real tool pusher.

"The first rigs I worked on were jackpost rigs. The drawworks weren't unitized. You had to take everything apart, and the bearings were poured Babbitt, which is a lead zinc compound. They didn't have roller bearings. And when unitization came along, they unitized the drawworks. The unitized engines drove it, and unitized water pumps supplied water to the boilers. The mud pumps were put on skids so you could pull one up to move them. Marvelous!

"The early rigs were powered by steam. Diesel and gasoline motors didn't come in until quite a bit later—in the mid-30s and early 40s. Of course, steam required lots of water and a reliable supply of fuel. If you had a long water line and something happened to that line and you couldn't pump water into your boiler, you could be in big trouble real fast.

"I've never seen a boiler blow up, but I've had old pot firemen tell me about it. They were very attentive to the water level gauge on the side of that boiler. . . . It was said when a boiler blew, part of it might be half a mile out in the field. I've seen lots of crew chiefs have to

(Left) An early day steam-powered rig (Below) Tents like these were home to rig crews and many times to their families.

O. G. McClain

'I might point out that in the early days geologists were the first ones thought of when a peculiar problem came up in a company. . . .'
—O. G. McClain

have a boilermaker out. Diesel motors were a godsend. They were just wonderful to have—even gasoline motors."

McClain worked for a year in the East Texas Field before moving to Southwest Texas to drill wells. His friend Ed Sellers talked him into working with him for Mills Bennett Production Company and JRB Moore, who were to drill ten wells in Duval, Webb, and Starr counties in the Government Wells and Mirando Trend. The Sullivan Ranch was the first one he drilled on in Starr County.

"The location was as far back in the woods as you can get. It was fifty miles from Hebbronville, thirty-five miles north of Rio Grande City . . ." he said. "Every morning I saw bobcat tracks on the hood on my car. I saw several panthers crossing the road at night. You didn't have to be told to carry spares and a survival kit.

"Roughnecking was a breeze. Roughnecking was fun compared to what I had been doing. In about our third or fourth well, we found the Lopez Field, which eventually produced twenty million barrels of oil."

Life on the rigs was rough. The crew lived in tents. It was difficult to move them, as roads were almost nonexistent. Water was scarce. There were blowouts and near blowouts.

"A water flow is pretty fearsome," he said. "You're drilling a surface hole. I've been out there on a workover, and a gas flow got away. I am partially deaf because of that experience of being in that jet stream of gas. We had to work right there and put the valves on."

He found another danger, totally unexpected, in a thunderstorm. Lightning struck the metal support pole in a nearby tent. A crewman and his wife were stunned, and Sellers received a terrible blow. McClain managed to get Sellers breathing by using artificial respiration he'd learned in a swimming class, saving his friend's life.

It took a month to drill a 3000-foot well, he said.

"We ate a lot of potatoes and eggs cooked on an open fire. We joked about going on a trip up the Devil River between wells, where we could camp out and cook over an open fire and get close to nature. . . . Instead we headed straight for Laredo and the Southern House run by Ed's aunt and uncle. We cleaned up, put on our best clothes, and called the girls."

They were accepted as eligible. Both married Laredo girls.

In 1934 he welcomed an order from the U. S. War Department to report to Randolph Field for training as a flying cadet. He had applied two years earlier and passed the physical.

In June 1934 "I experienced more excitement and sheer exhilaration than I ever had before or since," he said. "This came to a halt when my instructor decided to wash me out for having a 'natural inaptitude for flying.' He said he thought I would do better in the oil business."

About 100 out of the class of 156 were washed out after the first four months.

"I returned to work as a 'toolie,'" he said.

His first job as a geologist was with Mills Bennett in 1936.

Later he operated for some time as an independent before joining the staff at Southern Minerals Corporation, where he was employed as a geologist until the end of World War II.

"I might point out that in the early days geologists were the first ones thought of when a peculiar problem came up in a company, be it high density drill wells, how to complete most effectively—anything that was a special job there was a lack of experience in," he said. "In all likelihood, they'd go to the geologist and say, 'Look, what do you think about this? You figure out what we ought to do.'

"The geologist was credited with ability to handle those types of problems better than any other category of employee. Of course, this has been superseded later on because of all manner of experts and petroleum engineers and log analysts and everything else. The geologist once caught all these things."

He became a consultant in 1946 after the only consultant in Corpus Christi decided that Mississippi was the coming place.

"The consultant bases his utility value on the fact that many small companies, many individuals, just simply don't have funds to support a geologist on the payroll, so there was always an abundance of requests for reserve studies, for evaluating wildcat prospects," he said.

Financing wildcats often proved very difficult. Nobody would lend money on a wildcat unless there were proven reserves to back it up. It was not rare for a rig owner to give his workers a share of the profits if they would agree to accept lower wages.

McClain had a wide range of experience for a consultant's job. Foremost was his experience in Railroad Commission hearings. He had attended many and knew the commissioners and engineers in Austin as friends.

This well was drilled by the John F. Camp Drilling Company of San Antonio. As financing often proved difficult, rig owners sometimes gave their workers a share of the profits in exchange for agreeing to cut their wages in half.

One of the multitude of gas flares that lit up South Texas skies. It was said so much gas was being flared that automobiles did not need headlghts at night throughout much of the area.

"I had been a geologist, an engineer, a scout and did land work and heaven knows what else," he said. "It was not a difficult thing to turn a prospect. I had good success with attorneys, companies. I would go to friends and people who had resources, and I'd keep a small interest, an override. I didn't burden anyone. I'd take my profits modestly and let them have theirs.

"I didn't make much progress in many of the requests, but it was fun trying," he said. "If I couldn't do it, I told them so."

A delegation from Freer came to him and said the town was running out of water, as water from the Government Wells Field was drying up. He sent them down to the Hoffman Field, where water was available at 600 feet.

"They got all the fresh water they needed in five or six miles of line they had to lay," he said. "They were happy to do that."[12]

In the postwar period the Texas Railroad Commission continued to set policies and regulations that had profound effects on the industry. When Texas Eastern purchased the Big Inch pipeline after World War II and altered it to carry natural gas to the East, it was the dream Commissioner Ernest O. Thompson had been hoping for. As Republic and other companies started lines and independents began drilling for gas, prices rose and great markets opened for the industry, just as Thompson had visualized.

After Olin Culberson, who had been chief examiner and later chief of the gas utilities division, was elected commissioner, he served twenty years. It was Culberson who was responsible for replacing political appointees on the commission staff with graduate engineers.

In 1946 Gov. Beauford Jester appointed William J. "Bill' Murray, the first trained engineer ever to serve on the regulatory body, to the commission. In 1947 the commission, with Thompson, Culberson, and Murray as members, ordered wells shut down where operators were flaring large amounts of casinghead gas, gas separated from oil at the casinghead. Operators were to conform to new conservation rules and return the gas to the reservoir from which it was drawn, thereby maintaining pressures and increasing the yield from the reservoir. Some operators had already installed such a system voluntarily, but many lawsuits were filed before the flaring of gas was stopped.

The first order was directed at the Seeligson Field southwest of Kingsville. The companies got an injunction and took the case to the Supreme Court, but they lost. The first commission gas conservation order was law.

It was estimated that a billion and a half cubic feet of irreplaceable natural gas was wasted each day for years before the practice was ended. It took a long time, but the conservation machinery set up by Ernest O. Thompson once more saved the industry from itself.

Many oilmen resented the rules and restrictions the commission placed on them, but it was no piece of cake for the workers at the RRC, who were faced with mountains of unending paperwork.

Jim Herring got a quick indoctrination into the process. He joined the commission's Corpus Christi office in 1952 after working for a year for Magnolia Petroleum Company in West Texas. After two years the director, district engineer, geologist, and superintendent had all left, and suddenly Herring was district director.

'Now what do we do?' Owners and crew members of a Saxet rig ponder their next move after a blowout. Jim Herring took this photograph in 1954 when he was with the Railroad Commission in Corpus Christi.

"I didn't know what I was doing to start with," he said. But he learned in a hurry.

"The amount of oil produced was critical to the commission at that time, so we kept good records," Herring said. "We got a production report from the operator. By these he told us how many barrels he produced, how many barrels he sold. We got the reports from the pipeline that said, "Yes, I bought that many barrels from them, and I sold it over here either to another pipeline or to a refinery." Then we got a report back from the other pipeline and that refinery. So technically you could follow every barrel of oil that was produced from the ground to the refinery.

"It was lots and lots of hard work," he said. After a woman had spent the day with calculations, "there would be whole reels of adding machine paper on the floor. You could measure the work by the length of the tape."

Blake Fore handled the tenders each month. "They would be stacked up three feet high, and he signed every one of them. They wouldn't let us use a stamp."

In 1965 Herring was appointed examiner at the Austin office. There he "got a firsthand look at people asking for variances from the rules."

He also got a firsthand look at the politics involved in the offices of the commissioners. "Bill Murray was an engineer. He was more engineer than politician. The others were all politicians."

Jerry Sadler ran for the office, hoping to use it as a steppingstone to the governor's seat. "Luckily, he didn't make it," Herring said.

In 1980 he retired with thirty years credit, including time spent in the service. Herring suspected that during his tenure at the commission many operators were producing more oil than their proration permitted.

But one old gentleman was quite angry because he was ordered to do something about the flow of salt water from his well. He was compelled to drill an injection well.

"Several months later I saw him, and he said 'Man, I'm making more oil now than I ever did in my life. Greatest thing you ever did for me,'" Herring said.

That marked one small victory for a bureaucrat.

In addition to his conservation battles, Commissioner Ernest O. Thompson also had other victories. He led the successful fight to retain the Tidelands and thereby ensure state control of oil and gas leases three leagues into the Gulf of Mexico. He vigorously defended the 27-percent depletion allowance for oil producers, allowing them to compete against foreign oil imports.

In 1951 Thompson represented President Harry Truman at the World Petroleum Congress. Thompson's report anticipated problems that would arise from the mammoth pools of Middle East oil. He was convinced that the security of the world might lie in oil and gas

Desks go well with derricks

Don't sell these gals short— the women of Desk and Derrick, the club of professional women of the oil industry.

Without them, the oil business would be a mess.

They have been the ones who led you to information, who knew where people were, who were familiar with the business, who could tell you most everything you needed to know without troubling the boss, and who kept the office operation flowing smoothly,

The first club in Texas was the Houston club, one of the pioneering clubs of the association. The first club in South Texas was organized in 1951 in McAllen and the second the following year in Corpus Christi. The clubs formed a network that could recommend qualified members for jobs. The network also served to support members and help them learn to tackle various assignments.

Most of the members could speak with authority about what their company was doing, well status, and company personnel. Educational programs the club sponsored included field trips to drilling operations and lectures by industry experts.

Their membership dwindled in South Texas after drilling activity slacked off, but D&D remained a viable organization.

—Cecilia Venable

development in the Western Hemisphere. He believed in the sovereignty of each nation as he did in the rights of each U. S. state, and he thought the oil-producing nations of the hemisphere should be in the Interstate Oil Compact Commission as observers and participants.

In 1951 the American Petroleum Institute awarded Thompson, who had been promoted to two-star general in the National Guard, the Gold Medal for Distinguished Achievement, "petroleum's equivalent to the Nobel Prize. Only five had been presented prior to that time."

Thompson resigned from the commission in January 1965 and died June 28, 1966.

By the time of Thompson's death, the industry was facing hard times. As early as the late 1940s, the arrival of cheap foreign oil hinted at future problems, but drilling continued at a torrid pace. Over protests from independent operators, the rate of foreign imports continued to increase until the allowable production was cut in half in 1958. By 1960 the Railroad Commission cut the allowable to eight days, a move that brought great hardship to independent operators, many of whom were unable to survive such a drastic curtailment of their income. Numerous companies merged, and in 1969 Michel T. Halbouty, well known wildcatter and oil producer, estimated that three-quarters of the independent operators in the United States were put out of business.

Then the allowable was raised to 100 percent to help feed national petroleum reserves, and production hit a new high in 1972. Drilling was on the upswing as most price controls were removed. In 1973 the Organization of Arab Petroleum Exporting Countries, led by Libya's Muammar Gadafi, called for an embargo of oil shipped to the United States in retaliation for U. S. friendship with Israel. As a result, long lines of autos waited at service stations trying to find a pump that wasn't dry.

The next year OPEC, the Organization of Petroleum Exporting Countries, took control of the world price-setting apparatus of crude, resulting in higher worldwide crude prices. According to the *New Handbook of Texas*, the state's oil and gas industry had what might well have been its last boom during the 1970s and 1980s.

And that boom was short-lived. A number of factors combined to bring about a great slump in the oil industry. The Windfall Profits Act of 1980, which put heavy taxes on oil production, and the Economic Recovery Tax of 1981, which reduced most corporate and personal income taxes, made it more difficult to raise capital for ventures. The real slump occurred in 1986 when a drop in the price of crude to a low of seven dollars a barrel resulted in business and financial failures across the state. Many small refineries closed or were absorbed by larger operations.

Rigs were stacked, and displaced families moved from producing areas as more than a third of the workers in various jobs, including

By 1960 the Railroad Commission cut the allowable to eight days, a move that brought great hardship to independent operators, many of whom were unable to survive such a drastic curtailment of their income.

Jerry O'Brien

Jerome 'Jerry' O'Brien had a long, illustrious career as a wildcatter, oil company executive, and oil industry representative in national and international affairs.

office, technical, and field jobs, were cut. State and local governments saw their tax base diminished.

Corpus Christi, which profited greatly during the glory days of area discoveries, lost many district offices and residents when exploration and production offices moved away. Shell, Sinclair, and Superior had closed up shop in 1965. Gulf and Standard Oil followed suit, as did Mobil, Tenneco, Skelly, Texaco, and Continental. Coastal Corporation moved most of its eight hundred employees to Houston. Exxon was the last to leave, moving 250 production jobs to Houston in 1992. Its large exploration force had moved in the 1980s.

Although the declining production and increase in oil imports greatly diminished the importance of the Railroad Commission, the commission did play a part in bringing about a settlement of the LoVaca gas dispute of the 1970s.[13] As the twenty-first century began, the commission was a shadow of its former self, but it remained the model for intelligent conservation of natural resources. By that time the RRC was seldom in the headlines—OPEC got all the attention. This, however, was not at all strange, since OPEC had been patterned after the Texas Railroad Commission.

It was a South Texas oilman who introduced that working model to the oil-producing nations. That man was Jerome "Jerry" O'Brien, who had a long, illustrious career as a wildcatter, oil-company executive, and oil-industry representative in national and international affairs. Over the years O'Brien had become friends with Juan Pablo Perez Alphonzo, Venezuelan oil minister, who was upset over the extremely low prices of Mideast exports. O'Brien told him how the Texas Railroad Commission had solved a similar problem and suggested that the countries get together and set quotas. He gave Perez Alphonzo a copy of the Texas RRC regulations, and OPEC was born.

O'Brien, who was born in October 1906, had made an early start in the oil business. While he was a college student, he was a "petroleum transfer engineer" with Standard Oil in a swank Beverly Hills neighborhood.

"In other parts of town, you're pumping gas," he chuckled.

He had been recruited from his hometown in Spaulding, Nebraska, to play baseball at UCLA, a new university in Los Angeles, where he planned to major in medicine. After his bank failed and his savings were lost, he couldn't pay the fee for medical school. With hours in Latin, Greek and German, he changed his major to geology. When he got his degree, he applied for a job as geologist with Standard Oil.

The senior geologist said, "Son, please get out of here. I'm trying to keep my own job."

O'Brien went back to pumping gas and got a degree in petroleum engineering from the University of Southern California.

One day his professor came in and said, "Tomorrow we're going down to Signal Hill to witness an operation that's going to be real interesting. Shell Oil, which is a Royal Dutch subsidiary, is going to run a test—they think they're going to log a well electrically. We're going to witness that."

"Six of us in class went down there," O'Brien said. "Of course, there were no trucks then. It was just a little thing by hand. Shell had cored that well from top to bottom. He got the electric log out and laid it alongside that core, which showed shale, sand, shale. By god, it did show sand and shale. I'll never forget that. One of the drillers said, 'Hell, I could've told you when we were drilling in shale.'"

That test, on August 27, 1929, in Signal Hill Field, California, was the first Schlumberger electric log ever run on the American continent.

The History of Nueces County says that Marcel and Conrad Schlumberger ran the first electric log in the county on the Southern Minerals No. 2 Ocker on July 5, 1934, three years before this photo of a Southern Minerals gas plant was taken.

Derricks sprouted amid houses in Corpus Christi's Saxet area in the mid-1930s. While Jerry O'Brien was manager of Mid-Continent Operations for Sunset Oil Company, the company drilled 'a bunch of wells' in the Corpus Christi area, including six in the Saxet Field.

It was just a few years later that a similar exhibit was given in South Texas. *The History of Nueces County* says that Marcel and Conrad Schlumberger ran the first electric log in the county on the Southern Minerals No. 2 Ocker on July 5, 1934.

O'Brien worked as a mining engineer all over the West until he was hired by Sunset Oil Company, which sent him to Texas to evaluate properties owned by the Marine Corporation. He became manager of Mid-Continent Operations for Sunset and started Sunset's operations in Corpus Christi.

'All hell broke loose' all too often. Like O. G. McClain, many South Texas oilmen had vivid memories of such dramatic events.

They drilled "a bunch of wells at Corpus Christi"—six of them in the Saxet Field. In 1940 he was promoted to chief geologist for Sunset and transferred back to Los Angeles.

After Sunset got out of the exploration business during World War II and changed its focus to refining and marketing, Jerry and his boss, George McCarthy, formed Shamrock Drilling Company.

Their first drilling rig was a revolutionary new type—a steel cantilever rig that could be disassembled, moved to the drill site, and raised in just a few hours. They made a number of discoveries and had six rigs by 1946.

Another McCarthy, Houston oilman Glenn who was no relation to George, had a major impact on O'Brien's life. "Glenn McCarthy was always a character," he said. "Glenn and I were never close . . . but he allowed me to meet my lovely wife."

Shortly after O'Brien went to Corpus Christi, the Houston oilman called him.

"He said, 'I tell you what. You come over to Houston this weekend, and I'll take you to dinner.'

"I said, 'That's OK.'

"I got in my 1935 Model A Ford and drove over to Houston . . . I met McCarthy, his wife, and another lady, Mary Olga Payne, at the roof garden of the Rice Hotel," O'Brien said. "Six months later that lady was my wife.

"I lost her about five years ago," he said in 2002. "She was a really lovely person."

Jerry O'Brien told Venezuelan Oil Minister Juan Pablo Perez Alphonso how the Texas Railroad Commission had solved the problem of extremely low prices and suggested that the countries get together and set quotas. He gave Perez Alphonzo a copy of the RRC regulations, and OPEC was born.

In his retirement, O'Brien could remember a lot of things, but one of his most vivid memories went back to 1946, when all hell broke loose several miles west of Corpus Christi.

"I'll never forget it as long as I live. Lonnie Glasscock got me to move to Corpus Christi, and I had had Thanksgiving dinner at his house," he said. "We were driving out to his rig in the Agua Dulce Field to see how things were going on a well we were drilling. We got within three or four miles from the Sultex No. 1 Mary Eliff when it exploded. It was blowing wild, really wild. Two whole acres of cavity opened up. The rig with seven thousand feet of pipe went down in that hole. Everything went down in that hole.

"There has never been a blowout as big as this one. It is still a record for blowouts. Wells in the field were not properly cemented. Gas leaked and charged the upper sands, causing tremendous pressure.

"Ol' Lonnie said, 'Boy, I'm moving my rig out of here.' It was half a mile away. He took his rig off, and two months later his water well was shooting water up. The shallow sands all around it were pressurized. That was the famous blowout that set insurance history in drilling in South Texas.

"Here's a picture of it," he said, pointing to a photograph on the wall of his museum-like office in San Antonio. "Looks like the Grand Canyon."

After the drilling business began to slow in 1947, O'Brien became a consultant for the Jergins Company of Long Beach. He set up a Jergins office in San Antonio and eventually became vice president and general manager. In December 1949 Jergins sold out to Lehman Brothers for $32 million.

"We became Monterey Oil Company. . . . I had offices in Oklahoma City, New Orleans, West Texas. We had six floors in Midland. We were just operating all over and very busy," he said. At Monterey he was joined by Stanley Blanchard, Leonard Sayers, and Steven Blount. Walter and Arthur Buzzini and John and Bill Newman were drillers for Monterey Oil's Gohlke Field in Victoria County.

Arthur Buzzini, who was born in Manhattan, New York, and had graduated from Cornell University with a mechanical engineering degree, had worked as a roughneck after coming to Texas in 1933. In 1946 he and his brother Walter formed Buzzini Drilling Company, which drilled some five hundred wells in Wyoming, Colorado, and South Texas, including the Live Oak-McMullen counties Jacob Field that produced more than 2 million barrels of oil. Arthur Buzzini had also supervised rigs in southern Illinois and invented several devices that made rigs safer and more efficient. He died February 12, 2003, in San Antonio.

In 1957 Jerry O'Brien was president of the Texas Independent Producers and Royalty Owners Association. He was named Oilman of the Year that same year. In 1959 Monterey was sold to Humble Oil and Refining Company, and he became a vice president. He relinquished that position when he was named Director of Oil and Gas in the Department of Interior under Democratic Presidents John F. Kennedy and Lyndon Johnson.

Jerry O'Brien speaks to a gathering at the Mayflower Hotel.

Jerry, a Republican, and Johnson earlier had personal differences, but they developed "a working friendship." O'Brien attended NATO economic conferences and drafted a quota system on imports that helped save the domestic industry. When he resigned, he left a report recommending the consolidation of some sixty agencies that dealt with energy. Eventually this became the basis for the formation of the Department of Energy.

In April 1966 O'Brien incorporated the Colonial Production Company, and in 1975 he became chairman of McFarland Enterprises after it had merged with Seaboard Oil and Gas Company.

At one time he worked recovering heavy oil by heating it with a steam process. "You flow the pits and let the sand settle out," he said.

The process used to work well with 9 to 10 gravity oil.

"How do you tell if it's 9 or 10 gravity?" he said. "Well, you turn a rabbit loose. If they leave tracks, it's 10 gravity. If they don't leave tracks, it's 9 gravity."

He was drilling under the famous La Brea Tar Pits in California from which many dinosaur bones have been recovered. An expert on the subject, he also gave tourists lectures about the site.

"I would not recommend, if you do this type of work," he went on, "that you go home and walk on your wife's white rug."

He stepped down from the McFarland Board of Directors in 1990.

Douglas Weatherston

George Weatherston

Douglas Weatherston's friends estimated his work led to the finding of 150 million barrels of oil and more than a trillion cubic feet of gas. . . . His son, George Douglas Weatherston, and grandson, Doug Weatherston, were active with him in Discorbis Oil Company.

"I sold all my actual operations and now just have royalty checks," he said. "I'm getting tax bills. They tax the hell out of royalties. They can't be worth that much."

Jerry O'Brien was still going to work at his San Antonio office in the fall of 2002. Asked how a group of oilmen were living and active way up into their nineties, he replied, "I guess we just don't know any better."[14]

Another San Antonio oilman who scored high in the longevity category was Douglas Weatherston, who died November 14, 2003, at the age of 106. He was interested in the oil business to the very end. Five minutes before his death Weatherston made a call to ask the price of crude oil and the status of the Dow Jones Industrial Average. Weatherston's family had gathered in San Antonio just three days earlier to celebrate the birthday of Texas's oldest geologist, whose life could fuel a number of adventure stories.

He was born in Vera Cruz, Mexico, son of a Canadian cavalryman and planter. His mother was a Mexican lady whose family had mining interests and a coffee plantation. His father built the first hydroelectric plant in the Misantla area and developed utilities and roads. The plant powered milling and processing operations in the coffee-growing area.

Here young Douglas witnessed battles between rebels and the government of Porfirio Díaz and later against federalist soldiers. He was sent to live with relatives in Canada, where he attended high school and learned to speak English. He then attended Phillips Academy in Andover, Massachusetts. He entered the Massachusetts Institute of Technology but left to enlist in the British Latin American Volunteers in 1918 to fight in World War I. He was a Royal Air Force cadet in ground-school pilot training when the war ended.

He returned to graduate from the University of California at Berkeley with a degree in mining engineering and worked briefly for a copper mine in Sonora, Mexico, before joining Superior Oil Company in 1924 as a field geologist in Texas, California, and Mexico. In 1931 he was in McAllen as an exploration geologist for what was to become Standard Oil Company of Texas and later Chevron.

During the early years he bought Army and U. S. Geodedic Survey aerial photographs and U. S. Soil Conservation Service maps for surface geology clues, and his friends estimated his work led to the finding of 150 million barrels of oil and more than a trillion cubic feet of gas.

In 1945 he moved to San Antonio, and in 1950 he was associated with Edwin W. Pauley in making several discoveries in Mexico. In 1948, ten years after Mexico had expropriated most of the oil properties held by foreign interests, a childhood friend came to visit Doug in San Antonio. The friend was Jorge Cumming, a geologist for Petróleos Mexicanos, the Mexican national oil company. Cumming invited the Geological Society to hold its next convention in Mexico City. Antonio J.

Bermudez, president of Petróleos Mexicanos and host of the convention, arranged for Doug to introduce each of the delegates to Miguel Alemán, the president of Mexico. This conference was the first such exchange since the rift of ten years before.

"The wealth of good will and scientific exchange with new friends was felt across the industry."

Weatherston had met his wife, Ruth Warfield, while attending college. Both their son, George Douglas Weatherston, and grandson Doug Weatherston were active with him in Discorbis Oil Company, which Weatherston founded as Weatherston Oil Company in the 1950s. He served as president of the South Texas Geological Society in 1950.[15]

In addition to O'Brien and Featherston, many other individuals were prominent on the San Antonio oil scene. One who left a lasting legacy of scientific research was Thomas Baker Slick, Jr.

A native of Pennsylvania, Slick lived and attended school in San Antonio for a short period before his family moved back to Oklahoma when he was twelve. He attended Phillips Exeter Academy in New Hampshire and was Phi Beta Kappa at Yale University. After taking graduate courses at Harvard and the Massachusetts Institute of Technology, he came back to what he would consider his hometown—San Antonio.

His father, Tom Slick, Sr., was a nationally known oil producer in Texas and Oklahoma and a rancher who was known as both "King of the Wildcatters" and "Lucky Tom." He gained the latter title only after failing ten times before hitting in the great Cushing Field in Oklahoma.

"Two sons and a daughter could not be raised as close to the oil industry as we were," the younger Tom Slick said, "without being inoculated with at least a light case of the virus of excitement and romance of that industry."

Tom Slick, Sr., died in 1930, and his widow later married Charles F. Urschel, who had been a business associate of Slick and had been married to Slick's sister Flored before her death at age 38. Tom, Jr., formed various partnerships with family members and others. In February 1942 he with his brother Earl, Urschel, Urschel's son Charles, Jr., and members of the James. M. Hewgley family drilled the discovery well in Caesar Field in Bee County near where Slick had bought the Media Ranch. The following year they opened the Slick-Wilcox Field on the DeWitt–Goliad county line. In 1947 they found the Ruhman Field in Bee County. In 1949 Slick-Moorman Oil Company was an oil and gas producer.

The Slick brothers drilled on the plantation near Yazoo City that their father had owned and brought in the first substantial oil production in Mississippi, in the Tinsley Field. They also opened two small fields in Kansas. In 1947 Tom took over a well from the famous wildcatter Mike Benedum, who had stopped at 10,000 feet after starting a well in 1941 on

Tom Slick

Kidnap concern was justified

The fame of the elder Slick and news of his wealth caused concern in the family when there were reports that he could be kidnapped and held for ransom. After his death there were no more rumors.

So it was a shock when two armed men burst into Charles Urschel's home during a bridge game, kidnapped Urschel, and held him for ransom. The kidnappers were George "Machine gun" Kelley and Albert Bates. They held their victim for nine days until the ransom was paid off.

Urschel was able to reconstruct many details of the hideout, the directions of travel, estimates of distances, the weather, and sounds near the kidnappers' lair. He noted that an airplane passed overhead at exactly the same time every day.

FBI agents followed the leads and found the farmhouse where Urschel had been held. The kidnappers were later caught and sentenced to life in prison.

Ray Miles, *King of the Wildcatters. The Life and Times of Tom Slick.*

Tom Slick's research centers in San Antonio, the Southwest Foundation for Biomedical Research and Education and the Southwest Research Institute

a large acreage in Upton County, southeast of Odessa. With an interest in a large acreage surrounding the well site, Slick drilled to 12,000 feet and hit a tremendous oil sand, opening the Benedum Field.

In addition to oil, Tom Slick was successful in many business, land, and cattle enterprises. A Navy officer and transport pilot during World War II, he formed Slick Airways with nine surplus transport planes and created what would become the largest air transport company in the country.

Oil fields were being heavily produced in the postwar years, and new methods were needed to ensure efficient production from both oil and gas fields. In 1956 he formed the Slick Secondary Recovery Corporation to apply secondary recovery methods to semi-depleted and depleted oil fields.

Slick was an intensely gifted and curious man. He helped develop Brangus cattle and was co-inventor of the lift-slab type of construction. He was a world traveler and adventurer and a collector of modern art and sculpture. He organized expeditions in search of the Abominable Snowman in the Himalayas and Bigfoot in the Pacific Northwest. After a forced landing in a British Guiana jungle where he was on a diamond-hunting expedition, he found an Indian tribe so fascinating that he spent two weeks with them.

He was killed October 6, 1962, in a private plane crash near Dell, Montana, and buried in Mission Burial Park in San Antonio. He left a great legacy, however, not only to San Antonio but also to the nation in the establishment of nonprofit research centers around the country. By donating $2 million and 3,800 acres of land, he established the Foundation for Applied Research, later the Southwest Foundation for Biomedical Research and Education, and the Southwest Research

Institute, both of which continued to flourish in the twenty-first century. Both attracted research grants and remained self-sustaining, as Slick had planned. Tom Slick did his hometown proud.[16]

Among other prominent San Antonio independents were two who were still active on the oil scene in the twenty-first century, Louis Haring, Jr., and Wilford Stapp.

Louis Haring had based his successful career largely on following advice he had received from his oilman father-in-law. It could be compared to the idea of the baker's dozen a customer received when a store owner gave him thirteen doughnuts instead of twelve. In French, it's lagniappe. In the Mexican culture the same thing is called pilón.

In "An Interview with Louis Haring Jr." in the *Bulletin of the South Texas Geological Society*, Bonnie Weise described the idea—"a little extra."

The older man said he had drilled a well to total depth of 2000 feet. There it was plugged and abandoned. Some three years later someone came along and re-entered the well, drilled it to 2020 feet, and made a fine discovery.

"His advice was, 'Always drill twenty feet deeper,'" Haring said. "I've always drilled twenty feet deeper than I had planned. You never know when something might show up."

Born in Beaumont in 1916, Louis Haring grew up in San Antonio, showing a very early interest in geology.

Sometimes South Texas roads were dry and dusty, but on other occasions they were 'a sea of mud' for the men who traveled from oil field to oil field.

Louis Haring

'We cored a sand with good fluorescence and odor, but a drill-stem test of the interval tested salt water. My first baby was a dry hole. How disgusting!'

—Louis Haring

"When I was about seven years old, my mother, dad, sister, and I were wading in the Llano River near Junction," he said, "and I picked up brightly colored rocks from the river channel. I showed them to my mother, and she said, 'You're going to be a geologist.' That was the first time I had heard the word. Later in junior high school we were asked to write a paper on what we wanted to be or do as an adult. I wrote that I wanted to be a geologist. You never know what can happen to you."

In 1938 he graduated with a geology degree from the University of Texas. Jobs were scarce, he found, after trying for interviews in San Antonio, Corpus Christi, and Houston. He got a reception in Houston after the secretary announced, "Mr. Haring is here."

"The geologist's face fell when he saw me," Haring said, "He thought Pinky Herring was calling. Pinky Herring was a popular, well loved geologist. But they were nice enough to grant the interview. I met Pinky later. He was a very nice man. Later in my career he became a bank official in Houston, and through my connection with him, he loaned me some money for a development program."

After a brief stint with Petty Geophysical, Haring was hired by Trinity Petroleum in Corpus Christi. With chief geologist Robert Beatty, the crew was assigned to check surface geology in LaSalle and Dimmit counties.

"Every day we would leave the company Model A Ford at the service station to have thorns removed from the tires," he said.

In 1939 Trinity was drilling in McMullen County. A dirt road in to the rig turned to a sea of mud. At first he could set the hand throttle to set the wheels spinning. He would get out and push. When the tires found traction and pulled out of the mud hole, he had to run frantically to catch up with the driverless car. Then, near the Nueces River, it became hopelessly bogged. He walked some five miles to the rig, where he helped take cores until the well was plugged. The crew drove him back to his car. They found that five or six men had bodily lifted the car and placed it on the shoulder of the road.

He went to work for Stanolind and received valuable training under district geologist Gentry Kidd.

Called to active duty in 1941 as a lieutenant in the Army's field artillery, he was surprised to find himself riding a horse thirty miles a day in the cavalry. He transferred to the Army Air Corps and became a gunnery officer. After serving in the Orient, he was discharged a major on October 14, 1945. He remained a reserve officer and retired from the Air Force on his sixtieth birthday, September 27, 1976. He returned to Stanolind after World War II but resigned in 1949 rather than move to another city. He preferred to stay in San Antonio as an independent.

"In a few days I had subleased a 'huge' office—10 feet by 10 feet—in the Milam Building, purchased a desk, typewriter, small drafting table, and two file cabinets," he said. "I was ready for business."

Jerry O'Brien had a big office down the hall.

"I had known him back in 1939 in Corpus Christi," Haring said. "He called me in one day and said he was organizing a new company and wanted me to be secretary. The pay was two hundred dollars a month. Asked what the responsibilities would be, he said, 'None.' The money was a great help for a new independent. Jerry is a very 'updip' person."

Haring's first drilling deal was the Louis Haring No. 1 Altha G. Black in the northeast corner of Duval County. He said, "We cored a sand with good fluorescence and odor, but a drill-stem test of the interval tested salt water. My first baby was a dry hole. How disgusting!"

The next two years produced several dry holes and a few producers. Later he was active preparing prospects all over South Texas.

"I generated most of my drilling deals from my own geology, but through the years I had submittals from other geologists. I had the late Herb Davis on retainer for several years," he said.

They drilled in Refugio, Nueces, San Patricio, Bee, Goliad, Victoria, Karnes, Live Oak, and other South Texas counties.

"I never did consider myself a premium promoter," he said. "Fortunately, I had investors from the beginning, including Cecil Cox, Columbian Securities Corporation, Cyrus L. Heard, the Hawn brothers, and R. F. Schoolfield, one of the original organizers of the South Texas Geological Society. We had been partners for more than twenty-five years, and I couldn't have found more trustworthy people."

In spite of the potential for tragic accidents, agriculture and the oil industry usually coexisted in peace.

After 1955 Bay Rock Operating Company, owned by Paul Conly and Bill Richards, handled Haring's production and other petroleum-engineering business.

"We usually originated and drilled between ten and twenty-five wells a year," Haring said. "Kemp Solcher is an excellent geologist and has critiqued many geological prospects for me. Leonard Sayers is considered one of the best landmen in South Texas. He negotiated and purchased most of our oil and gas leases. We have been meeting for lunch one day a week for more than forty-five years."

Things went well during most of his fifty-two years in the business. Among other honors, he served as president of the South Texas Geological Society. However, in 1977 a farmer's father combining a crop of grain ran over the Christmas tree of one of his wells in Bee County. It burst into flames, destroying the farm equipment and fatally burning the man. The ensuing lawsuit resulted in a substantial payment.

"Regardless of that judgment, the cost of the well control, or anything else to do with this tragedy, I still deeply regret the death of the farmer's father and pray for him," he said.

Haring's wife, Isabel, died in 1986. He finished the well he was drilling and never drilled another. He closed his office in 1999 and did business from his home.

He said he would like to see the industry have a steady time all of the time. "This boom and bust is not good for anyone. In the old days you could see lights of forty to fifty rigs at night driving from San Antonio to Corpus Christi, and you knew from the scout tickets who was drilling what. Most of them were independents. Now when you drive to Corpus Christi, you might see only two or three rigs running.

"Those were fun days, but, of course, we drilled a lot of dry holes," he said. "Now they've got more seismic and are more sophisticated, which is probably better."

Haring encouraged young people to enter the field of petroleum geology.

"There's still the thrill of finding something. . . . Every time you bring in a well, you're excited," he said. " . . . I never got rich, but I had fun."[17]

While Haring grew up knowing he would be a geologist, Wilford Stapp had other interests. Throughout his life he enjoyed what you might call a harmonious conflict between geology and music. As a young man, he had a problem deciding which one would determine his future. As it developed, they both did.

Born in 1918 to Baptist missionaries in Bahia, Brazil, the youngest of four brothers, he received a varied and unusual education. His father worked at building schools and other buildings, creating lakes for growing fish, installing roads to allow passage of trucks, and engaging in other activities to improve the lives of the people.

Louis Haring said, 'There's the thrill of finding something. . . . Every time you bring in a well, you're excited.'

The family lived on school grounds, and Wilford learned Portuguese before he learned English.

"Because my mother was ill, I was raised by a couple of great women that I still think about," he said.

He also studied French and German and later learned Italian while spending two years in Italy.

"Languages have been easy for me," he said in 2000. "I was very fortunate in that sense. I also knew some Spanish, but when I had to report in Spanish to the Dominican Republic government officials whom I was working for in Santa Domingo, they all laughed because I had a Brazilian accent."

He was a singer all his life and had a dance band when he attended Baylor University, where he majored in geology and French.

"The reason for the French was that I had to do something to make an A+," he said, "because otherwise I was having such a good time that I kept making poor grades." He felt guilty about that until years later, when the expertise came in handy in Morocco and later in France. In college he was in the a cappella choir and took voice lessons, "and there was some ambition to get me into opera."

However, one of his brothers had taken a course in geology at Baylor "and he kept talking about it. I plunged into it and found it fascinating," he said. "When I was a senior about to graduate in geology, Martha Barkema —of the school of music—had arranged for me to go to the Eastman School of Music on a scholarship, which I turned down. She quit speaking to me."

His actual start in the oil business had been in Brazil when he was eleven years old, carrying firewood and translating for an old drunk who was drilling on the beach with a cable-tool rig. "He was keying off the oil seep; and now, just offshore from where he was working, are a bunch of fields."

After completing work for his degree at Baylor in 1938, Stapp worked as a geological "flunky," as he put it, for Shell Oil Company in Houston but took a leave of absence in 1939 to attend graduate school after "they told me that I could not get anywhere unless I had a master's degree."

When his funds ran out, he left graduate school at the University of Texas and went to work for the Drilling and Exploration Company of Hobbs, N. M. He got the job because he spoke Portuguese and they were trying to entertain some Brazilians.

The company was developing the Wasson Field in Yoakum County, Texas. Through "blundering and misunderstood orders" they found oil.

"It taught me a lesson that you find oil by trying," Stapp said. "It's like what they say about baptism of children in a church—you don't have to understand to enjoy the blessings."

Continued on page 208

Barrier islands posed a challenge

Drilling on the barrier islands before the development of offshore drilling barges was a challenge. This from the *Corpus Christi Caller* of August 21, 1938:

"A new phase in South Texas oil development is in the offing. The Texas Co. this week announced location for a wildcat test to be drilled on Mustang Island, which separates Corpus Christi Bay from the Gulf of Mexico. Only one attempt has heretofore been made to drill on the numerous islands along the Texas Coast and as far as we know that was not successful.

"The Mustang Island test will be highly interesting, not only to the oil fraternity but to outsiders as well. It will be interesting to note the methods used to move the heavy drilling equipment to location. It is doubtful that the causeway or ferry to Port Aransas will support the large trucks necessary to transport the drilling rig and material. It will be necessary to determine if the sand along the island beach will sink under the heavy load.

"Much future development hinges on geological information to be uncovered by the test. Will sands that are known to produce inland pinch out entirely or will they increase in thickness that far down dip? The test will be a true wildcat as geologists will have little or no previous information to go by.

"If the Texas Co. test is completed as an oiler, the field will more than likely extend into the Gulf. Also, information obtained by the well will probably have a lot to do with whether Corpus Christi Bay will be drilled. The bay has been 'shot' by almost every major company and a number of the larger independents."

Billy Blake, "The Dope Bucket," *Corpus Christi Caller*, August 21, 1938.

A barrier island rig overlooks the Gulf of Mexico. While with Sunray Oil Company, Wilford Stapp was involved in early Mustang Island activity.

The company ran out of money in New Mexico but managed to find funds in Connecticut. As geologist and assistant to the drilling superintendent, Stapp ran samples, inventoried the rigs, and surveyed locations, all for $187 a month. That was for an 18-hour day.

"A year later their situation was much improved, but I didn't get a raise," he said. This was in 1941, and military service loomed, as he had been in the National Guard at Baylor. "They knew where I was going and saw no point in giving me a raise, but they treated me well," he said. "They promised me at the end of the war they would take care of me. . . ."

Stapp said that when he realized he'd be recalled "and become a dogface again in the infantry," he "made desperate moves to become a cadet in the Air Force and get a commission." He joined the Army Air Corps the week before Pearl Harbor and served as a staff officer in communications and radar in B-24 bombers in combat in Europe. The toll was heavy but he survived. He was even luckier that he did not stay with the Army unit. Of 170 men in it, only 13 came back alive.

"At the end of the war [the company] gave me a $50 raise," he said. "The manager told Mr. Nelson, president of the company, 'You promised this boy you'd take care of him. He made us maybe $50 million, and you gave him a $50 raise?'

"'Well,' he said. 'That's why I hired him—to find us some oil.' That field has produced two billion barrels of oil."

Logging or perforating. Note tubing blown out.

Stapp received his degree from the University of Texas in 1946 and joined Sunray shortly before that company took over Transwestern Oil Company, which had headquarters in San Antonio. In his new office in the former Transwestern office, an upset young lady appeared at his door and demanded to know, "What are you doing in my office?"

He didn't know it, but he was talking to his future wife. He and Margaret Clarke were married in 1948. Within two years they had adopted four children at one time, the youngest about two and the oldest five and one-half.

While with Sunray he was involved in Seeligson, Slick, Yoward and other South Texas fields, including Mustang Island. He was a geologist in San Antonio for four years, then an exploration manager, in Abilene for three years and in Corpus Christi for three more years, ending in 1956.

In Corpus Christi he had eight geologists, a couple of seismologists, and the land department. Sunray had seven or eight rigs working most of the time and usually five seismic crews. "And," he said, "I didn't see my kids in the daylight for a long time."

From 1956 to 1959, he did well in Louisiana, as both a consultant and an independent. Then he was chief geologist for Phillips Petroleum in San Antonio until 1964, when he started Stapp Drilling Company. That turned out to be an unfortunate decision because he went broke, largely because he could not collect funds owed him. It took eight or nine years, but he managed to pay his debts without bankruptcy.

An unfortunate drilling incident also contributed to his financial problems. When he was drilling around Corpus Christi, the driller disobeyed orders and went deeper than he should have.

"I had wanted to go into it with great care, but he went recklessly into it, and it cost me a fortune. . . . I rushed up there and it was really rough. I mean they had a stream of mud coming out over the pits and over a highway with cars running over it. That was in one direction. Back in the other direction was a school full of children."

The school superintendent came out and asked about the noise. Stapp told him the well kicked but was under control. Then the land owner came out there, and said, "You know, I had wondered where that old well was that my daddy told me they had drilled way back when he was a kid."

"And there I was," Stapp said, "looking out to a column of muddy water rising into the air where somebody had drilled a water well two or three generations ago, and this thing was popping up at me. But the worst part of it was that the person that I was drilling for came out, and I had to order him off of the property. You see, not only did I have this turnkey contract, but I also had a one-eighth interest in the well, and I said, 'I am in control here, and I'm telling you that I don't want to be responsible for your injury or death. So please get the heck out of here.'"

It finally bridged over in three days at a cost of many thousands of dollars. "I went in there and perforated my drill pipe at 3800 feet," he said, "and made a nice test. . . . I was able to make a settlement with them. . . . In other words I left a whole bunch of money in the hole, but at least I came out alive. . . . The well lasted eight or ten months and then died."

He began working as vice president for exploration for Meridian Oil Corporation in 1971, a job that led to scary adventures in such places of unrest and revolution as Angola, Morocco, and South West Africa. He also faced natural dangers. On one occasion in South West Africa, a rogue bull elephant nearly caught up with their jeep, which was slipping in sandy soil. In 1972 he returned to San Antonio as a consulting geologist.

Stapp continued to devote his efforts to his passion—music. He worked to keep the San Antonio Symphony alive and to promote an opera hall for the city. He also supported a multitude of other civic causes.

"It doesn't take everybody agreeing on everything," he said, "just a few people who feel that it is important. Without the sports and entertainment, there would be more crime, more divorce, and more everything. When the Spurs were playing for the championship, 911 calls were down by 50 percent.

"You can't dispute the numbers," he said. "If you want to reduce crime, what you need to do is hire ballerinas and quarterbacks."[18]

Wilford Stapp

Wilford Stapp began working as vice president for exploration for Meridian Oil Corporation in 1971, a job that led to scary adventures in such places of unrest and revolution as Angola, Morocco, and South West Africa.

Back-road cafés like the Ruff Neck sometimes gave scouts and geologists a chance to stop for a 'cool one.'

Corpus Christi geologist Jay R. Endicott, Jr., a scout for Sunray Oil Company early in his career, recalled Stapp as his teacher.

"At times when I was not involved in scout work, Wilford Stapp took me on as a student. I would sit in his office, and he would lecture about what this South Texas area was made up of.

"He had a large surface map in colors on his wall, and each band of sediments on the surface was in a different color. The importance of the map was to show the configuration of each formation on the surface and give a reasonable idea of how it would appear in the subsurface thousands of feet below," Jay said. "He led me through the early days of exploration in South Texas and the zones of extreme importance. He had a lecture and quiz session for the six months that I was scouting."

Endicott said that he himself had been confused by his first exposure to the oil business. It came on a visit to his Uncle Cass west of Ponca City, Oklahoma. Uncle Cass had married a widow with a 320-acre farm.

"Uncle Cass leased the land in the early 1920s. It turned out to be right on top what became Three Sands Oil Field. I was about six or seven

years old. My grandfather was walking with me down the road right down the middle of the field. The place was brightly lit by flares from all the oil wells. As we passed some buildings, young ladies on a balcony were waving handkerchiefs at my grandfather. I couldn't understand. When I asked Grandfather why they were so friendly, he wouldn't answer."

Endicott didn't realize until years later that he had experienced something of an oil boom.

Jay Endicott graduated from Arkansas City High School in 1938. Jobs were scarce, so he and a buddy, Ken Russell, joined the Army Air Corps in October 1940.

After a year overseas in the African Theatre, he returned to the states, entered flight training, and received his second lieutenant's commission as a navigator/bombardier on November 11, 1944. He was sent to Lincoln, Nebraska, for flight assignment on B-29s, but a problem with his sight was discovered and he was medically retired in April 1945.

Among the first ex-servicemen to attend college under the GI Bill, he majored in geology at the University of Oklahoma. He was hired by

The doggy was a doodlebug

Sometimes oil-patch yarns are so farfetched they almost seem true. Apparently it came from the *Blowout*, produced for an oil-field celebration in Alice, Texas, by way of another industry sheet, *Historic Legends of Western Oil.*

It seems that a barroom bum named Lucus Miller appeared at a boomtown saloon in company with a mangy-looking dog, begging for coins for drinks and treats for the animal. He often got both, plus tobacco, to get him to leave.

After the boom had begun to slow, Lucus approached an oilman and told him he could talk to his dog, Clyde, who had the knack for sniffing out oil. The wildcatter was fresh out of luck, but he had some faith in doodlebugs. Besides, he was more than a little bit superstitious, so he agreed to put $10 on the dog's nose.

They walked out into the countryside. Clyde seemed totally disinterested. His master sat on a stump and mopped his brow. Immediately Clyde ran in a big circle, sat down, and gave a mournful howl.

"This is the place," Lucus said.

And it was. The oilman drilled, and oil fairly squirted out of the ground. Clyde's reputation was assured. There were offers, and the price of a sniff went to a hundred, then a thousand dollars and much more.

Finally, one day Lucus came to town, took a bath, bought a new suit of clothes, wired $100,000 to an out-of-town bank and left on the next train. The oilmen were puzzled, then panic-stricken without their canine doodlebug. Then it occurred to them. They didn't need Lucas. They would go out to his shack and find Ol' Clyde. They went to his shack and found the dog starving and dying of thirst.

They took Clyde out, but he wouldn't give his usual performance. They cajoled and begged. Nothing. Finally one of them sat down on a stump in disgust and wiped his brow. Ol' Clyde took off in a big circle, then sat down and howled mournfully.

Then it became painfully clear. Lucus wasn't much for talking to dogs, but he was a master at training them.

Sunray and joined an exploratory staff that included Marion Moore, Stapp, Forrest McClain, and Glen Sandberg, geologists, and Leroy Cockrill, district landman.

He was assigned as a scout for six months. "My job was to get information from all sources as to wells that were being drilled, leases recorded in county seats, and geophysical activities. Scouts would meet once a week at different districts. We would be assigned counties to check, wells being drilled that did not have scouts in the check," he said. "I would leave San Antonio on Monday, gather the latest information, and go to Alice for a scout check. Then we would come to Corpus Christi in the basement of the White Plaza Hotel on Tuesday [for a scout check]. On Wednesday it was back to San Antonio for another scout check for the northern counties of the district. The scout checks themselves were

(Above) Hitchens # 1 in Saxet (Right) A drill-stem test. Note the turnbuckle.

handled by a bull scout, who took charge of the list of wells in the district that were read off. Whoever had that area would give a report on what they had found."

Back in the office there were telephone calls to be made for information from oil operators.

He rode the road with scouts from other companies, and they became personal friends.

"Once a group of us stopped at a café in Oakville. We noticed an old feller at the counter. He looked like a rancher. He asked if we were in the oil business. Then he pulled out what looked like a bob on a string.

"He said, 'I have something here that will tell you where oil is and how big a field it will be.'

"We asked how it worked.

"'If it turns this way, it shows oil. When it turns that way, it shows how deep the oil is.'

"When we asked him what was in the gadget, he drew himself up and said, 'Sir, I would rather you see my wife naked than show you my secret.'

"He stomped out and was long gone, and we were still laughing," Jay said.

In 1952 Sunray moved its district office from San Antonio to Corpus Christi. Jay and Bill Easley were unable to sell their houses, but Sunray bought them. The office was on the ground floor of the Wilson Tower.

"On January 1, 1949, I was no longer classified as an oil scout," Endicott said. "I was a geologist. I was sent out with Glenn Sanders to take sidewall cores in Starr County."

Endicott gained varied experience in subsurface geology all over South Texas. He also learned core analysis.

"Wilford Stapp put me through the ropes on well sitting. I accompanied him at different times on different wells. This was primarily the deeper zones, the Wilcox," he said.

It was in this sand near Huisache in Goliad County that they completed a good well with fifty feet of productive sand. The landowner was ecstatic. His fortune was assured. His brother-in-law, who had the adjoining lease, was rubbing his hands in anticipation because a well was to be drilled on his land, thinking a well so close to the discovery would surely be a winner, too.

A log was run, and there was no oil-bearing sand. The leaseholder was extremely agitated by the letdown. A few days later his brother-in-law went over to check on him and found him dead from a self-inflicted gunshot wound. He'd left a note saying, "I do not understand the oil business."

Endicott noted, "It's a good thing geologists do not take a situation like this to heart and follow that poor fellow's example."

Jay Endicott

The Porter Field was Jay Endicott's first subsurface job. The contractor was Lucien Flournoy.

He remembered another test in the same area where a drill-stem test failed to show much pressure. There was enough gas to spew water over the top of the pipe. Then it would subside and another stand of pipe was removed. As each stand was pulled, water spewed over and settled down.

"Finally pipe was pulled, and water didn't spew over. The driller, a Polander with a terrible temper, walked over to see if he could hear anything. He peered down the pipe. Just as he put his head over the pipe, a powerful blast of water hit him in the face, knocked his hard hat off, and doused him good with water. He ran to the toolbox and grabbed an axe, ran over, and beat the metal windshield to pieces. He looked up sheepishly to find a considerable audience had been watching," he said.

The office changed considerably when Sunray bought Barnsdall Oil Company. Among the many changes was the transfer of Stapp to Abilene. A number of other geologists also moved off to other districts. Among the acquisitions was Richard "Red" Houser, brother of Laredo

Sunray had good production in Redfish Bay, where Jay Endicott ran logs and evaluated wells. As this White Point rig illustrates, bay drilling was under way by 1942.

independent oilman Monk Houser, as landman. Endicott said Red had been driver for O. W. Killam before he joined Sunray after college.

"He told me of some of his experiences driving for Killam," he said. "They drove out into the countryside in Webb or Jim Hogg counties, and as they drove along, Killam would yell, 'Stop the car! Stop the car!' Killam would jump out of the car with a shotgun, shoot the gun off into the air, and listen with one hand cupped over his ear. Then he'd get back in the car and say, 'This is not the place. Go on.' This went on for a number of stops before he said, 'This is the place!' and got out and marked the ground. He would later move in a rig and drill a well."

The Porter Field area was Jay's first subsurface job. The contractor was Lucien Flournoy.

"He had this rig that was a large workover rig [Flournoy built it himself] that could drill at reasonable depths," he said. "The well was drilled to the projected depth. No problems. Schlumberger ran the log. That's where the project came to nothing. A dry hole. My first attempt was not very rewarding."

He was "sitting" a well in McMullen County when a turkey flew up from the fence line and the crew's chief felled it with a single shot from a .22 caliber rifle. "We had barbecued turkey for several days."

The incident stimulated the hunting urge among the crew, since the nearest restaurant was miles away.

"Oscar, the tool pusher, and Charlie Baker, production foreman, decided to spotlight a deer. I went with them. Across an open field we could see eyes shining in the dark. Charlie took a shot at the animal. We went over and found nothing there. It was a lucky miss Charlie made because the 'deer' was a prized Brahma cow," Endicott said. "He could have made the rancher very unhappy."

On another occasion bulldozers digging pits to prepare a well site in southern Hidalgo County encountered bones of a very large animal. They sent a sample to a university zoology department to find out if the bones were those of a dinosaur or some other exotic creature.

"Then the report came back," he chuckled. "We had a dead elephant on our hands. Investigation showed that many years before, the area was the wintering grounds for a small circus. It had a very old elephant which died and was buried at the site."

Endicott declined an offer from Doug Weatherston to join him in a venture financed by Edwin Pauley to work in two areas in Mexico. He found Corpus Christi more appealing. Exploration of the bays was well under way. Sunray had good production in Redfish Bay, and Jay ran logs and evaluated wells there. Then, in 1954, he decided to go independent. About this time he learned that Cosden Oil was looking for a district geologist. He reached an agreement with Cosden to open an office, which was located on the sixth floor of the Wilson Building.

One way to move a mountain

Steanson Parks, a Dallas engineering consultant for a number of companies faced with upgrading their acquisitions, said, "Some of the people came in and didn't do the maintenance they really should have done. Then they couldn't figure out why it's catching on fire or blowing up or releasing bad stuff in the atmosphere. Pipes were eaten up and equipment was worn out."

He told how the company that bought Southwestern wanted to move a battery of tanks across the road.

"It's quite a job when you dissemble a tank, move it, and put it back together," he said. "What we did was build a dike around the tanks, flood it with water, float them across the road with the help of a bulldozer. Saved a lot of money."

2003 Interview at Greene & Associates in Dallas

Like Lee Durst, Jay Endicott, and the San Antonio examples, most other South Texas oilmen of the postwar period had served in the military during World War II. Among these were Charles M. 'Chuck' Forney, Oscar Sherman Wyatt, Jr., David 'Tex' Hill, William C. 'Bill' Johnson, and Hewitt Fox.

Endicott was instrumental in arranging the acquisition of the Pratt-Hewit production. The Hawn Brothers owned the Hewit part of the holdings, and Pratts were investors out of Nebraska. Eventually Cosden traded 123,000 shares of stock for the holdings, which included half interest with Atlantic in the new Refugio Field. Cosden joined with Texas Crude and Plymouth Oil Company for a prospect in Louisiana, and Jay was elevated from district geologist for the South Texas Coast to division manager for Cosden. He was involved with Cosden in building an offshore rig.

In the late 1950s W. R. Grace gained a controlling interest in Cosden. He had tax troubles, and the company was taken over in 1963 by American Petrofina. This company reneged on promises made to Cosden employees, and Endicott quit in 1966 to become an independent.

He joined Norcal Petroleum that year and worked for a year in California doing property valuation, steam flooding, and secondary recovery of oil of very low gravity. After finding that there was no way to make the endeavor profitable, he returned to Texas in 1967 and started a consulting business with a Fort Worth operator, Duer Wagner. Most of their work was looking for additional reserve in areas where fields had been discovered. The first was in the Boyle Field area south of Rincon Field in Starr County. They also got production in the West Rincon Field and then moved up to the Greta Field in Refugio County.

"Independence was good but was not lucrative," Endicott said. "I presented a proposal to Bill Volk. He turned it down but offered me a job with Corpus Christi Oil and Gas as their onshore geologist. They were starting their offshore projects. I worked with Emmet Wilson. I worked with them until 1985 until I lost enough sight that I could not perform any more."[19]

Like Lee Durst, Jay Endicott, and the San Antonio examples, most South Texas oilmen of the postwar period had served in the military during World War II. Among these were Charles M. "Chuck" Forney, Oscar Sherman Wyatt, Jr., David "Tex" Hill, William C. "Bill" Johnson, and Hewitt Fox.

Charles M. "Chuck" Forney entered the University of Texas to get two years of college credit so he could enter the Army Air Corps. At first the courses he took—anthropology and geology—were of secondary importance. Then he took a real interest in geology, making his highest grades in the subject, which was to fashion much of his life. Just as he finished the two years, the United States entered World War II and he was in training to become a B-17 bomber pilot. His combat experience was limited by the Germans, who shot him down over Germany after only three missions.

"We had to parachute from 23,000 feet," he recalled. "The Germans were standing on the ground with their rifles pointed at me, saying, 'For you the war is over.' I don't speak German, but I could tell what they were saying."

The crew, he said, all got out of the plane, but one man died on the ground.

He joined 2,500 other prisoners in a camp for Air Corps officers in Poland. He spent nineteen months and twenty days in captivity.

"We depended on the Red Cross for a considerable amount of aid," he said. "In the winter of 1945, we were moved on a ninety-mile march to a German staging-camp area, and we were put in boxcars and transported to a camp in Austria. Toward the end of the war, [Gen. George S.] Patton's army happened to be close by, and their tanks liberated us. The Germans had all left. Life in prison camp was an education. I really don't regret it because I came through it."

After that, college was a breeze. He finished his degree in geology at the University of Texas with the GI Bill. Then he learned subsurface mapping, paleontology, and lab work. He liked the work, but the pay wasn't much, so he was easily lured away from Sun Oil Company and West Texas by La Gloria Corporation, which offered him $150 a month more and a move to South Texas—"an offer I couldn't refuse," he said.

He worked for Champlin Oil & Refining Company, which was purchased by Chicago Corporation, and was moved from the Valley to Wichita, Kansas, to Oklahoma City, to Fort Worth, to Corpus Christi.

"None of my kids ever graduated from the school they started in," he said. He thought his youngest son would graduate from Ray High School, but he was one of the seniors transferred to the new Richard King High School. The Forney record was intact.

When the company, then Champlin, wanted him to move back to Fort Worth, he quit to become an independent.

"They gave me a year's severance pay," he said. "I thought, I'll find oil before a year is over. I did, but I quickly learned I would have to work up more prospects to be a success."[20]

Fortunately, he did.

Oscar Sherman Wyatt, Jr., who was a bomber pilot in World War II, got an early start on his flying career but almost missed out on military aviation. He took up flying at age fourteen, when he and his friends built a glider and hooked it onto a car to get airborne. He got his pilot's license at sixteen. At Navasota High School he was the heaviest player on the football team, at well over two hundred pounds. He was the scourge of other teams, intimidating and frequently penalized for holding. He graduated from high school in 1941 and played as a lineman for the Texas Aggie freshmen.

After the United States entered the war, Oscar tried to enlist in the Army Air Corps, but the size that served him so well on the football field proved a handicap. At 240 pounds he was far over the 180-pound maximum. He started a very strict diet, worked on two farms, exercised, and ran miles and miles. In a few months he had lost sixty pounds and qualified as an Air Corps cadet.

Charles M. "Chuck" Forney's combat experience was limited by the Germans, who shot him down over Germany after only three missions.... He said, 'The Germans were standing on the ground with their rifles pointed at me, saying, "For you the war is over."'

All lit up. This is the type of view that may have given Oscar Wyatt the idea of developing pipelines to market the natural gas that was being wasted in the skies of South Texas.

He flew many combat missions in the Pacific Theater and received a number of decorations as a B-24 pilot. As the war was winding down in 1945, an engine failed on takeoff on Okinawa, and he crash-landed his plane, fracturing both legs, his jaw, and his skull. He was able to crawl out and help his crew to safety. All survived.

Oscar's toughness reflected an unhappy childhood. He was born in 1924 in Beaumont to a big, boisterous, heavy-drinking father and a pious Southern Baptist mother. Oscar frequently left the house at night with a blanket to sleep under a tree to escape family fights and abuse from his father. The family moved to Navasota, where a neighbor described Oscar as a street fighter and a hard worker who would finish any job, even under a broiling sun. A local doctor, who also had endured an unhappy childhood, took the boy under his wing. The doctor had a tough, brusque manner about him, a persona Oscar adopted. His classmates nicknamed him "Chink."

After the war he attended Lamar College in Beaumont on the GI Bill until he and a friend made enough money to return to A&M, where he graduated with a degree in mechanical engineering in 1949. He supported himself at A&M by student teaching and selling used cars.

Oilman Lee Durst said, "I had classes with Oscar at A&M. He'd sleep all through his classes and make straight A's. I had to work like everything and stay up all night studying to try to make C's and D's. Oscar was a very smart, fast fellow."

He was also rather truculent. Years later he named his 18,000-acre Duval County ranch "Tasjillo Ranch" for a cactus that grows in that area. Tasjillo cactus is somewhat like a porcupine. Anyone who messes with it is going to get stuck with its spines. The description, his enemies would agree, sort of fit Oscar himself, for he never backed away from a fight, even if it appeared to be a losing cause.

His life was filled with victories and some pretty serious setbacks as he fought his way through a series of acquisitions and hostile takeovers, yet he nearly always seemed to come out ahead in the rough-and-tumble corporate dogfights. Though he took his lumps, he probably inflicted more on the competition. He rubbed a lot of his fellow oilmen the wrong way, but almost to a man they said, "Whatever you think of Oscar, he was damned good for South Texas."

Their reason was that his success contributed to their success. By taking small spur pipelines and combining them, he created a major distribution system that opened markets to many producers.

After graduation from Texas A&M, Oscar started out with Kerr-McGee, working as a drilling engineer on an offshore rig near Freeport. For a short time he was a sales engineer for Reed Roller Bit Company, and he quickly became the firm's top salesman in the Gulf Coast area, selling all types of drilling bits.

Durst was one of his customers. "He was a very capable fellow, but he talked a little above my head. He said right away he was tired of what he was doing and was going to get into the other end of the business. He said he was going to get some oil and gas," Durst said. "He wasn't kidding."

Wyatt started out slowly, using his old Ford as collateral for an eight hundred dollar loan dated October 11, 1950, from the then-Citizens State Bank. Years later Henry Ford II congratulated him on his billion-dollar company.

"It was the first time I ever heard of a company getting started with a Ford," he said. He sent Oscar a gold model of a Ford with the message.

Wyatt used the eight hundred dollars to start the whimsically named "Hardly Able Oil Company," which was hardly a success. In 1951 he joined A. A. "Bus" Moore to form Wymore Oil and Gas Company. They drilled several dry holes before getting a significant discovery in St. Joseph Field in Webb County, which got him out of debt.

Jay Endicott described an incident involving the Wymore partners that occurred shortly after Endicott joined Cosden.

"Our first prospect was near Robstown," Endicott said. "We took a deal with Lawrence Hoover, an independent geologist who had just quit Pontiac Refining. Our drilling contractor was Wiley Singleton. We drilled down to the Frio to a sand around 6500 feet. It had good oil shows. We set pipe and perforated, and the well started flowing oil. Everything was

Oscar Wyatt

Oscar Wyatt rubbed a lot of his fellow oilmen the wrong way, but almost to a man they said, 'Whatever you think of Oscar, he was damned good for South Texas.'

Wyatt opposed war, helped free hostages

Oilman Oscar Wyatt received international notice when he and former Texas Governor John Connally flew to Iraq in 1990 and convinced Saddam Hussein to release a group of Americans held hostage there.

Wyatt fiercely believed the United States was wrong to go to war against Iraq in Operation Desert Storm in 1991 and said so in a speech to a large audience at the annual Corpus Christi Chamber of Commerce Banquet. He explained he supported the U. S. troops but not President George Bush's decision to go to war. The audience reaction to his stand was angry and loud. Many walked out.

The next day *The Corpus Christi Caller-Times* reported that Mayor Betty Turner said she was "disappointed, appalled and embarrassed" by Wyatt's remarks.

After reading of her statements, Wyatt returned the Key to the City that Turner had presented to him after his speech.

Turner, who said she had known Oscar for thirty years, said, "I fully respect Mr. Wyatt's right to having and giving his opinion on the Persian Gulf situation. However, I do strongly disagree with the forum which he chose to express those opinions."

In his letter returning the key, Wyatt said that he, too, was disappointed—at the published reactions of the mayor and the Chamber.

He said news reports had falsely indicated that he did not support the U.S. service personnel and pointed out that he had stated that "our troops deserve and should have the full and complete support of all Americans."

He said, "I accept the fact that you and the Chamber may not agree with my personal views on this or any other subject. That's fine. What I cannot, and will not, accept is the idea that contrary opinions are not appropriate and should not be heard.

"... Thought control does not belong in my hometown."

"Wyatt to Mayor: 'Take Back Your Key to the City,'" by Jim Steinberg, *Corpus Christi Caller-Times*, January 26, 1991

going fine. Wiley and I were standing at the pit as the well was cleaning up. Wiley said, 'Don't turn around. Look at the filling station.'

"I saw a couple of heads peeking around watching us. One was Oscar Wyatt, and the other was Bus Moore, his partner. Then the well started making water. Oscar and Bus started walking away. They had leased the little tract around the filling station. They were disgruntled. It was decided that our cement job might have failed. The well was squeezed, reperforated, and oil flowed. We drilled a second well, which was a good well. And the third test made a well. Later I ran into Oscar. He wanted info on our wells. I told him we couldn't give it to him."

Someone else drilled the little tract across the tracks. It was abandoned.

Wyatt bought out Moore in 1955 and formed Coastal States Gas Producing Company. By the end of the year, the company showed a net profit of four thousand dollars.

Selling oilfield equipment had given Oscar an opportunity to study the layout of nearly every gas well and field in South Texas, and he was credited with having a near photographic memory. Perhaps he saw

a view of his future one dark night in the early 1950s as he flew over the Orange Grove Field and looked down on hundreds of great gas flares. It occurred to him there was money to be made if all that wasted natural gas could be piped out.

At that time large producers paid operators for acreage in proportion to their share of the whole field, giving the small operator less. Wyatt managed to break the monopolistic control of the large producers and gathering firms by signing the small operators to contracts that gave them more money by paying them specifically for the gas from their wells. The next step was gas gathering.

His first purchase was a 68-mile line from Hidalgo Field to Elsa. By the end of 1956 Coastal States had nineteen gas-gathering systems in operation. His holdings continued to grow and included oil and gas production. His successful consolidation of small stringer pipelines tying into larger pipelines leading to markets was profitable to him and even more important to many producers in remote areas who would otherwise have had to wait for connections. Before this consolidation, producers often were forced to sell out to larger operators.

Laying gas lines in the South Texas brush was a hot and dirty job. This crew was working in August 1937, well before Oscar Wyatt began his operations.

Flying Tiger ace also found successful role in Oil Patch

David Lee "Tex" Hill was a Flying Tiger ace during World War II and holder of many notable decorations, including the Distinguished Service Medal, second only to the Medal of Honor. He also had a fair amount of success in the oil business.

"Most of the time, success comes from other people's mistakes," he laughed, and recounted one of his own.

"I bought a lease from Ralph and Dale Rowden where four counties come together. You had to turn off at Cotulla to get there. It was primitive as hell. I made a deal with Nueces Land & Cattle Co. It had an oil sand at less than 400 feet. I talked to them and figured and muddled.

"I had an idea one day in the Manhattan Café in Hebbronville," he said. "If I can drill that well, I might be able to make it better. People then drilled with cable tools in holes like pistons in cylinders. I knew the earth. I took three-inch pipe, slotted it. I knew where the sand was. I set pipe. Didn't cement it, put a bottom pump on it, and the well flowed over the top. That was the beginning of the Rodriguez Field, which had thirty-seven wells.

"That convinced me I had to have secondary recovery. I sold to Bob West. I didn't keep residuals. They sold the field to Texaco. They fire flooded it and took 1.4 million barrels."

Which goes to show your mistakes can make success for the other guy.

Author's interview with Gen. "Tex" Hill, 3 June 2003, in San Antonio

By 1965 Coastal was a conglomerate, boasting 4,500 miles of pipelines; fifteen subsidiaries; banking and land holdings; and crude oil, gas, and petrochemical lines stretching along the Gulf Coast. Holdings also included the newly purchased Sinclair Refinery at Corpus Christi.

Louis Beecherl, Jr., of Texas Oil and Gas, spoke of his company's competition with Coastal. "One of the companies that we emulated and later became competitors with was Oscar Wyatt's Coastal. Oscar Wyatt was one of the first people to strike upon the idea of gathering small packages of gas and taking them to market. Major pipelines did not want to fool with small packages of gas."

By 1965 Coastal was supplying natural gas for Central Power & Light at a number of points, gas for electrical and domestic use to the City of San Antonio, and gas for the City of Austin's electrical and gas system. Wyatt had moved his base of operations to Houston.

Natural gas was selling for 20 cents a thousand cubic feet in the 1970s until the supply dwindled, the price rose, and LoVaca Gathering Company, a Coastal subsidiary, was unable to fulfill its contracts. On June 5, 1973, Coastal's stock plummeted from a 1972 high of 55 to 7. Furious South Texans called Oscar a "robber baron" and worse, and by 1975 Coastal faced more than a billion dollars in lawsuits.

In January 1980 the lawsuits were dismissed; State District Judge Herman Jones signed an order, which the Railroad Commission approved, dissolving LoVaca and providing for the formation of a new company called Valero Energy Corporation. In a very short time Valero grew to giant proportions, with LoVaca's CEO, William Greehey, as president.

The settlement was considered the largest corporate spin-off in history at the time. The *San Antonio Express* reported, "Valero was formed under terms of a settlement agreement between its former parent company and its customers. The plan called for each Coastal States shareholder to receive one share of Valero common stock for each Coastal share held."

Five years later, Wyatt's biggest hostile takeover was the acquisition for $2.5 billion of American Natural Resources Company, which was called a "company maker" for Coastal. The takeover gave Coastal a pipeline system in the Midwest and provided coal and trucking assets. A key 1973 takeover was the purchase of Colorado Interstate Gas Company for $182 million, giving Coastal a pipeline and three refineries.

Wyatt was not so successful in another hostile takeover battle in 1990. He lost a litigious battle with Panhandle Eastern Corporation as Coastal tried to take over Texas Eastern's pipelines that serviced the Northeast.

From the early 1970s Wyatt conducted business with Iraq, and he drew criticism for dealing with governments out of favor with Washington. The government dropped an inquiry into a December 1990 trip he made

Rig with General Motors Detroil diesel engine-driven pump. Stewart & Stevenson was one of the largest distributors of General Motors generators.

to Baghdad after he came home with more than thirty hostages who had been held by Saddam Hussein as the First Gulf War loomed.

His blunt and earthy attitude did not win many friends. Though he owned the mansion Hugh Roy Cullen built in plush River Oaks near the Country Club, he was not invited to join, even though his wife, Lynn, was a legend in Houston society.

One member said that some oilmen in the club are tough and rough around the edges, but "Oscar is too damn mean." Another compared him to "a bug zapper in the back yard" in attracting controversy.

But he had his defenders. One executive said Oscar would add to the club because "he's not living on five generations of inherited wealth."

Another said, "He's the meanest, crudest, foulmouthed son-of-a-bitch in the world, and I like the guy because you know exactly what he thinks."[21]

Another South Texan with industry connections considered Wyatt a friend. He was Richard N. "Dick" Conolly, Sr., of Stewart and Stevenson, who said, "I watched Oscar Wyatt from his beginning and watched him grow. He has been my friend all these years."

Conolly, who also served in the Army Air Corps during the war, did not seem to consider himself an oilman.

"I was never in the thick of drilling wells where they made millions and lost millions. We just sold them diesel engines and turbine engines," he said. " I never got into the big money like many of my friends did."

But he voiced appreciation for the industry.

If a well was starting to blow, nobody would want to stick around and turn a valve.

"The oil business, to my way of thinking, over the past one hundred years has been a great benefit to South Texas," he said. "The land, the port, which came along at the same time as the oil business—it was a long and happy marriage.

"When I think of the early days, I think of people like Bob Kirkwood. A number of the early drilling contractors lived in Alice," he said. "So much happened in the Alice area. Then there was Jimmie Storm and Gus Glasscock who built the first drilling barge."

Others he mentioned were Sam Wilson, Frank Zoch, Paul Turnbull, Jim Wilson, retired from Shell, and Dan Pedrotti.

"He's published several books on wildlife. Bought the old Glasscock Ranch, you know," he said.

"David 'Tex' Hill was at A&M when I was there," Conolly said. "He's best known as one of [Gen. Claire] Chennault's Flying Tigers—he shot down twenty-one Jap planes. He got into the oil patch as soon as he got out of the service. He worked all over South Texas."

Although Miriam "Ma" Ferguson, elected Texas governor to fill the seat of her impeached husband Jim, was accused of selling prison pardons, profiting from land deals, firing Texas Rangers, and other various malfeasances, Conolly had reason to think well of her.

Because he had a friend who was in her good graces, she got him a job so he could attend Texas A&M University in 1933. He graduated in 1937. After his military service, he moved to Corpus Christi to open the local and area offices of Stewart and Stevenson before it entered the oil-field equipment business, a field in which it became preeminent.

The man responsible for that achievement was Fred Koomey, who had been reluctantly hired in the firm's shipping department. As a fill-in he was transferred to sales, where he became the company's leading salesman in the nation. He tried to interest Cameron Iron Works, one of his customers, in a new idea he had for a blowout preventer for drilling rigs.

"Preventers in use at that time," Conolly said in a 2002 interview, "had a large valve that had to be closed by hand. If a well was starting to blow, nobody would want to stick around and turn a valve. He said an air motor could shut off the well in three seconds."

Even though Cameron wasn't interested, Koomey wanted to build a prototype to make his point. He built one, but before he could show it to Cameron, a contractor bought it to put on a new rig. He built another, and the same thing happened. Stewart and Stevenson decided to market his device, and thereby the Koomey blowout preventer was the genesis of the Oilfield Equipment Division of Stewart and Stevenson, the company's premier division in 1957.

Conolly retired around 1980 as vice president in charge of offices and dealers worldwide.[22]

Ordinarily a geophysicist studied geology, engineering, and a host of other studies related to the oil business. William C. "Bill" Johnson took another route. An uncle who worked for Gulf Research and Development Company, the geophysical arm of Gulf Oil Corporation, gave him a job as a rod man with a geophysical crew when he was discharged from the Army after service in World War II.

"One day they told me, "'Billy, if you are ever going to get any place in this business, you are going to have to have a degree, even if it is in religion.' I had two years at Texas Tech. I went back and got a degree in mathematics and entered a training program for Gulf," Johnson said.

After three years he joined Shell Oil Company and then Forest Oil Company, which transferred him to Corpus Christi in 1955.

"It was a family-owned business, and the manager was Richard Dorn, who was pretty much a family man. When other companies started centralizing and moving to Houston or other places, he decided to keep the office here open."

Dorn said the money to transfer the office and employees "would buy a lot of airplane tickets."

The company had been founded by Clayton Dorn, who drilled water wells and injected water into oil and gas formations in 1926 and started secondary recovery in Pennsylvania.

Firm generated hurricane help

Stewart and Stevenson had started in 1902, when horses provided the power for a shop owned by James Stewart, a blacksmith who made carriages and later automobile bodies. Stevenson was a woodworker who also made carriages and auto bodies. The two turned to automobile engines, and their company later was one of the largest distributors of General Motors generators. It became a Fortune 500 company while Dick Conolly was with it.

Conolly was especially proud of company's relief efforts after Hurricane Celia struck Corpus Christi in 1970.

"I had been through several hurricanes," he said. "I was in West Palm Beach when it was nearly wiped out. I was on Okinawa and went through a typhoon there. We got worried that the hurricane was going to hit Corpus Christi. I called every branch of the company in the area and asked them to send every available generator they had to Corpus Christi."

The company furnished generators to Central Power & Light Company, Southwestern Bell Telephone Company, the Blood Bank, the Port of Corpus Christi, and hospitals that had no auxiliary power.

But one point bothered him.

"We got absolutely no credit for providing such a service in a time of crisis," he said.

Interview with Dick Conolly, October 2003

"He had five sons involved in the operation of buying abandoned oil leases and recovering millions of barrels of oil. Clayton established an exploration business. Four of the sons opened district offices. One was also named Clayton. Others were Gail, David, and Richard. Offices were located in Bradford, Pa.; Midland; Houston; San Antonio; and Corpus Christi. The Corpus Christi office took care of the Gulf Coast, offshore, Southeast Asia, and Australia."

Johnson worked wells in the Gulf Coast and offshore and traveled extensively in Southeast Asia and Australia.

On one trip he visited a hospital in the New Guinea jungle operated by a doctor and his nurse wife from New Zealand. They had been caring for the natives for five years.

"I asked him how long he was going to stay," Johnson said. "He said, 'I suppose until I die.' He said he had lost touch with what is going on in medicine. He will have to live with that decision—being out of the mainstream—the rest of his life. I would think that would apply to geologists and geophysicists. If you do not study and stay up, you will be in the same position until you die."[23]

Another World War II veteran, geologist and independent oil operator Hewitt B. Fox, served with the combat engineers who helped free the troops trapped in the Battle of the Bulge in France in 1944. A native of Tennessee, Fox attended the University of Texas. After the war he was with the headquarters of Atlantic Oil and Refining Company at Dallas.

"It was a marvelous opportunity to see how a major oil company operates," he said. But as a geologist he preferred to be in the field. He worked in various areas and had been in Corpus Christi about three years when the company asked him to move back to headquarters. As he didn't like the way things were handled there, he quit. In 2004 he said he'd been "unemployed" ever since.

Fox prided himself on finding something that seemed unsalable or unusable and making something useful out of it. He believed in doing the same thing to oil wells. For example, he had taken over a well that had not been doing well. Backers gave up on it. He got finances from some California investors. The well refused to be completed. It produced and then fell off.

"I suddenly found out that those coyotes out there owed me about thirty thousand dollars in back bills for operating costs and whatnot," he said. "They just quit paying. If they were not going to get anything out of it, they weren't going to spend any money on trying to rehabilitate the well."

He offered them a deal. He would forgive the thirty thousand dollars and maybe try to make a well out of it, and he would pay the bills. They agreed, happy to be free of the debt.

(Top) A 1941 blowout at a Humble operation in Flour Bluff
(Bottom) Killing operations at the blowout

"Not too long thereafter—I thought it was there all along, but I hadn't gotten up to it," he said. "I got up to it and perforated it and it came in for about 300 barrels a day."

That was in 1973. The well was still producing in 2004.

In 1957 Fox formed a consulting partnership with Bill Miller.

Miller had been with Humble Oil and Refining Company and later was geologist for Oscar Wyatt at Wymore Oil Company and Coastal States Gas Producing Company.

Miller and Fox prepared a book to inform potential investors of the advantages and pitfalls of financing exploration projects and to serve as a textbook on how the oil patch runs. Had it had widespread distribution, the honest presentation could have prevented many investors from making costly mistakes.

They were partners in a gold mining venture in Mexico. Miller was keen for the project. Fox wasn't. He traded his share in gold for his partner's share in oil, which proved to be a wise choice.

Fox was highly vocal in his opinion of the Texas Railroad Commission and its motives. He blamed the commission for the failure of many independents when it cut producing days from thirty to eight a month in 1960 because of the heavy importation of foreign oil.

"How are you going to operate a lease on eight days?" he asked. "It was marginally economic because back then oil was five dollars to ten dollars a barrel. This was all the way from the sixties to the seventie s. They literally cut us off."

In addition, he pointed out, the Windfall Profits Tax ranging as high as 90 per cent made it very difficult to raise money for exploration.

"I survived, but it was tough in those days. That was a bad period," he said. "All this had to do with the State of Texas. I don't think the rest of the country realized how bad it was."[24]

Two more South Texas oilmen shared the name Bill Miller.

One of them had aviation experience, both military and commercial. He didn't expect to get rich when he went from commercial airline pilot down to truck driver in 1949, but the pay for the latter job turned out to be much better.

"I was flying as first pilot for Eastern Airlines when they laid off all the young pilots," he said, "I had an engineering degree at Texas A&M College at College Station, and I applied for a job with Halliburton Company at Houston. They hired me as a petroleum engineer. They sent me to Alice, Texas, in February 1950. Mr. Halliburton was quite active in the company at that time. He was building a cement plant . . . in Corpus Christi, and he started all engineers driving cement trucks so they would learn the basic fundamentals in the oil fields."

He was hired as an open hole-logging engineer. The company training school involved driving a huge truck that was a big pumping unit with high-pressure-powered pumps.

Sharing the Wealth

Galloway Selby operated oil and gas wells in South Texas as Selby Walker Oil Company. His wife was Mary Jo Walker, a member of a family operating oil and gas wells out of the Dallas-Forth Worth area. Galloway earned a law degree from Southern Methodist University, played football there as an undergraduate, and served as a trustee on the board of the university for several years.

Galloway lived near Ocean Drive and liked to fish in Corpus Christ! Bay near his home.

Almost weekly, when fishing was in season. Galloway would get up in the morning, walk the short distance to the bay, and catch fish. He would then return to his house, clean up, and take the fish to the Driscoll Hotel, which was a favorite meeting place at the time.

Cooks in the hotel would cook the fish, and oil and gas men had a standing invitation to come to the hotel and enjoy the fish Galloway had caught that morning. He shared dove and quail with men in the industry in the same way during those seasons. Galloway loved fishing, hunting, and sharing his good fortune with others in the industry.

The Maude Traylor No. 1, drilled by Layton Brown, which opened the Maude Traylor Field

"My job was to handle the heavy iron and hoppers and the equipment to rig up so we could pump cement into the well to isolate formations," he said.

One of his first assignments was an emergency call to a blowout in the Rincon Field in Starr County. "They said, 'Go south from Alice to Encino, turn right, and go fifty-five miles, and you'll see the fire.'"

It was a Continental Oil Company rig.

"Mr. Myron Finley, one of the world's finest and notable fire fighters, was in charge of the operations for Continental. His son-in-law, Red Adair, was working with him, and two of their swampers were Boots Hansen and Coots Matthews."

Their highly technical services were not needed. Under the direction of Finley, "Most of the work was done by persons like myself and other Halliburton workers, along with tool pusher B. D. Lee, riggers, and roughnecks from the rig. We pumped enough mud and tagging material into the well to stop the blowout," Miller said.

Continental went on to develop one of the better fields of South Texas, he said.

"One of my first wells was with Layton Brown Drilling Company No. 1 Maude Traylor in Calhoun County. Wallace Graham was one of the local geologists, and it was his prospect," he said. "We logged about thirity feet of the prettiest Frio oil sand that I've ever seen. This particular well established the Maude Traylor Field. That was in 1958, and to my knowledge it is still producing."

One weekend he was logging a well with Gene Garvis, drilling superintendent, and Bob Wilson, supervising geologist. The operator, A. O. Morgan, "had some unkind words to say to them. Mr. Morgan fired them both. Monday morning I went down to the office, and they were both working doing their normal duties.

"I said, 'I thought you were fired.'"

"'Aw, this happens all the time when things didn't go right,' they said.

"They were both effective, competent employees," Miller said. "It didn't seem to bother them. It bothered me more than it did them."

In 1952 Miller was promoted to head Halliburton units in five South Texas cities. There were many rigs drilling in the division, from San Antonio to Brownsville. Humble had forty-four of them.

When the Railroad Commission cut production to eight days in 1958, he said, "I was required to cut about 40 per cent of our employees due to the fact that the drilling industry had cut back."

Things improved after that until 1968, when his combat crew, the 67th Military Airlift Squadron, was called into active service for two years. He was promoted to squadron commander in Vietnam. In 2003 he was still a bit angry at the reporting of the Tet offensive. "It was one of the greatest victories in recent times, and the newspapers in this country took it as a defeat."

He returned in 1973 and served as division sales manager for Halliburton for two years. Then he joined J. W. McClellan, brought in a number of wells, and in 1991 became an engineering consultant.[25]

Much of the search for oil and gas in South Texas goes on in arid, hardscrabble brushland, like that pictured here in earlier days.

Dan Pedrotti

'My theory is that oil is where it finds you. Therefore, you have to go to a lot of places to give it a chance.'

–Dan Pedrotti

The third Bill Miller was William B. Miller, who had worked for Superior Oil Company and later formed an alliance with Corpus Christi oilman Daniel Pedrotti.

When Pedrotti was growing up, his father was training him to be a grocer. Then his father decided to send him to school to be a lawyer, but a seismic crew came into the grocery in Del Rio, and Dan asked how he could get into the oil business.

"If you want to have it made, be a geologist," the crew chief advised.

"That was my total counseling," Pedrotti said. "I made a deal with my dad and took off for college, not knowing the slightest idea of what a geologist was going to be."

Although he wasn't a star in college, he finished in the middle of the pack and continued as an Air Force cadet and pilot. A professor talked him into getting a graduate degree.

"He said, 'It's worth six dollars a month more in pay,'" Pedrotti said.

After seven years with Texaco, he went into business for himself.

"Early in my career as an independent," he said, "a geologist from San Antonio, J. B. Means, came to my office and was looking at some of the things I was doing and said, 'Dan, you are trying to find an oil field. You don't need an oil field. You need an oil well.'

"I said, 'What do you mean?'

"He said, 'Well, an oil well here and an oil well there, an override, and finally you'll have enough to live on and you can go do something big. Quit chasing these elephants because they don't work most of the time.'

"I took that to heart," Pedrotti said, "and sure enough, I had four or five different small wells and had an override."

He had enjoyed a working relationship with Miller, and soon they formalized an alliance. It would be a long-lasting relationship. They drilled wells and made some profit. Then Miller figured out a new pay in Charamousca Field in Duval County.

"We had three farmouts there, but nobody wanted to make a deal. Then Scott Manley had a pile of money he needed to spend."

At Charamousca, Pedrotti said, "We had been up all night waiting on this log, and Bill was out in the car asleep. I was in the logging truck and saw the 100 feet of pay on the log. I went out and woke Bill up and said, 'Man, we're millionaires. We have just found 100 feet of pay and we have a quarter interest in it.'

"He cranked up his car and took off for town. Before he left, I said, 'Now look, this is pretty significant. We need to keep this information tight. We need to find out what else is available out here. We can make our play before anybody knows about this.'"

Pedrotti stayed on the well, getting cores and information together. That evening there was a margarita party at Vaughan Plaza.

"I got back in town about 7 o'clock and stopped by the party," he said. "The first thing everybody said was, 'Hey, congratulations.' The word was all over town before I got back."

They had found a million-barrel oil field.

In 1970 Hurricane Celia made an impression on them, too, but not such a favorable one.

"It wrecked downtown Corpus Christi office buildings, turned my house trailer over in Live Oak County, damaged my mother's house in Del Rio, and went all the way to Comstock and damaged my sister's house there," Pedrotti said.

Because his side of the office building was damaged, he took his maps, logs, and other records and put them on his clothesline to dry. Miller wasn't so lucky. His office was devastated. Before he could look it over, a salvage crew had bulldozed his desk, files and everything, out the window, nine floors down. Everything was gone.

"We got back to work and started finding oil," Pedrotti said. "I told him, 'Bill, we got rid of all your crappy maps. Now we're starting over and doing pretty good.'"

They had some luck during the seventies and had money when the collapse occurred. Then Sun Oil Company wanted to drill a deep well and offered them a farmout on one of their leases.

"We said, 'We will give you seventy-five percent of the deep rights, and we will take seventy-five percent of the shallow rights,'" he said.

"One day Bill gets a call from the Sun geologist, who said, 'Guys, I forgot to tell you we've already spudded the well and are already down and logged it.'

"Bill said, 'Why. you no good outfits. You did all this without even telling us?'

"The geologist said, 'The good news is we have 300 feet of pay.'

"Bill said, 'You can do that any time you want to.'

"We named it Suemaur Field. It provided us with income right up to 1990 to buy Harkins Drilling Company—prospects, rigs, leases, seismic data base, plus the guys who were running the exploration and were looking for jobs. So we formed Suemaur Exploration and Production Company."

When thinking of a name for their company, they thought of their wives, Maureen and Sue. "Maursue sounded like a Japanese freighter," so it was Suemaur, a name that stuck for more than thirty years.

"My theory is oil is where it finds you," Pedrotti said. "Therefore, you have to go to a lot of places to give it a chance. You can't become like the wildcatters of old who took a hundred percent of their deals and went for broke—rich today and broke tomorrow. We were always more conservative."[26]

However, old-fashioned wildcatting paid off for one South Texas oil company, when Reese Rowling's TANA Oil & Gas Company, unable

W. L. 'Wally' Popejoy

What's in a name: Part 2

W. L. "Wally" Popejoy moved his petroleum lease broker office from Midland to Corpus Christi in 1967, when practically every office building in the city was occupied by oil companies. The following year he began a long association with Reese Rowling in obtaining leases in South Texas.

"I was checking out a tract of a potential lessor in a farming area near Falls City soon after I moved here," he recalled. "I went to the courthouse and determined the mineral owner was a man whose last name was Moczygemba. One of the people in the clerk's office said it was a Czech name and told me how to pronounce it. I practiced saying that name all the way out to his farm.

"I walked up to him and said, 'Mr. Moczygemba, my name is Wally Popejoy.'

"He said, 'Who?' So I repeated my name.

"He slapped his leg and roared with laughter. 'That's the funniest name I ever heard,' he said."

Author's interview with W. L. Popejoy

Reese Rowling

'My friend Reese Rowling was a throwback to the old wildcatters. For Reese, the fun was in the hunt.'
-Wally Popejoy

to get financing for a promising prospect, decided to drill it anyway and ended up with a successful well.

Rowling, a graduate of Texas Western University at El Paso, had worked for Evans Production Company until 1972, when he became an independent. He hit his first big producer after taking a risk on a prospect in Live Oak County.

Wally Popejoy, Rowling's friend and business associate, said, "My friend Reese Rowling was a throwback to the old wildcatters. For Reese, the fun was in the hunt."

Geologist Ernie Easley agreed, "He instilled something in me that is with me today—that the search is like a treasure hunt."

Mike Popejoy, Wally Popejoy's son and one-time president of TANA, said that Rowling "was right more than he was wrong. He was deeply involved and a deep thinker. When he made a discovery, it was likely a new field or a new trend."

TANA, which was formed by Rowling and associates, was named for a discovery in Webb County about thirty miles north of Laredo—at that time out in the middle of nowhere, so far as the oil and gas business was concerned.

In 2003 Wally Popejoy said, "We leased over 100,000 acres of land before we drilled the first discovery well. That was more than thirty years ago, and there are still at least a hundred wells producing in that field."

The company name came from the formations that produced in acreage in which they had an interest in Webb County: TA from Taylor and NA from Navarro, to form TANA Oil & Gas Company.

Shortly after Reese's son Robert graduated from the University of Texas with degrees in business and law, the younger Rowling decided he was in the wrong business. He was talking to clients, most of whom were in the oil business when, he said, "My desire was to be on the other side of the table."

After entering the oil profession in 1981, Bob Rowling threw himself wholeheartedly into the business, learning sales, the field, pipelines, everything. TANA became one of the most successful oil and gas exploration companies in the state.

When the industry collapsed in the '80s, TANA was free of debt, with cash on hand and a promising prospect in Hidalgo County, but it was hard for an independent to get financing for a prospect at that time.

"I'll bet I showed that deal twenty-five times and couldn't get anybody to buy it," Bob Rowling said. "The leases were expiring, and we believed in it enough that we drilled it ourselves.

"It's funny how things work out. Instead of owning 25 percent of the deal, we owned 100 percent."

It was a very successful well. TANA joined forces with Bud Treadaway and formed pipeline company TECO. In 1989 the two companies merged to form TRT Holdings. Before long the partners were considering buying another oil company, but then their plans changed.

"Suddenly we decided we were on the wrong side of this equation," Bob recalled. "We needed to offer our property up for sale."

Up to the time the papers were signed, the elder Rowling was not convinced that he wanted to go through with it.

"I may not sell," he threatened, but Texaco offered them a deal that even Reese couldn't refuse—nearly half a billion dollars in stock and cash.

They immediately looked for opportunities to diversify. They bought the Corpus Christi National Bank for $20 million and sold it two years later for $131 million. They then bought two Omni Hotels in Corpus Christi. Across the country hotel owners heard that the Rowlings had paid cash for hotels, and offers poured in. And thus the Rowlings made a highly successful transition to the hotel industry.

"In the next three or four months we spent about $150 million, buying eight hotels in the Southwest," Bob said.

Jim Caldwell, Omni Hotel president, joked that the only thing they knew about hotels was that they had slept in them but said "the

Corpus Christi's Omni Marina hotel (left) and Omni Bayfront (right) flank One Shoreline Plaza on Shoreline Drive. TRT Holdings bought the Omni chain in a diversification move after a highly successful strike in Hidalgo County.

'My dad loved the energy business. He stayed an oil guy until the day he died.'
—Robert Rowling

hotel business, like every other business, requires the same business principles."

TRT bought the entire Omni hotel chain for half a billion dollars in 1996 and spent millions bringing the properties up to four-star level. The company gave up millions more in revenue by taking X-rated movies out of the hotels, but Bob Rowling said that at the depressed prices of the period, "we bought the company for a song. We were getting really priceless real estate in places like New York City, Boston, Chicago, Houston, Dallas."

Although it was not an easy decision, the company moved its headquarters from Corpus Christi to Dallas. In mid 2003 TRT had forty-one hotels and resorts and was negotiating for more.

Reese Rowling died in 2001, proud of what his company had accomplished. However, Bob Rowling said, "My dad loved the energy business. He stayed an oil guy until the day he died. He had a really, really significant discovery six months before he died. In fact, we are still drilling wells."[27] Likely Reese would have been prouder of that than anything.

Wallace Graham's family was another in which the son joined forces with the father. Wallace got his start by majoring in geology at Texas Christian University after serving in the Navy during World War II, and two of his children followed him into the field.

In 1952 Wally took a job with Pure Oil Company, which was drilling with core rigs. He then joined Republic Natural Gas, which was conducting an aggressive exploration program in the Corpus Christi area.

"Red Kelsey was chief geologist for Republic," Graham said. "He was a real character but a great teacher. Once Red was lining up a block for a lease and had reached agreements with all of the landowners but one. Red went out there, and the guy said, 'Hell, no. I won't give you that lease.' Red said, 'Okay. I'll just contour you out.' He pulled out the map and an eraser and wiped out the contour lines on the man's land that indicated the geological view of the lease."

It was as if Red put a hex on the man. As it developed, the man's property was in a low, and when production was found on the rest of the lease, "the guy never did make a well."

Wally left Republic in 1955, much to his good fortune.

"I had worked up a prospect in Calhoun County. Company brass came down and looked at it. It was 100 feet from a Humble well. We sent a landman over to get a farmout from Humble. He came back with it and showed it to the company president. Sam McCord, who was with him, said, 'That's a graveyard.' That was the end of the deal right there. The interview was over."

As an independent Wally asked if they wanted the lease. They told him he was welcome to it. Layton Brown drilled it for him. It was the No. 1 Maude Traylor, the well Halliburton's Bill Miller cited as "30 feet of the prettiest Frio oil sand that I've ever seen." The discovery opened the Maude Traylor Field "that supported me for fifty years," Wally said.

"When I got the word about the well, I jumped in my car and got a speeding ticket right in front of Spohn Hospital," he said.

"That well was just 100 feet from a Humble dry hole. They drilled another one exactly one mile to the north. It was a dry hole, too. Just 100 feet to the north and they would have had a new field," he said.

Layton Brown drilled most of the wells on the 1,200-acre lease, and F. William Carr took over operations and drilled the rest of them. Wally estimated the field produced six million barrels of oil.

He made other discoveries, but the Traylor was the most significant. They weren't all that easy.

"We were up coring wells on the Cataleen lease, and me and the tool pusher decided to go into town and get something to eat, but everything was closed. It was Christmas Day. The tool pusher had a sack of pecans. That was our Christmas dinner—pecans and oranges we found growing there."

Then there was the saga of a drunk roughneck who came out to the well site looking for trouble.

"He pulled Frank Robinson out of his bunk and whipped him good. He beat everybody else in the crew. We were sitting around talking later, and everyone said this guy had whipped their butts. The derrick man said, 'He didn't get me. I was up on the derrick. But he did whip mine before I climbed up there.'

"It took all of the sheriff's deputies to put that guy down, but he whipped a few of them, too. 'What about you, Wally?' they asked. I was in the top bunk just as quiet as a mouse. I was the only one he didn't whip up on."

There was little doubt that Wallace's son Robert would be in the business.

"We'd take him out to the rig. When he was about three years old, he would sit out on the run and watch the crew," Wally said. "He could tell you what every man in the crew did."

"I remember as a little kid," Robert said, "there was a bunkhouse, and there was a bay on either side of the well site. We fished at day and floundered at night. It was a picnic."

When Robert wanted to be a roughneck, Wally was a little leery.

"He really didn't want me to, but he took me over to talk to Roy Jindra at Harkins & Company because they were real conscious about safety," Robert said.

Wally remembered trouble with a well he had drilled for Clayton Williams. With all sorts of bad luck, everything went wrong, including drilling costs. It was completed but was not much of a well.

Twenty-six years later, Robert, who also became an independent, had a go at the lease.

Wallace Graham

'When I got the word about the [Maude Traylor] well, I jumped in my car and got a speeding ticket right in front of Spohn Hospital.'
–Wallace Graham

"Let me tell you about that lease," Robert said. "A lot of geologists claim they aren't superstitious. I don't. That Clayton Williams well was started on October 31, 1978, Halloween night. It was a mechanical nightmare. They had two side-track holes and were on the job for more than a year. Then twenty-six years later they spudded a well on the same lease. Would you believe it was October 31, Halloween night, at 10 o'clock? They could have waited two hours, and it would have been a different day. A jinx. It took twenty-six years to sell that deal, and it was a four million dollar well contaminated with hydrogen sulfide and carbon dioxide. Am I superstitious? You bet I am."

The family connection didn't stop with Robert. Wallace's daughter, Leslie, also became a geologist—and married another, David Kirk.[28]

Mestena Oil and Gas was another family operation, and few families have had greater influence on the all phases of development in South Texas than members of the Jones family, on whose property the company was founded by Clarence Hocker, a San Antonio oil and gas attorney. Hocker was a son-in-law of W. W. "Dick" Jones, one of the largest individual landowners and financiers in South Texas.

The first member of the Jones family in Texas was Allen Carter Jones, born in South Carolina in 1785. He arrived in Nacogdoches in 1826 and received one of the earliest land headrights for his service in the Texas Revolution. "Dick" Jones, who was born in 1884, was his great-grandson.

Making up a joint with a spinning chain

The first county judge of Jim Hogg County, "Dick" Jones developed the family properties in several counties and, with the rich oil and gas discoveries, consented to Hocker's formation of the Mestena Oil and Gas Company in 1935. As president of Mestena Oil and Gas, Hocker negotiated a number of leases that were still producing sixty years later.

As one example, the Alta Mesa Field, developed by the company, produced 80,000 barrels of oil a day during World War II. The field produced almost eight million barrels.

Hocker died shortly after the company was founded, and his wife, who became Mrs. Donald Alexander, was chairman of the board and president from 1938 until she was succeeded by a nephew, Benjamin Eshleman, Jr., who died in 1972. He was succeeded by W. W. Jones II, and he was followed by George Tanner, who was married to Jones's niece.

When Tanner graduated from college with degrees in mechanical engineering and business with a specialty in power systems, he found there was no demand for engineers, so he went to work for a large international construction company called Chicago Bridge & Iron Corporation.

"Before coming to Mestena, I had built practically everything ... I built some large gathering facilities and large surface facilities for Exxon," he said in 2004. "I was actually involved in filling stations and everything in between—refineries, petrochem and gas plants,

Like the Jones family, the Welders combined ranching and oil operations. This scene is from the Welder Wildlife Refuge, established by Rob Welder to show that wildlife preservation could coexist with the two occupations.

pipelines. . . . I was involved in construction and engineering, manufacturing, and sales and marketing.

"I was also with them in developing offshore Gulf of Mexico rigs. . . . I ended up selling seven large structures that are in the Gulf of Mexico. When they decided to get into the manufacture of these, I recommended that they choose a site at Ingleside. Actually the Gulf Marine facilities at Ingleside were originally purchased by Chicago Bridge & Iron."

He and his wife lived in New Orleans but wanted to come back to Texas, so he left the company and they moved to Houston. He was doing well there, but when W. W. Jones II invited him to join Mestena, he considered the Houston traffic and a few other factors and accepted the offer in 1982. He already knew all the family.

When he first came to Corpus Christi, he went to the ranch one day a week so he could learn the location of everything and know what was involved. At that time "we were a lessor and everybody else was lessees and operators," he said, "so I had all these operators, and I started going down and making all the logging runs and going on every rig and going by every piece of production equipment and learning every aspect of the business."

He closely observed drilling in the Queen City, a very complicated formation. By watching a number of operators, he learned "there is a right way and a lot of wrong ways to drill, complete, stimulate, and produce the Queen City."

He organized an operating company, because, he said, "part of the Mestena is really good pay. We started picking up and acquiring stripper operations that were on the ranch because I could take an eighth interest and operate in an economic climate that an operator could not do."

He got 3-D seismic interpreted, and "then I knew exactly how I wanted to drill it and how to complete it and how I wanted to stimulate it."

He found a partner as an operator "and the first well worked, and we ended up drilling about fifty plus Queen City wells and had almost 86-percent success rate."

Predominately the wells produced 3.5 to 4 million cubic feet of gas a day with 42 barrels of condensate per million.

In 1986 W. W. Jones II died, and Tanner became president of Mestena, Inc., the corporate-held family group that owned a substantial amount of acreage in Jim Hogg, Brooks, Starr, Hidalgo, and Duval counties. He was also president of Mestena Operating, Ltd., which managed over 200,000 net mineral acres in Brooks and Jim Hogg counties. The managing general partners were called Mestena GP. All were owned by Mestena.[29]

Jones family had ranching roots

Capt. A.C. Jones, the son of Texas pioneer Allen Carter Jones, fought in the Civil War and was sheriff of Goliad County. As a banker and railroad promoter, he was known as "the father of Beeville." His son, William Whitby "Bill" Jones, born in 1857, did not become a merchant or banker as his father wished. Rather he drove cattle to Kansas City and became a rancher, acquiring thousands of acres of rangeland in Bee, Jim Hogg, Duval, Brooks, Starr, and Hidalgo counties, which remained the heart of the Jones properties into the twenty-first century.

"Bill" Jones was an imposing figure, his six-foot four-inch height magnified by western boots and a large broad-brimmed hat. In 1905 he moved to Corpus Christi, where he established businesses and banks, built one of the first multi-story office buildings, and owned the famous 300-room Nueces Hotel. He assisted in the establishment of the Port of Corpus Christi and was associated with banks in Hebbronville, Alice, Corpus Christi, Houston, and other towns in the area.

Corpus Christi, The History of a Texas Seaport, 138

Casing stacked up on location

Another member of the Jones family, William T. Vogt, Jr., was a latecomer to the oil business.

"My mother told me to major in geology," he said, "but I didn't listen to her. I majored in English and journalism. I have been learning geology ever since."

His grandmother was Alice Jones Eshleman, W. W. Jones's youngest daughter. She married Benjamin Escheman, Jr., who came from Philadelphia in 1969 to take over as president of Mestena Oil & Gas and died at an early age in 1972. Vogt was exposed to the business by working three years for the Guardian Oil Company in Houston. Guardian, run by Buck Rogers, had drilled some wells on the Mestena. Vogt moved to the ranch in 1982 when the company was sold.

"When I got here, I thought I was going to be in the oil business," he said. "They had just discovered Mestena Grande Field, and there was quite a bit of drilling on the ranch. By 1984 it was clear the industry was taking a nosedive."

He found he could make more money with quail leases until the 1990s, when he put together his own company with Andy Crews and called it Black Diamond Exploration. Later he was appointed to the board of Mestena.

"One guy we dealt with was called Leslie C. Texas," Vogt said. "He had been a geologist for the Hungarian National Oil Company until he hijacked a plane and escaped the Communist country."

"He operated an elevator in New York City until he learned English, and he migrated to Houston where Getty took him on. He had a hard-to-pronounce name, and he always said he was going to change his name to the place that got him back in the oil business. So he had his name legally changed to Leslie C. Texas. By 1980 he had his own drilling rig and flew his own airplane, and he owned a tiger and a ranch. His company was the El Texas Petroleum.

"He floated his offering on the English Stock Exchange, and the company was worth about $80 million. He wore safari clothes and

A secret of Hunt's success

Jerry Clark recalled working on a joint exploration program with Hunt, Tidewater, and Getty.

"We were drilling a 22,000-foot well in the Delaware Basin. For some reason I had to call the Hunt geologist," he said. "It was like a Wednesday at 2 o'clock in the afternoon and nobody ever answered. Then a voice came on the line and said, 'Mr. Clark, this is H.L. Hunt. How are you today?'

"I said, 'I am fine, Mr. Hunt.'

"He said, 'Today is the day of our annual picnic, and I'm the only one in the office.'

"I gave him the well number and told him what I was after.

"What can you say? He let everyone go to the picnic and he stayed in the office. He was still the worker and interested in what was going on in that well."

—Jerry Clark

always had a big knife in his belt, and he wore big gold bracelets. He was really very colorful.

"Guardian Oil Company took an option from Mestena Oil and Gas Company and had drilled half a dozen shallow wells. Most of them weren't any good. Then in about 1981 they decided to drill down to the Queen City. Then they pulled the log. Queen City logs look like crud. Guardian had sold its interest in the well, and there were a couple of company guys and some other people there monitoring the well. The company guys looked at it and said it didn't look like anything.

"But Leslie was enthusiastic and finally sold the company guys on setting pipe. After all, it wasn't their money. So they perforated and fraced [pronounced fracked and short for hydraulic fracturing] it. The last time I saw it, that well had produced about 70 billion cubic feet of gas and is still producing. Leslie had about a quarter interest in it.

"He did pretty well for awhile, but it all toppled over finally. The ranch, the airplane, the rig—all gone. But the fact that he was on location with that log really made a difference. Otherwise the company would have walked away from it. Somebody else might have come in behind and found it, a field of about 200 Bcf."[30]

Unlike Vogt, Richard C. "Dick" Wilshusen knew he wanted to major in a petroleum-related field. This gave him a problem. He lived in Fort Worth and always dreamed of going to Texas Christian University. Then he discovered that TCU did not have the petroleum engineering courses he wanted, so he attended Southern Methodist University, the place he had always considered the "enemy."

Wilshusen worked with Core Industries from 1955 to 1962 and learned the intricacies of analyzing cores and performing other tests. He came to Corpus Christi as a reservoir engineer and formed a consulting partnership with geologist Dick White. His specialties were log interpretations and core and rock analysis.

Wilshusen told of a time when geologists would gather when logs were being run. "Sometimes three or more companies would be involved in a project. There would be a fascinating interplay when they were getting to the end of the well. One would come in and look at the log and shake his head and say, 'Oh, no. This doesn't look good at all.'

"You could see the spirit in the whole trailer go down.

"Then the next geologist would come in and say, 'Oh, my gosh. Did you see this?' And the trailer would get all excited. You might get another reaction from a third geologist.

"The tug of war was more emotional than technical, and the strongest personality would usually win. Watching it all play out was fun. Of course, if I was directly involved, it probably wouldn't have been as much fun."[31]

Another Corpus Christi oilman who spoke of an especially enjoyable period of his job was Jerry P. Clark, who got an early start in geology while he was still in junior college. He answered an ad for a position at the Geophysical Department of Tidewater Oil Company in Corpus Christi.

"It was my first job," Clark said. "The fellow who hired me was Ray Lowe, a geophysicist. In 1955 Tidewater was moving up the street from its office in the Driscoll Building. They needed somebody to carry a bunch of geophysical boxes down the street to their new offices."

Actually, it was a good contact, because the early connection paid off after he got his geology degree. He was hired by Tidewater, which became Getty Oil, in 1958 and worked for the company twenty-five years.

Clark was born in Laredo. "My family was oilfield trash," he smiled. "My mother and father were married and went to Refugio for the boom there. Then they moved to Hebbronville. I was born in Laredo because there was no doctor in Hebbronville."

Although he did not realize it at the time, one of the highlights of his career with Getty occurred in the early 1960s when "I was the rookie on the block on a well in Lee County, New Mexico.... I didn't realize until many years later that I had the privilege of sitting on the discovery well of a 100-million barrel oilfield."

Some of the productive pay zones in the Vacuum Abo Field were 600 feet thick and still producing more than forty years later.

In 1983 he quit Getty and joined Edwin L. Cox in Corpus Christi. When he left Getty, he was exploration manager for the Lower 48 but took a $15,000-a-year salary cut to work for Cox.

"I quit because I was totally dissatisfied with the bull stuff in the company," he said. "They were more interested in buying insurance companies and TV networks than they were in looking for oil and gas."

Cox was the son of Edwin M. Cox, who, with Jake Hamon, drilled some two hundred wells in a field in Webb County in the twenties and thirties. Hamon split with the elder Cox when the father wanted his son to join the firm after he came home from World War II.

"The twenty years I spent with Cox were the most productive years of my professional life—certainly the most enjoyable," Clark said.

"It's real interesting when you look at the successful independent operators. Invariably they take care of their employees through overrides or some other reward. Ed's philosophy was, 'If I'm going to make money, you're going to make money.' I made lots of money working for Ed Cox."

Clark left the company to become an independent in 1990, and Cox closed up shop about two years later.[32]

While family connections, fortuitous circumstances, or childhood ambition brought many into the field, it was William Wordsworth's fault that Lawrence Hoover became a geologist. In his first year at the University of Texas, he fully intended to become an English major.

Lawrence Hoover

'I've known people who were perfectly efficient as geologists but just didn't enjoy the timing. My wife and I remind each other just how lucky we have been.'
—Lawrence Hoover

Humble's Stratton Camp. Lawrence Hoover remembered sharing meals at Humble housing 'in the remote sections of the brush country.' The companies built camps for the families because the operations were so far from towns.

"I loved the Romantic poets and was in a literary class reading and studying Shelley, Keats, Coleridge, and Wordsworth. I really enjoyed it," he said. "Then one day my professor asked me to read this passage from Wordsworth. I read 'There was a time…,' and he stopped me and said, 'Mr. Hoover, what is the significance of that?'

"I had never seen it before, so I didn't know I was supposed to know that Wordsworth had used the same words in another poem. I hadn't been exposed to it," he recalled. "He gave me a "C" in that course. About the time I was required to take some science courses so I could qualify for the Bachelor of Arts Degree. I made A's in physics, chemistry, and geology. I said, 'I'm in the wrong line of business.' So I switched my major around and got my degree in geology."

Hoover was born, appropriately, in a town named Hoovertown near Paris in Lamar County. The Hoovers came to Texas in 1850 from Tennessee because the steel plow had been invented and a man could plow 160 acres of Texas land by himself. Raised in Dallas, Lawrence graduated from UT in 1948 after serving in the Navy. He worked for Humble Oil and Refining Company, Gulf Producing Company, and Pontiac Refinery and was involved in uranium-exploration work before becoming an independent.

His first oil production was almost a poor-boy operation.

"In 1956 Wallace Singleton, a drilling contractor, drilled a prospect I had worked up. I had a working interest in the well. We tested, and it looked favorable, so we set casing, perforated, and the well flowed oil," he said. "I felt like celebrating, so I took the whole crew to lunch at Joe Cotton's Restaurant in Robstown. We got back to the rig, and it had gone to water. I used every bit of money I had to buy the cement to seal off the flow.

"Then the flow changed back to oil. That was one of the high points of my career. It was a good well, or I might have been out of the business. I still hesitate to buy anybody lunch," he smiled.

Hoover credited new seismic and other technologies for the great increase in natural gas exploration after 2000, but he appreciated some of the older and simpler improvements, like communications.

"Geologists sometimes had to stay at the well site in the remote sections of the brush country," he said. "There were no telephones, and conditions weren't good. I remember Humble built houses and had a dining room in the area. I used to eat with them."

He had been blessed with good fortune all these years, he said. "I've known people who were perfectly efficient as geologists but just didn't enjoy the timing. My wife and I remind each other just how lucky we have been."[33]

Other military veterans, including Albert T. "Al" Harris and Dr. Ray Govett, entered the field after service in the Korean War.

A job as statistician for Chance-Vaught Aircraft didn't sound at all promising to Harris, a newly released Marine Corps captain, who had been in charge of 150 artillery batteries in Korea. He had been a high school teacher, but he wasn't interested in returning to that, either.

Driving into Dallas, Harris looked up at Pegasus, the Flying Red Horse, symbol of Magnolia Petroleum Company, on top of the company building. The highest point in downtown Dallas at that time, it was like an omen.

"I walked into the lobby of the Magnolia Building and asked about job openings," he recalled. "They sent me to the office of Howard Brooks, chief scout for Magnolia. We talked for an hour, and he asked 'How'd you like to be a scout for Magnolia?' The pay would be twice that of statistician."

"My career in the oil and gas business was just beginning," Harris said.

At the time Magnolia operated its own rigs, and Harris was initiated into the new world as a roustabout in West Texas, catching all the dirty jobs until the crew accepted him. They proved it by holding him down, raising his leg, and pouring ice water down his pants.

"I cleaned the drawworks, painted and washed the rig, raked the location, laid pipe, tore down, and rigged up," he said. "I posted wells

A Matter of Degree

Gary McAtee was ready for a career change. After a stint as a junior high coach, he had moved to Mary Carroll High, where the golf team he coached finished fourth in the district. One day a newspaper want ad for a company named Gemini caught his eye.

Gemini had been founded in 1964 by Mike O'Connor and his partner, Bert Guthrie. O'Connor had invented and distributed a small natural gas compressor that proved extremely popular in the oil fields. It became a national success. Later the firm would be sold to Universal Compression Company, but in 1976 the company needed someone "with a PE degree." McAtee a Physical Education graduate, jumped at the chance.

He was a bit chagrined to learn that the "PE" meant Petroleum Engineering.

But all was not lost. According to the Carroll High School yearbook, Coach McAtee left coaching that year "to enter private business." He wasn't a petroleum engineer, but his outgoing personality made such an impression on the partners, they hired him to sell their compressors.

Jon Spradley; Carroll High School *Anchor*, 1977

Ray Govett

After serving as an Army pipeline-engineer officer in the Korean Conflict, Ray Govett earned master's and doctorate degrees in geology at the University of Oklahoma.

on company maps and learned a lot about oil-field trading. I also worked with Magnolia's geophysical department and with the district landman."

He became bull scout of the North Texas Scouts Association, and one of his achievements was ending the squabbling among scouts who suspected others were making more money than they.

"I had each of them write down the amount of money showed on their last pay stub and drop it in a hat," he said. "I made a composite list from high to low and passed it out to the group so each scout could see how he stood with respect to the others. I don't think we learned much, but it stopped the salary talk. Actually, we were all probably being paid a heck of a lot more than we deserved."

He was promoted to landman for the Houston District Office. "From 1966 to 1969, I was area landman for Mobil Oil Corporation's Corpus Christi office," he said.

His most exciting moment occurred when Clinton Manges ordered all gates to the Duval County Ranch locked. Mobil had exclusive rights to all the ranch.

"He let me know his men would patrol the ranch," he said. "I told him I had a 30 ought 6 rifle and I personally would shoot the locks off. He had his men remove the locks."[34]

Unlike Harris and many of the other South Texas geologists, Ray Govett grew up around the smell of oil. As a child he sat on the front porch of his home in Bartlesville, Oklahoma, and heard the sound of wells pumping and smelled the earthy odor of crude oil. Frank Phillips, founder of Phillips Petroleum, and Mr. Arutunoff, founder of the Reda Pump Company, lived a few blocks away. Phillips and Cities Service Company were headquartered in town, as were several early oil operators.

Growing up in Bartlesville was indoctrination in itself. Govett worked summers on seismograph crews for Phillips and earned a geological engineering degree from Colorado School of Mines. After serving as an Army pipeline-engineer officer in the Korean Conflict, he earned master's and doctorate degrees in geology at the University of Oklahoma.

He came to South Texas with Sunray Oil Company, which was started by a Mr. Sun and a Mr. Ray. Through a series of mergers and acquisitions, the firm's growth was so phenomenal that its officials did not always know the extent of their new properties. Barnsdall, Greta, Flour Bluff, Mid-Continent, and Transwestern Oil companies were among those merged into Sunray.

Before World War II Sunray had the huge Keeran Lease in Placedo Field in Victoria County. The company had drilled six very productive wells on it but, because of wording in the lease and wartime limitations on available casing, was forced to relinquish much of that lease that later was successfully drilled by other companies.

Sunray transferred Govett to Corpus Christi, where the company owned an interest in the Suntide Refinery. His mission was to find oil as stock for the refinery.

During this period there was a 94-year-old Eastern investment banker who worked with Sunray on acquisitions and frequently visited and talked with workers in the Tulsa office. Even at his age, he had a "Charles Atlas-type physique" and had no trouble walking up and down twenty flights of stairs, but he died shortly after returning from one visit.

At the time some fifteen geologists worked for Sunray in the Corpus Christi area, and there was confusion as to what the company owned. For example, it had an interest in a Flour Bluff Field that included the Corpus Christi Naval Air Station, but Sunray landmen were not aware of it until after bidding for the leases.The company plugged wells in some fields that other operators successfully reentered. Soon Sunray, with assets greater than the company's value, mergd with Sun Oil Company.

Because of the impending takeover, working conditions became undesirable, and Govett went to work for Superior Oil Company for a couple of years before going independent. Much of his work involved the rugged brush lands in northern and western Webb County. In the area there was a crude road called the Old Mines Road, which paralleled the

The Suntide Refinery was under construction in Corpus Christi in 1953. Sunray Oil Company transfered Ray Govett to Corpus Christi to find oil for the refinery, in which the company had an interest.

Brian O'Brien

Brian O'Brien was born in England. The family lived on the outskirts of London, and he remembered the Blitz of German bombers and buzz bombs.

Rio Grande from Laredo, Texas, to Eagle Pass, a distance of about one hundred miles.

About a 50-mile stretch of it had been closed during World War II for use as a firing range for aircraft from nearby military airfields. Some ranchers who owned property along the road did not want it reopened after the war, even though the southern end was designated as Farm-to-Market Road 1472 and the northern end Farm-to-Market Road 1021.

"One of my clients was drilling a well near the road and needed to get logging and cementing vehicles to the location in the middle of the night," Govett said.

"They found a locked gate across the road. The Webb County judge was called, and he called the rancher and told him he would go to jail if he was ever called in the middle of the night about a locked gate."

Later the Scibienski family donated land near the south end of the road at the Rio Grande so the United States and Mexico could build a toll bridge there. The new highway replaced the southeasternmost part of the Old Mines Road and connected with IH 35 north of Laredo and U.S. 59 east of Laredo.

Govett's client worked with Bill Sutton of San Antonio, who was associated with Joseph Kennedy, father of President John F. Kennedy, in a firm named Kenoil. This was in addition to Kennedy's association with Corpus Christian Jack Modesett in Mokeen Oil Company.

Sutton had become acquainted with Kennedy at a hospital where they each had a child under treatment. After the deaths of President Kennedy and his father, the partners worked through the Kennedy Foundation. It was apparently controlled by Ted Kennedy, who was not friendly to the domestic oil industry. Eventually the holdings were sold.

Govett consulted some for Edwin L. Cox, of Dallas. Joe O'Brien was the geologist in charge of Cox's Corpus Christi office. On one occasion, when Cox was drilling a well west of an existing field near the Webb-Duval County line, O'Brien told Govett to "take some sidewall cores."

"I took cores from the Miranda and another sand and took them to Hebbronville. They showed oil," Govett said.

The well definitely had a split personality. The electric log indicated salt water, no oil, but sidewall cores indicated an oil show at 1900 feet.Govett couldn't figure it out. The office couldn't figure it out, either.

"Bring in the cores, and we'll go from there," he was told. Finally the oil shows won out.

(Left) An early geophysicist with his field equipment. (Below) The J.K. Culton Baldwin No. 1 was spudded October 6 and completed November 1, 1935, at a depth of 4082 feet. It was 9200 feet west of Corpus Christi and 5400 feet north of the Tex-Mex Railroad.

A driller on another rig asked what depths they were drilling pipe. They told him they were running casing to test a sand near 1900 feet.

The driller gave them an explanation "That's the depth where my drill pipe stuck. I used oil from a lease tank to pump into the well to free the drill pipe."

Cox set the casing and used the well for saltwater disposal.[35]

Tony Sanchez was not one of Govett's clients but worked with him on several projects. He had a typewriter store in Laredo but developed a more successful sideline—buying and selling oil leases. He teamed up with geologist Brian O'Brien in Sanchez-O'Brien Oil and Gas, which became very successful.

O'Brien's name was Irish and his accent was Texan, yet under it all he was a one-time Englishman whose family roots went back to the earliest days of South Texas history.

Here is how this came about.

His mother, born in Sinton, met his father when they were working for the same company in Venezuela in the late 1920s. They married and moved to England before the outbreak of World War II. Brian was born in England. The family lived on the outskirts of London, and he remembered the Blitz of German bombers and buzz bombs.

"As soon as the war was over, we moved back to Corpus Christi," he said. "According to the English, if you are born in England, you are always subject to the Crown." Just to be safe, he was naturalized as an American citizen after he returned home from the Army in 1955.

Actually he already had the best of credentials as a Texan. His great-grandfather worked for the White family across Corpus Christi Bay at White's Point before the Civil War. He returned from the war and worked for rancher D. C. Rachal after the Whites perished in a yellow fever epidemic.

Landon Curry

'You and/or people who trust you put out quite a few dollars to drill a well. If you're right, what a thrill and what a reward. If you are wrong, the family nest egg is gone, and there's an empty feeling in the pit of your stomach.'

—Landon Curry

Brian graduated from Corpus Christi College Academy and went to a junior college in Oklahoma, where he studied animal husbandry.

"I didn't make good grades and decided I didn't like animal husbandry. I came back to Corpus Christi, and my dad told me he wasn't going to spend any more money on college until I decided what I wanted to do," he said.

In the meantime he became a meter reader for Central Power & Light and became acquainted with a great many backyard dogs, most of whom were not friendly. His next move, to the Korean War, was decided by the local draft board. After his discharge he recognized Ray Govett's Army Pyle jacket as that of another Korean War veteran, and the two became acquainted. At that time Govett was a graduate geology teaching assistant at the University of Oklahoma, and he urged O'Brien to major in geology.

O'Brien earned a B. S. Degree in 1958 but found no market for geologists, so he worked in oil fields in Texas, Louisiana, and New Mexico. He returned to O.U. and received his master's degree in 1963.

He worked on seismic crews for Atlantic Refining, then was transferred to geology and sent to Houston, where he worked both onshore and offshore. He was sent to Corpus Christi in 1968. The following year he went to work for Mesa Petroleum, a T. Boone Pickens company.

"I met Tony Sanchez in '72 or '73," he said. "He and I became good friends. He came to Mesa one day and showed them a deal near Laredo. I showed the South Laredo Play to my supervisor. Mesa was not interested. They were interested in offshore, and that's where they wanted me to go."

He split with Mesa, and in 1973 Tony Sanchez, Sr., Tony Sanchez, Jr., John Blocker, Joe Thomas of Sinton, and O'Brien formed Sanchez-O'Brien Oil and Gas.

"We put together a 10,000-acre deal and showed it to a lot of people. Nobody was interested in Webb County. There was one pipeline to Mexico, and nobody was interested in gas. There was very little activity. Pemex was drilling right across the Rio Grande in Mexico. Land was leasing for one to five dollars an acre. We did a lot of borrowing," he recalled.

After they drilled a well on the river, gas was selling for 11 cents a thousand cubic feet. Then events changed. The winter of '72-'73 was very severe. There was a shortage of intrastate natural gas. The University of Texas was forced to shut down, and the price of gas jumped to $1.80. Then a pipeline was opened in 1974 from Laredo to Agua Dulce, opening the Laredo Pay to the interstate market.

The first Lobo Sand well was drilled in September 1974.

"We hit it big," O'Brien said. "In the first go-round there were sixteen wells drilled. Since then maybe three thousand wells have been drilled, and many are still producing. We drilled with one string of pipe. The

South Laredo Play merged into one giant field of close to three trillion cubic feet."

He credited Sanchez with the company's success in leasing the large tract. Sanchez could deal with the small Tejano landowners honestly. They trusted him, and hundreds profited by his deals.

The senior Sanchez died in '92, Thomas died in '90, and Blocker also passed away. O'Brien and Tony Sanchez, Jr., split before O'Brien made another astounding discovery, this time in Maverick County. His company was Saxet Oil, a nostalgic name that stirred the memory of his youth in the Saxet Heights area of Corpus Christi. It also recalled the historical Saxet Company that opened oil production in the area as Saxet Field.

Maverick County was another area that had not been worked a great deal. Again, it was not a matter of luck that he discovered one of Texas's biggest oil producers of the latter part of the twentieth century.

"We 3-D'd 200 square miles, starting at the river and going east. We started drilling seismic prospects. The first effort was a dry hole. The second was an oil well. That was March '92."

Saxet sold out in June of 2003, and the Glen Rose sand at 8,000 feet had produced more than one million barrels of oil by the fall of 2004.[36]

Like Ray Govett, Landon Curry was a Corpus Christi geologist who worked for Sunray early in his career. Curry studied engineering for about two years at the University of Texas, then took a job as a seismic computer for North American Geophysical Company.

"In those days you would shoot and record returns at one shot point and do the same at another location three or four miles away. The computer—a human being—would try to correlate energy returns from one record and then the next.

"I worked at this through East Texas, Arkansas, and North and South Louisiana for about two years, during which time the company became Shaeffer Geophysical Company. It was my first exposure to structural geology and stratigraphy."

He went back to the university, enrolled in geology, and "crammed all the geology courses into the next year and a half or two years."

To Curry subsurface geology was much like a crossword puzzle.

"In a crossword puzzle you put in a word you think is right. Then you work the next two or three words. You see you were wrong and make changes," he said. "With subsurface geology you map an area with all the data from wells previously drilled. You say, 'Aha, now I've got it right. This is the way it is from now on. Drill here.' So you drill, and the well—like the next word in the crossword puzzle—will prove you wrong. You make the necessary corrections and say, 'Now I've got it right. This is it from now on.'

"Working crossword puzzles is a challenge and a thrill if you are right. If you can't fill all the blanks, no big deal. There'll be another paper with another puzzle tomorrow. Creating oil and gas prospects from

Contraband Cores

Stampede Energy Corporation drilled a well in North Weslaco Field in the spring of 1979. The well was drilled and logged with no problems. Sidewall cores were taken and sent to Hebbronville for analysis. Some of the sands had shows of gas in the sidewall cores, so more were taken and sent to Hebbronville for analysis. Some of these also showed possibilities of gas, so more were taken.

Larry Teague, the Core Service analyst in Hebbronville, made the three runs to the well to pick up and analyze the cores. Several investors in the well were there for the log run and had been on location for a couple of days by the time the last core run was made. Larry was told to call us at a motel in McAllen when he had the final analysis.

After waiting five or six hours for the call on the final cores, we called the office in Hebbronville, with no response. After several more calls, we called Jim Brewster, with Core Service in Corpus Christi. After another four- or five- hour wait, the sidewall cores were found in jail in Falfurrias.

The Border Patrol suspected Larry of running drugs when he went through the checkpoint three times, and they thought the sidewall cores were some kind of dope. Core Service had to have the company attorney, Benny House, get Larry out of jail.

—Ray Govett

(L to R) Landon Curry, Bob Staewin, Jay Endicott, and Bill Shelton on a field trip to Mexico in 1962

subsurface geology entails higher stakes. You and/or people who trust you put out quite a few dollars to drill a well. If you're right, what a thrill and what a reward. If you are wrong, the family nest egg is gone, and there's an empty feeling in the pit of your stomach."

After graduation from the university in January 1950, Curry worked briefly for Southern Geophysical Company in North Texas. He joined Sunray Oil Company in San Antonio as a "well-sitting" field geologist in the summer of 1950. He was hired by Marion Moore, who had also been a geophysical computer.

"I was the first applicant he interviewed for the job," Curry said. "We hit it off well from the start, probably from our similar backgrounds and temperaments. He was meticulous, well-ordered, a very sharp geologist."

Sunray had just merged with Barnsdall Oil Company, which had many undeveloped or partially drilled leases in fields scattered from Nueces County to Victoria County in the Frio Trend. Sunray was drilling development wells to help pay for the merger, and Curry's job was to supervise the logging, coring, and open-hole testing of these new wells.

"In those days we cored almost all the objective sands and ran open-hole drill-stem tests on them. That's something that is rarely done today. We would core during the day and usually find sand we would need to test by afternoon. We would circulate and come out of the hole in the wee hours of the morning to rest at daylight. That way a well could be cored one afternoon, another well logged that evening, the first one tested at 5:30 a.m., and the second one cored in late morning, and so on, like juggling plates on the end of sticks. In the midst of this, the service and supply companies had bedrooms in their stores where you could clean up and take a nap before going back out again."

This was a time of great change in personnel at Sunray. Marion Moore became division geologist at Tulsa. Harold Picklesfimer took his place but soon resigned. Forrest McClain became district geologist, and about that time Wilford Stapp left San Antonio to open a new district office for Sunray in Abilene. McClain quit, and Billy Easley became district geologist.

"Billy was the only employee from Barnsdall to come to Sunray in San Antonio," Curry said. "He had a couple of years experience well-sitting in Houston but had not yet made the contacts or established the financial resources to get better positions or to go out on his own as his senior employee-comrades had done, so he moved to San Antonio within a week or so of the time I did.

"He had been caring for Barnsdall's Frio development wells, and he showed me the ropes. In return I did much of the contouring. We were low men on the totem pole, so we gravitated to one another. We worked well as a team. Though Billy was relatively inexperienced, he was soon the district geologist. I—also green—was his right-hand man."

Soon the district office was moved to Corpus Christi, where Curry and Easley were exposed to an environment of small companies, large brokers, promoters, and independent geologists. There were also major oil companies whose landmen would farm out a tract or two to get a lease block evaluated.

"I thought about quitting, and Billy had thoughts about quitting. We each realized that our future lay as independent geologists, not as company executives.

"Billy came to me one day and said, 'I've been thinking of going to work for Herman Heep.' I told him I'd been thinking of quitting, too. He said, 'Good, let's keep the team going. I was afraid you'd quit before I could get a deal worked out.'"

Billy Easley

"[Billy Easley] was an administrative type, outgoing, a wheeler-dealer, sociable. He would have been a great landman, and he knew his geology well."
—Landon Curry

They quit Sunray and went to work for Herman Heep, who had been quite a wildcatter early in Refugio. Heep had a half interest in Conroe Drilling Company in Corpus Christi, which he owned with Jack Modesett. They went to work for Heep Oil Corporation and officed with Conroe Drilling.

"We drew salaries greater than Sunray's and overrides on anything acquired on their recommendations," Curry said.

Their first prospect resulted in an oil discovery in Starr County, followed shortly by the discovery of West Rockport Field in Aransas County. They kept Conroe's three rigs busy and made "many good wells."

Later they left Conroe and formed Easley and Curry, Petroleum Geologists. They worked on retainer for a year or two for Edwin L. Cox. During that period they discovered North Texana Field in Jackson County.

"We had a lot of fun," Curry said. "If I found something promising, I'd take it to the office next door to Billy. My part was finding places worth drilling. Billy handled all the rest. He was an administrative type, outgoing, a wheeler-dealer, sociable. He would have been a great landman, and he knew his geology well."

Curry knew Oscar Wyatt from the early 1950s.

"Sunray drilled a well, promoted by Wyatt and Joe W, Thomas, at Round Lake in western San Patricio County," Curry said. "It was shortly after Wyatt and Bus Moore had formed Wymore Oil Company, and the well was probably drilled with Bus's rig. The well was dry. We all caught the flu but Oscar. He called a doctor and nursed us back to health."

Curry said Wyatt was a down-to-earth sort of person. When Curry and his wife were living in Clairlaine Apartments, "on hot summer nights they'd come by and pick us up to go to Snapka's to get a hamburger."

He said that Wyatt was very sharp. "He had a small office in the Wilson Tower. Once when Wyatt's secretary got sick, Curry's wife went down to his office to help him out. She told how Wyatt would buy and sell leases and make trades over the phone with slide rule in hand, figuring percentages and costs.

"His mind was quick," Curry said. "The Railroad Commission had rules, but they always had exceptions. Oscar obeyed the rules and parlayed them and the exceptions to his advantage."

The Round Lake fiasco also started a close relationship between Curry and Joe Thomas. Their contacts were infrequent for a number of years until one day when Easley and Curry were working for Cox. Thomas came in and said an old classmate of his and new president of Royal Oil and Gas Corporation wanted to renew their participation in South Texas Lower Frio exploration.

"We gave Joe a four-well package in San Patricio that Royal then drilled. Three were producing, the best being in North Midway that had over 200 feet of pay. This led to the drilling on a number of wells in

A rabbit runs through it

Sometimes the women know as much about the oil business as the men.

O.G. McClain recalled when he was geologist for Southern Minerals Corporation. A map was hung up in a storage tube, and he was having trouble getting it out.

"A girl who was working for me as a draftsman watched me for a while and asked, "Why don't you run a rabbit through it?"

I said, "Oh, my goodness. What does this girl know about running a rabbit through a pipe?"

He explained that when used pipe is used, a collar, smaller than the pipe, is run through it to make sure there are no obstructions in it. The collar is called a rabbit.

She explained. "My daddy was a drilling contractor, and I used to work for him sometimes when he was shorthanded. When we had to make a trip, I was a backup."

McClain said her father's name was Swift. "He invented a core barrel. The Swift core barrel was popular for quite a while."

Letter to author from O. G. McClain

North Midway, South Midway, Gregory, Geronimo, and Portland. Most were productive."

Curry felt the business was much more personal when geologists spent more time at the wells. In 2004 he expressed this opinion on the industry's technological advances: "All this is good, but I'm afraid something vital is being lost by not having the human, the personal touch."[37]

During both World War II and the Korean War, when most of the steel capacity was going into armor, a shortage of steel created problems throughout the industry. However, another South Texan, Bob Buschmann, was keeping production flowing by making casing available in the oil patch. At that time Buschmann was operating an office for National Supply Company in the Milam Building in San Antonio after moving from an office in Corpus Christi.

"I became very popular as a provider of casing, and I didn't have any trouble getting into anybody's office," he smiled. "Most of the independents officed in the Milam Building, and the coffee shop next door was a popular meeting place for talk about wells and leases. Gorman, George Coates, and other famous independents were there."

Hank Harkins (left) with President Gerald Ford

In 1955 he became manager of National Supply's largest store in the world, in Houston.

As a salesman in 1952 he had sold Field Drilling Company, named for Field Davis, its first land drilling rig. Davis had two longtime independents as partners, W. Earle Rowe and Bret Northrup, who put up the financing. In 1957 Buschman heard that Field and his drilling superintendent were planning to organize a new company.

"Having known Earle Rowe quite well," Buschman said, "I called him up the minute I heard about it and negotiated with him to come over and talk about a place in running the company. I succeeded in doing that. That's how I got in with Field."

The firm, which had grown to four drilling rigs in a short time, went into expansion and soon had nineteen land and inland barge rigs in the United States and five offshore rigs scattered around the world.

"When we went into offshore, we went into partnership with the Rutherfords in Houston—Pat and Mike. . . ," he said. "That was in the late '60s, and we put together Seagle International Drilling Company."

Seagle drilled in Borneo, Rhodesia, Angola, and Trinidad. That was some distance from projects he drilled with Earle Rowe in South Texas.[38]

While some other oilmen from the area also plied their trade in international circles, not all of them operated from the glass towers of San Antonio or Corpus Christi. Some of them, including Henry Burton "Hank" Harkins and Lucien Flournoy of Alice and Bill Carl and Dan Hughes of

'Burt' Harkins

By 1985 Henry Burton 'Burt' Harkins, Jr., had taken over as president of Harkins & Company. Like his father, 'Burt' Harkins played high school and college football and graduated from the University of Texas.

Beeville, were content to distance themselves from city life for the more leisurely pace of small-town living.

Harkins began his career in San Antonio but became a highly successful drilling contractor, wildcatter, and producer in the "Hub City" of Alice. Such a future did not seem likely when, with an accounting degree in hand, he applied to Robert L. Kirkwood, the successful San Antonio drilling and producing contractor, for a job in 1946.

Hank's only experience in the oil business was pumping gasoline in front of his father's general store in the village of Scottsville in East Texas. Born in 1921, he had worked in his father's store for two years as a soda jerk for 15 cents an hour. He played football before the days of platooning, on the field the full sixty minutes of the game. He was offered a contract with the professional Cleveland Rams after the war for $175 a week. Football and the Navy had both taught him the value of teamwork and the will to win.

Kirkwood, a well known wildcatter, wasn't too impressed that the recently discharged naval officer had a distinguished war record aboard the aircraft carrier Cabot or that he had been an All-American player on the University of Texas national championship football team, with his picture on the front of *Life* magazine. He wanted someone running his office who knew something about drilling an oil well, so he set a condition for Hank.

"I'm going to start you as a roughneck," Kirkwood said, "because I don't want no goddamned bookkeeper working for me who doesn't know a Kelly joint from a rotary table. After a couple of years on a rig, you might be worth something to me."

Hank started working in the hot South Texas sun in 1946. Within six months he was promoted to tool pusher and not long after that promoted to vice president in charge of four rigs.

His friend Wilbur Steen described an incident that illustrates Hank's work ethic. Steen, who was Kirkwood's drilling superintendent at the time, told of going out to a well fire.

"I went up there to talk to them about what we could do, and when I got near the well site, there was this fellow lying on the ground. He had a three-day growth of beard and looked like a bum all curled up in an effort to get some warmth out of the jacket he was wearing and trying to get a little shut-eye. I didn't even recognize the man, so I asked someone, and they told me it was the young fellow from the office who was a bookkeeper. I looked again, and sure enough it was Hank Harkins trying to get a few minutes rest. He had only been with the company a couple of years, and no one expected the bookkeeper to be out in the field fighting a fire day and night for seventy-two hours, but he was there, and he was working damned hard, too."

He was moved up to president of the company in 1949. With Hank's promotion to president, Kirkwood voluntarily took the position of vice president. Under Hank's direction the company drilled one hundred wells in 1954.

In 1956 Harkins decided to leave Kirkwood and strike out on his own. He formed Harkins & Company and, with Kirkwood's approval, hired Kirkwood's drilling superintendent, George Standard. Ray Butler would succeed Standard after his retirement. Kirkwood sold out to Holland American.

Hank never pretended to be an expert in exploration, but he surrounded himself with men who were experts, and he listened to them. When a prospect looked good, he would offer to drill a well at a reduced rate in exchange for an interest in the well. This helped him become a producer as well as a drilling contractor.

When Harkins split with Kirkwood, Hank had roughly one million dollars with which to start a new company. In 1976 he opened a big field, Greens Creek Field, in Mississippi.

In 1956 there had been twenty drilling contractors in Alice. By 1984 there were only three, because of the industrywide slowdown of that decade. Harkins, anticipating that the boom of the seventies would not continue, had curtailed his expansion and built up his cash reserves to help him during the lean years of the eighties.

He had an explanation for his success. "As I am listening to a deal, I sometimes get a strange feeling for what I am hearing, even though it may seem to be a bit wild at the time, and when this happens, I am inclined to go with my gut feeling. Since I don't believe much in ESP or clairvoyance, I call it good old gut feeling. Of course, working like hell to make the feeling correct has to have something to do with whatever success comes about."

He was selected as a charter member of the All-American Wildcatters in 1968 and honored in 1969 at the Explorer's Club annual banquet in New York City. He was active in other professional organizations and also served as mayor of Alice, city council member, and president of the school board.

Harkins' wife, Mary McGrath Harkins, and his mother died tragically when the family home burned December 31, 1971. Hank Harkins died February 8, 1990. By 1985 Henry Burton "Burt" Harkins, Jr., had taken over as president of Harkins & Company. Like his father, "Burt" Harkins played high school and college football and graduated from the University of Texas. His degree, however, was in petroleum engineering.[39]

Unlike Harkins, Lucien Flournoy achieved his success without the advantages of a college degree.

As the youngest of five children and the only boy, he knew poverty on a farm near Greenwood, Louisiana, a small town between Marshall, Texas, and Shreveport. Perhaps that was the reason that—for all his honors and wealth—Flournoy never lost touch with the common man.

Lou Flournoy worked as a roustabout and rig builder. Riggers, who built and dismantled the derricks, had the reputation of being the toughest 'athletes' of the oil patch.

Throughout his life he took pride in keeping a connection with his dirt-farm beginnings. In almost every interview he gave, he included this statement: "I feel more at home in the kitchen of the country club than I do in the dining room.

"My father was born in 1876. He died in the flu epidemic of 1918. I was born in 1919," he said. "To supplement our income, my mother had a boarding house in Greenwood. That gave me the opportunity to see a lot of people from the oil fields."

His first job was carting lunches and water from the boarding house to a crew that was laying a pipeline across their property. "I was sort of a water boy until they got three or four miles from our place and the job faded away."

He sometimes skipped school to help on the farm, with its four mules and a wagon. He worked on an ice truck delivering ice, and he drove trucks. In later years he bragged he could still drive any kind of truck, even the monsters.

He learned the value of hard work, patience, and persistence from his mother. With only a small blackland farm and the boarding house,

Flournoy Drilling Company celebrated the drilling of its 1,000th well for Exxon Company, USA, in 1977.

she managed to send all four of his sisters to college and insisted that he go, too.

He was seventeen in 1936 when he hitchhiked to the oil field at Rodessa in northern Louisiana to work as a roustabout. It was his first full encounter with the oil business, and the sights of a boomtown opened his young eyes.

"I set out to be a petroleum engineer at Louisiana State University," he said. He did all sorts of odd jobs to stay in school, including becoming a competent butcher. Even so, after two and a half years he was running out of money.

"At the time I was nineteen, and I had the highest average in my class in calculus and the lowest in English," he said. "One of my professors had a friend in Corpus Christi who had an oil company. It was Tom Graham Oil Company. He gave me a letter to Mr. Graham. I sold some of my belongings, went by and told my mother goodbye, made another loan of about four dollars, and headed for South Texas. That was 1939. I got a place to sleep in Corpus Christi. It cost me 35 cents a night. I cleaned up at the bus station, put on the best clothes I had, and went out to meet Mr. Graham and give him the letter."

Flournoy had been on the road for a week, and "the letter had been rained on a bit." Graham had an operation going at Robstown and put young Lou to work as a roustabout and rig builder. Riggers were the athletes of the oil patch, with the reputation of being the toughest of the bunch.

"You can see pictures in the museums of guys hanging in the air bolting the steel derrick together," Flournoy said.

Lou didn't see much of his boss, but he did get a look at Robstown and Sinton, both of which were booming. After eight months in Robstown, he heard of a job opening up in Louisiana across the Mississippi River from New Orleans. He hitchhiked again.

"The last car we had in the family was before my daddy died, before I was born. My older sister married a driller, and that was the first wheels in the family," he said. "It was nice."

Planning to save enough to return to college, he worked in Louisiana as a roughneck and roustabout, then as derrick man for Loflin Brothers Drilling Company, drilling a well for the Texas Company.

"That was just before the war started. A service company called Baroid, owned by the National Lead Company, had bought the patent rights and was experimenting with what was to be a well-logging unit to analyze drilling fluid for particles of gas or salt water.

"We cored. Back then the crewmen and the technicians played a part in the process. They had these portable laboratories and they were experimenting with gas detectors, and I helped them with one operation for a man who didn't know how. The word got out among the scientists and those doing the work, and they hired me."

Lucien Flournoy and his politics

Flournoy said he was a Democrat "because the party is more compassionate. I'll have to confess, my wife is a Republican, because of the abortion issue."

Maxine Edmondson Flournoy, who married Lucien on June 29, 1946, was not only an oilman's wife but also an achiever in her own right. A licensed pilot, she served during World War II in the Women Air Force Service Pilots (WASP) program. After the war she searched all over the country for a job flying and ended up in Alice. She quit flying when their children—three girls—were born but took it up again twenty years later, flying to attend WASP conventions and to visit their daughters.

"My kids and grandkids are all Republicans," Lucien said. "Sometimes I don't know what they think of me."

He had pictures of himself with President Jimmy Carter on Air Force One and with other Democrats. He had actively supported Democrats on the national and state levels but produced letters showing he had also contributed to Republican candidates.

"[A geologist friend] just hates Democrats. When he comes by, I put the picture of [former Vice President Al] Gore in the drawer," he laughed.

Interview with Lucien Flournoy, 19 Nov. 2002

In this Suemaur Exploration photo Lucien Flournoy is seated with business associates and some of the geologists he credited with helping him achieve success. (L-R) seated: Jim Harmon, Flournoy, Bill Miller; standing: Dan Pedrotti, Edward Elliott, Bill Maxwell, Owen Hopkins, Dave Webster, and Jim Devlin.

Lou marked that as the turning point in his life. The company sent him to schools at Houston and in Kansas and Bakersfield, California. He spent a year working for Baroid in California.

"I finally had enough money to go back to school, but I didn't get back. . . ," he said. "World War II came along, and I enlisted in the Air Corps."

During training he suffered a severe fracture of the ankle on the obstacle course and subsequently received a medical discharge. Back home he was hired by Arkansas, Louisiana Gas Company, a division of Cities Service "which is now CITGO," he said in 2002. "Sold all that to the Venezuelan government. If those [gas station] signs said Pemex, nobody would buy it. But it's Venezuela. Same thing. But they do pay taxes."

In 1945 he was transferred from Morgan City, Louisiana, to Alice as drilling engineer for Arkansas Fuel Oil Company.

"I had a good job, was single, had a company car, but I still wanted to do something," he said. "I wondered why nobody had thought of designing a hopper to mix dry mud at well sites. With Ted Sage, a welder, we bought a blacksmith shop. We made a low pressure mud hopper that was successful, and we sold quite a few. Unlike H.E.B. [the grocery chain], my market was limited; and once everybody had one, their problems were solved. We made steel benches and catwalks. They were popular, too, for a while. That's when I built my first rig. It was a small rig. I formed Flournoy Well Company. I didn't patent it because I probably would have got into lawsuits."

His other equipment included an old surplus Navy truck bought at auction and picked-up used equipment and iron. He explained that if you just have to do something, you can do it. He could change truck flats, start dead engines, and operate trucks and heavy equipment. He was the tool pusher on the little rig for four years.

"Then I worked with Dresser Industries—now [part of] Halliburton—to design a future rig. You hear a lot of stories about roughnecking and how glamorous it is," he said. "It really isn't. It pays more money than any other common labor. In the early days it paid big money for the time."

Roughnecks don't need a union, he said. "I'm not opposed to unions. If I worked in a refinery, I'd probably join one if I didn't like the way they treated my family."

He said he had always held a real appreciation for geologists. "One of the first ones I ever got involved with was a geologist with Magnolia Petroleum Company. He was out of San Antonio, and his name was Dan Freeman. He moved to Corpus Christi, and I officed with him. I was the contractor drilling a 2500-foot well. It wasn't the first well I drilled, but it was the first one I made a million dollars out of. It was in the vicinity, and nobody else wanted to deal with it."

From that point Flournoy Drilling grew into one of the biggest independent drilling firms in the country. It wasn't always easy. There were fires, blowouts, and other troubles, but success quickly overcame obstacles to the point that Lou was termed "the statistics maker."

It happened first in 1964 when, with the oil business in a downslide, he set a national record with six rigs drilling 281 tests for a total of 1,408,911 feet or 288 miles. His rigs remained active during the great slump of the 1980s. Much of this work was done for Humble, Amerada Petroleum, Sun Oil Company, AMOCO, and Coastal States.

Dan Hughes said, "Lou just loved iron. Built his first rig himself. When everybody else went broke in 1980, he kept his rigs going, and he'd go to these sales where they'd sell all this equipment for five cents on the dollar. He'd go buy a rig and stack it in his yard. He knew it would be worth a lot in the future. When the boom came, he was ready."

Through the years Flournoy drilled some 1,800 wells for Humble/Exxon on the King Ranch and a total of more than 7,000 wells altogether. Drilling that many wells would have been impossible if he had not been bitten by the wildcat bug. He drilled his first wildcat in 1950 in Refugio County.

Lucien Flournoy

'I feel more at home in the kitchen of the country club than I do in the dining room.'
—Lucien Flournoy

Bob Wilson

Bob Wilson,who worked for Lucien Flournoy for more than thirty years, said that many oilfield workers wouldn't work for anyone else.

Wildcatting was exciting, he said, not so much for the money in it but rather for the challenge of the battle against the odds. The secret to winning that battle was finding the best people and treating them right. This policy paid off in success, efficiency, and longevity of his employees. And the key to it all was good geology.

"You've probably noticed, you drill a well where the stake is driven. A lot of guys take all the credit [for a successful well]—the mothers, the drillers, the roughnecks, the banker, and what have you. The guy who drives that little piece of stake—it's a one by two about 18 inches long—is out there first. He tells them where it's going to go. I've always trusted geologists," he said. "And for some reason they have trusted me."

Independent geologist Lawrence Hoover was one of his first collaborators.

"He was a geologist for Ed Singer," Flournoy said. "He came by in 1958 and wanted to know about a location in Starr County. That opened the North Rincon Field. He made me look like a smart guy with my two rigs."

He pointed to a photograph on his office wall of himself with a group of men.

"I didn't put that up just to look at it. There. He's a geologist. He's a geologist. He's a geologist. He's a geologist. He's a geologist, and he's a geologist. And he's a landman. And there I am—a college dropout. I have a magnetic attraction for geologists."

He credited them all with making him a success, but Hoover cited additional reasons. He said that Flournoy had a passion for the business and because of that adapted his procedures to be competitive. He added that Flournoy Drilling kept a reliable supply of roughnecks.

Lou was very loyal to his employees, he said, and one reason people liked to hire him was that his crews had "been with him for so long." He added that Flournoy would have preferred a new mud pump to a swimming pool or set of golf clubs, that his way of expressing himself was devising some new time-saving device.

Bob Wilson, who worked for Flournoy for more than thirty years, said that many oil-field workers wouldn't work for anyone else. "He'd give you the shirt off his back. When times were good, he gave big bonuses. He could irritate the stew out of you, but in the end, he earned everyone's respect."

On the local level Flournoy was an Alice city councilman, mayor, and school board member, generous with schools, hospitals, parks, and other charities. In 1983 *Texas Business* magazine named him one of the twenty most powerful Texans.

In 1997, six years before his death in March of 2003, Flournoy sold his drilling company to Houston-based DI Industries in a stock transaction valued at $38 million. He remained a major stockholder on the DI board of directors, and the company continued to employ Flournoy personnel.

"The reason we're buying Flournoy Drilling Company... is precisely because of the fine people they have operating there," said Tom Richards, president and chief operating officer of DI, "and we're not about to change that."[40]

Like Flournoy, William C. "Bill" Carl appreciated his geologists, giving them credit for much of the success he had as a wildcatter.

Carl, a semi-retired independent who moved from offices in Corpus Christi to his ranch just south of Beeville, fell in love with geology at Williams College, a small Massachusetts liberal arts school. During World War II he was in the V-12 officer-training program and attended midshipman's school at Columbia University. He served on a PT Boat in the war and returned to Williams College to complete a degree in geology. In 1948 he started in the land department at Stanolind Oil and Gas Company, later AMOCO, in Houston.

William 'Bill' Carl

'I just loved the geology, and I loved the exploration and the sense that you can conceive of an idea, put together the lease block and raise the money to drill a well and test your own theories. . . .'
—William 'Bill' Carl

Storm Rig # 10. Bill Carl used Glasscock barges in Laguna Madre and worked with Jimmie Storm's drilling company.

'Years and years ago you went with the rig and stayed there. As a geologist it was fun. I used to say we did a better job on the back of an envelope out on the drilling rig than we did in the office.'

—Bill Carl

"I started out the first four months as a scout. That's the way Stanolind started out all their geologists," he said. "You'd ride several weeks with the scout from another company to see what it was all about."

"They were good PR people for the company. They would bring a geologist or a landman from one company together with another to exchange information. That was their purpose for a long time. If you wanted information from Humble, you called the scout or the landman. They took the load off the other professionals and made it easier for everyone. I thought it was a good idea.

"[Now] I don't know of anyone who uses scouts," he said in 2003. "They might have them in some other form but not the way they used to work. It was more of a fraternity back then. Today it is cold and rather impersonal, I'm afraid. Today you don't have to go far in Houston—you can find any company in the world there. But you don't have any personal contact.

"Some of the scouts weren't college graduates, but, boy, they were street smart. A lot of them moved up in their companies. Scouts had a big barbecue down at Laredo every year. It was one of the most popular barbecues in the industry. It wasn't just for scouts. It had geologists, landmen, engineers, and the scouts. The industry lost the personal touch when all this passed. But isn't this what's happened in all industries in this age?"

He first drilled in Corpus Christi while he was based in Houston. He remembered that the skies were bright at night from flares of hundreds of wells.

"Years and years ago you went with the rig and stayed there. As a geologist it was fun. I used to say we did a better job on the back of an envelope out on the drilling rig than we did in the office. We'd look at samples and figure which was low and which was high and why. And the other companies' individuals would be there and share the knowledge, which you'd be real 'tite' about in your office.

"I just loved the geology," he said, "and I loved the exploration and the sense that you can conceive of an idea, put together the lease block and raise the money to drill a well and test your own theories—usually unsuccessfully because we mainly are in the wildcatting business. Of course, the techniques are improving every day, and the success ratios are much more attractive today than they were. Maybe there aren't as many big fields to be found, but there is still some activity or rather prospective areas that will pay off over the years, I think."

In 1951 he was moved to Corpus Christi. Later he left Stanolind and worked for about a year with Midland independent Ralph Lowe. In about 1953 he and Jeff Carr formed a company called Carrl Oil Company, combining both their names. In the beginning most of their drilling was in South Texas.

"We were heaviest in Brooks County and opened the Loma Blanca Field down near Falfurrias" in the early sixties, he said. "We drilled on our own and had a relationship with General Crude. . . . They were partners. We had a drilling company, Frio Drilling Company, . . . and used Glasscock barges in Laguna Madre. We got offshore and did some work with Jimmie Storm's new drilling company.

"A well we were drilling with Killam and Hurd in Zapata blew out. I flew down there and it was burning real good. Red Adair was there, and I asked him, 'Red, what do you think?'

"Red answered, 'Well, for the shape we're in, we're in pretty good shape.'

"I've often thought that applies to an awful lot of things in life," Carl said.

"When we started our little company, all the major companies had offices in Corpus Christi. As a small independent you take farmouts. I'd say we carried more farmouts than most companies. We did more original geology than any of them. We always had a lot of geologists on

Sometimes the 'corduroy road' provided the best—or only—means of travel through the South Texas brushlands.

Joe McCullough

Joe McCullough led Carl Oil & Gas in an aggressive, successful drillling program that resulted in discovery of many new fields and extensions of others.

the payroll. At one time we had as many as twenty on the payroll—I mean good geologists. We are proud of them. We did mostly origination. A company is only as good as its geology," he said.

Carl said that sometimes the biggest discoveries are not the best days in the oil business. "Sometimes it's so many relationships or restructuring some new concept of financing, or whatever."

"I never had a super-duper discovery," he said. "I wasn't lucky, though Loma Blanca was very good. The problem was that gas was thirty-two cents and we had an old contract we couldn't get out of. Just before gas moved up to a dollar plus, our partners wanted to go with Houston Natural Gas, who sold the contract to AMOCO. It could have been the biggest thing we ever hit, but dollarwise it wasn't."

He and Carr terminated Carrl Oil Company in 1970. Then he operated independently as Carl Oil & Gas. "We had offices in Denver, New Orleans, Lafayette, Houston, and Corpus Christi and worked the whole Gulf Coast from 1975 to 1984," Carl said.

Joe McCullough became exploration manager for the company soon after it started, McCullough, who grew up in San Antonio, graduated from Texas A&M in 1951. At that time all A&M students were members of the Corps of Cadets, and Joe entered the Army upon graduation. After serving the required two years on active duty, he went to work as a geologist for Lion Oil Company and was soon transferred to Corpus Christi.

Joe left Lion in 1958 and worked with Carl until 1995. He led Carl Oil & Gas in an aggressive, successful drillling program that resulted in discovery of many new fields and extensions of others. Joe was responsible for discoveries in the Cretaceous, Wilcox, and Frio trends and extended the company's operations into South Louisiana.

When things slowed in the eighties, Carl merged with MacFarland Energy from California.

"I stayed on as vice chairman," Carl said. "We had some of our offices in the Gulf Coast. We kept the office in Corpus Christi. It was a slow time for the oil business, and we sold out in 1990. When we sold, we did pretty good, but it wasn't a happy time. California was a different world from Texas, and it wasn't good. We are a small company, mostly me. My daughter works for us. We are not active. My low point was California. Those people come from a different part of the globe. Companies do things that you'd blush about ethically. They are a different breed of cat."[41]

Although these oilmen enjoyed the pleasures of small-town life, they didn't exactly cut themselves off from the outside world.

"I keep a jet out at the airport here; and when I need to run up to Houston for a meeting, I can be back in time for lunch," Beeville independent Dan Hughes said.

In a few hours he could be in San Francisco, but he didn't have to leave home to oversee his operations in the Rocky Mountains, Colombia, and Australia.

Most oilmen recall entering the business after their military service. Hughes did them one better. He took the plunge while he was on active duty. He and his twin brother, Dudley, worked summers during World War II for United Gas, building pipelines when they were sixteen, and later worked as roustabouts for Magnolia Petroleum Company. The brothers graduated from Texas A&M with degrees in geology during the Korean War, and immediately Dan was called into the Army.

Dan A. Hughes

'I asked to be transferred [from New Orleans] to some place in Texas, and they sent me to Beeville. I've been here ever since.'
—Dan A. Hughes

Lt. Hughes spent his spare time at El Paso's Fort Bliss mapping the geology of New Mexico and managed to develop prospects, buy leases, and drill a discovery well, which opened Saladar Field. He then spent a year in Korea. After his discharge he joined Union Producing Company, which sent him to New Orleans.

"I worked there for about a year," he said in a 2003 interview. "It was a great place. I was single, and the three thousand dollars I had saved in the Army was gone. I was broke. I was getting $325 a month. I asked to be transferred to some place in Texas, and they sent me to Beeville. I've been here ever since."

At the time Union Producing had some 150 employees in Beeville, where several companies had offices, as there was heavy activity in a number of large nearby fields. He resigned in 1961 to go independent.

He was a consulting geologist with Caddo Oil Company, which drilled some 450 shallow wells on his geology in old fields—Bar Creek, Von Ormy, San Caja, Somerset, and Taylor Ina. He also developed friendships and connections with scouts from many other companies, some of whom later headed large corporations.

"When I first moved here," he said, "I started out as an oil scout. I worked as an oil scout in New Orleans and for a while in San Antonio. There was a lot of camaraderie among the scouts. We'd have scout checks on Monday nights at Laredo, Tuesday at Del Rio and other towns, then go back and fill out the dope. We don't get the information we used to get. A lot of them now just do without the information. They think it's all been published. It really hasn't. We still use the old records. The industry is so money oriented, it has lost a lot of the personal touch."

In 1965, after five years as an independent, he and Dudley formed the Hughes and Hughes Oil and Gas Company as a partnership. Later it would become Dan A. Hughes Company, one of the leading exploration and production companies in the country, and Dudley would head an oil exploration and production company in Mississippi.

Dan Hughes had a great many discoveries, but one in South Texas, Las Tiendas Field in Webb County, was the biggest. It started with a strange geological clue.

In the Beeville office of Dan A. Hughes, a geologist checks two computer screens for the latest data.

"In 1967 I was deer hunting with Judge John C. Beasley on a ranch that had a few small wells and some dry holes," Hughes said. "John C. had invited me down for the weekend. We were driving around drinking beer and such, and he showed me an old well casing sticking up. He lit a match and dropped it in. It went 'SWOOSH!' and flared up.

"I went back and looked up the record on that well and found it was right in the middle of a prospect I was working on. I did some subsurface and got some old seismics, and we were able to use it to some extent. I got a broker and we ended up with 15,000 acres—most of it was two dollars an acre.

"On the first well at 2800 feet we had a good gas flow of a million or so (MMCFD). Got down another hundred feet, more gas. We went on down through four sands and logged the well. We had gas down to 3600 feet."

The field covered 11,000 acres. Hughes had the whole field and drilled some thirty wells on a 320-acre tract.

"We made a deal with Houston Natural Gas, later Enron—in those days it was a good company. It was very aggressive. They built a pipeline from Mt. Lucas in Live Oak County 114 miles to our field. Those wells are still producing. We sold them five years ago."

That was a high point. A potential low point occurred with a well he drilled with Hank Harkins, a gas well in the Wilcox Sand.

"The well came in good," he recalled, "We kept it choked back to about half a million a day because we didn't want water to come into the sand. This went on for two or three years, and all of a sudden they called and said, 'There's some uranium wells around your wells out there blowing out gas around them.'

"We found out that the $2\frac{7}{8}$-inch casing had pulled apart below and was charging all the pipe. We thought this couldn't last long with this gas on water. We couldn't bleed the pressure off. We got the thing up to 25 million a day and still couldn't do it. We finally produced 4 billion feet of gas out of that well before it died and watered out. Apparently we weren't gas on water. We just thought we were."

Actually that wasn't a low point at all. He got a high price for all that gas. And then:

"After it was all over, the insurance company paid us another million dollars for insurance on the well. We came out smelling like a rose."

Hughes attributed his company's success to many talented and loyal employees and co-workers.

"We've had several investors who stayed with us through the years, and we've shared some great times. It has been my pleasure to work with some great South Texas people, too, like Lou Flournoy, Pete Murphy Drilling, Hank Harkins of Alice, Bill Carl, and many others."

He credited Flournoy with "getting me started."

After his stint as consulting geologist for Caddo Oil Company, Hughes started drilling deeper wells on his own. "I got with Flournoy. He operated the wells, and I was the geologist. He had the whole operation. He'd say, 'If you can give me ten thousand dollars, I'll carry you for a quarter interest in that well. I'd sell half of that to somebody else, and he'd drill it. Lou was always very generous."

At the end of 2002 the Texas Railroad Commission reported that Hughes ranked 37th among the top one hundred natural gas producers in Texas. His stature was even greater when his worldwide operations were considered. Dan Hughes opened major gas fields in Canada and Australia. His son, Dan A. Hughes, Jr., who joined the firm in 1974, was operations manager. He was also in charge of operations in Australia, which the elder Hughes found a good country to deal with.

In 2002 Hughes was also having great success in Colombia, while many hesitated to brave the dangers there. The Colombian government, he said, "lets you recover all of your investment before they come around for their share."

He said he had never been there himself but directed the operation using native labor.

"I don't go there," he smiled wryly, "because I have no desire to become a hostage."[42]

'It has been my pleasure to work with some great South Texas people, too, like Lou Flournoy, Pete Murphy Drilling, Hank Harkins of Alice, Bill Carl, and many others.'
—Dan Hughes

The Move Offshore

The Republic of Texas didn't worry about its offshore waters except when invaders threatened. When it was admitted to the Union in 1845, it enjoyed some of the privileges it had enjoyed as an independent nation. A map drawn by Gen. Sam Houston and approved by President Andrew Jackson and later by President James K. Polk recognized the Texas boundary as extending three leagues–10.34 miles–into the Gulf of Mexico.

In those days there was no thought of mineral rights under the sea. But in 1946 the United States set off a legal firestorm by filing suit against California, claiming federal control of offshore lands. Although Texas claimed to be exempt from the ruling, the Supreme Court denied the claim. The battle raged back and forth and influenced national politics. Congress passed laws giving coastal states the rights to their offshore lands, but President Harry Truman vetoed them. The controversy was a major issue in Texas during the 1952 presidential campaign. Dwight Eisenhower, who promised to sign a bill supporting the states' claims, carried the state and nation and, soon after his inauguration, signed the promised measure.

However, there was one final struggle for Texas when Attorney General Herbert Brownell filed suit to limit the state's lands to three miles instead of three leagues. On June 1, 1960, the Supreme Court ruled in the state's favor, and the matter was finally settled.

Between that time and 2004 the schoolchildren of Texas benefitted by approximately $2 billion from oil and gas recovered from the Gulf of Mexico.

Crewmen aboard a helicopter make a dangerous inspection of an offshore well blowing out.

An offshore rig is being outfitted before being towed out into the Gulf of Mexico.

(Above) Humble drilled this well on the edge of a beach in July 1941, but it would be years later, after World War II, when men like Gus Glasscock, James C. Storm, and others would take their rigs into the bays, like Glasscock # 4 (left) and then to the deep waters of the Gulf of Mexico.

Offshore drilling provides bounty

If it takes nerves of steel to play the oil game, the Glasscock family was well prepared. They were used to walking the tightrope. They walked the high wires as headliners in the circus, entertaining audiences in vaudeville, on Broadway, and in Madison Square Garden.

Gus Glasscock ran away from home at age fourteen to join his oldest brother, Alex, who owned a small circus. He and brothers Donley, Mike, and Foster were crack performers. The circus moved from town to town in wagons. Gus first watered the animals, substituted for a clown, and learned to be a high-wire walker with his three brothers. After six years the act broke up with the start of World War I.

For all his nimbleness of foot on the high wire, Gus Glasscock was rejected for service in the Army. He had flat feet. He worked in construction during the war and operated taxicabs in San Antonio. In 1917 he met and married seventeen-year-old Lucille Freeman. Then he went to Ranger where, he said, "a man can start from the bottom."

He took his young wife and baby to the booming oilfield, where she set up housekeeping in a tent. And Gus used his nimble feet atop a derrick for eighteen dollars a day. He lost his savings twice, investing in dry holes. They moved to Wichita Falls and repeated the process.

Then the Glasscock brothers, Gus, Donley, Mike, and Lonnie, "got their act together," this time not in show business but in the oil business. Their first investment was for a lease in the Somerset Field, a few miles south of San Antonio. This, noted Lucille, "was the first lap on a lifelong breathless chase." She added, "This was a four-brother act on the shakiest wire they had ever walked."

Their first effort was a dry hole. The second was a grandly successful producer of ten barrels a day. In a year and a half they brought in five other producers. Gus and Lonnie formed their own partnership, spent ten thousand dollars, and were immediately broke again. The other brothers dropped out.

In this group of circus performers, Gus Glasscock is at the right.

Gus Glasscock, Sr.

Gus Glasscock, Jr.

Gus and Lonnie Glasscock developed an oil reclamation system that trapped the errant oil and treated it. Magnolia Petroleum Company gave the Glasscock brothers the contract to retrieve the oil.

The next stop was Lytton Springs near Luling, which Edgar Davis had put on the map with his mystical discoveries. Lonnie Glasscock's wife, Luna Edmondson Glasscock, told how the move came about:

"One night we heard a knock on our north door, the one we never used," she said. "A stranger in a black slicker stood there in pouring rain. 'I know you're looking for a place to invest your money,' he said . . . . The man said he had just come off a rig in a new oil field at Lytton Springs. The core was dripping with oil.

"My husband went to the field and took a franchise on a spring near the field. He piped water to the field, three miles away. You had to have water to drill, and it was too expensive to haul. So it was from water, not oil, that we made our first real money."

Oil from completed wells soaked the soil and fouled the streams. A huge gusher was polluting streams and getting the attention of the State Game and Fish Commission. Gus and Lonnie developed an oil reclamation system that trapped the errant oil and treated it. Magnolia Petroleum Company gave the Glasscock brothers the contract to retrieve the oil.

This paid them enough to buy their first steam rig and new wooden derrick. They drilled their own well, which came in with a roar and a shower of crude. Soon Gus and Lonnie owned drilling rigs, drilled wells, and invested in leases.

It was a time of boom and bust. Luna Glasscock said, "It was a time when you could go to bed a pauper and wake up a millionaire or the other way around."

The careers of the two brothers took an upward turn. They had success in West Texas near Big Spring and finally made some good hits in the big East Texas Field. But it was in South Texas that Gus Glasscock would win his greatest distinctions. In 1939 he and Lonnie moved to Corpus Christi. They dissolved their partnership and each formed his own drilling company.[1]

The costly and difficult job of constructing well platforms in the inland bays had long limited exploration there. Gus Glasscock had drilled a discovery well in Shamrock Cove, but it had taken more than two months to build an island on which to move the rig. He had heard that barges were being used in Louisiana, and he decided to try it in the Southwest Texas bays. Aided by his son, Gus, Jr., and James C. "Jimmie" Storm, at that time his son-in-law, he designed the first rig, which was to be built by Bethlehem Steel. It slid down the ramp and hit the water July 29, 1949, Glasscock Rig No. 6.

It was a strange sight, a floating house with a huge jackknife derrick. It was 155 feet long, 52 feet wide and weighed 1,100 tons. The derrick weighed some 400 tons and could support 15,000 feet of drill pipe, powerful mud pumps, and a huge storage area for mud. Drilling crews were amazed to find a fully equipped kitchen, with refrigerators, a

Before the days of offshore drilling, pipelines crossed South Texas rivers and Nueces Bay. This scene is from August 1935.

This series, taken by Corpus Christi photographer Sammy Gold, shows the collapse of Mr. Gus No. 1. It appeared stable. Then it began to tilt, tilted more and more, and collapsed on its side. Most of the machinery was salvaged.

deep freezer, tables, and chairs. There were a stateroom and quarters for six men, a locker room, showers, and clean sheets—all the comforts of home. Roughnecks had never had it so good.

Soon Glasscock rigs dotted Corpus Christi Bay, East White Point, Nueces Bay, the Flour Bluff fields, Laguna Madre, Riviera Field, Matagorda Bay, Nueces Bay, San Antonio Bay, Bird Island, and other bays and inlets along the coast. Each new rig was averaging a completion a month, a small revolution in South Texas.

In 1957 the C. G. Glasscock-Tidelands Oil Company drilled twenty-three wells, fifteen field discoveries and the rest wildcats. That same year Glasscock Drilling Company drilled fifty-six wells and worked over twenty-seven existing oil and gas wells for twenty-five different companies.

But there was a bigger challenge, across the barrier islands, far out in the Gulf of Mexico. That was the year Glasscock moved into deeper water, launching Mr. Gus I, an offshore rig capable of drilling in 150 feet of water. The beginning was not auspicious, for Glasscock's first big mobile offshore drilling platform sank as it was setting up off Padre Island south of Corpus Christi. The Gulf floor collapsed under the supports, and the rig toppled over on its side.

It was not possible to save the entire rig, as it rested in thirty-nine feet of water, but the drilling equipment and foundation hull were salvaged and reconditioned. Heavy seas delayed the salvage project. Men with torches cut the legs from Mr. Gus I into 40,000-pound sections and removed them by barge. Engines that powered the drilling equipment were cut free by deep-sea divers equipped with torches. After five months of tedious work, the hull itself was lifted above the water's surface and moved to a dry dock in Corpus Christi, where Glasscock maintained a shipyard. He also operated a helicopter rental service.

Gus Glasscock's Mr. Gus 1 toppled over on its side when the Gulf floor collapsed under the supports.

In April 1965 Storm Drilling Company Rig No. 1 drilled the deepest hole drilled by an independent contractor up to that time when the Phillips Petroleum Company A-1 Nueces Land and Livestock Company reached a total depth of 24,220 feet in McMullen County.

On the bright side, an even larger and more sophisticated Mr. Gus II was already launched at a cost of $6.6 million.

In 1958 Gus Glasscock sold a two-thirds share in the C. G Glasscock-Tidelands Oil Company to Murmanill Corporation of Dallas. Gerald C. Mann, chairman of the board of Murmanill, had served as Texas Secretary of State and three terms as state Attorney General.

The Glasscocks were likely influenced by critical reductions in the number of days a well would be allowed to produce, a glut on the oil market, confused federal controls of natural gas shipments, and the uncertainty of pending Tidelands legislation, which would increase the state claim to offshore waters from three to ten miles.

The sale included oil and gas holdings, producing companies, a shipyard, and other subsidiaries.

Doyle C. Miller was the new president and James C. Storm was the executive vice president of Tidelands Drilling Company and subsidiary companies. Jimmie's brother, Ralph, was put in charge of foreign operations, handling drilling in Cuba, Chile, and Argentina.

"We had a thousand-well contract in Argentina, but there was such a risk of not getting paid that we turned it down," Ralph said. "Bill Clements took the risk. He made about thirty to forty million out of it."[2]

The new owners immediately announced movement of the executive offices to Dallas, but it would not be long before the Storms would be back in Corpus Christi.

Jimmie Storm didn't get to be an All-American Wildcatter by playing it safe. He rose from roustabout to vice president of a successful drilling firm, then resigned, took his money, borrowed a bundle, and bet it all on a new venture.

Jimmie was born in Goodnight in the Texas Panhandle, where he rode a buggy three miles to a one-room schoolhouse. The family later moved to South Texas. Ralph, Jimmie's younger brother, said he saw his first rig near the family farmhouse near Premont.

"I was just a kid, about seven or eight years old," he said. "I'd go over there and sit and watch a drilling rig from a distance. Finally they said, 'Come on up here,' and they'd let me sit between the rotary and the drum and the drawworks. It was a wooden type rig. They told me to sit there and don't move. That thing started bubbling over, and they said, 'Run for your life! Here she comes!'

"I looked back over my shoulder, and it was hitting the crown block. It was blowing out. Then they always seemed to blow out. It was gas and water and mud. It finally bridged over, and they moved away."

Jimmie graduated from Alice High School and attended the University of Texas for a year before returning to South Texas in 1934 to work in the oil fields. He worked his way up from a Magnolia Petroleum Company roustabout, gauger, driller, and toolpusher on steam land rigs to becoming an operating partner with Glasscock Drilling Company, operating bay and offshore rigs. "In the 1950s he was instrumental in the design, construction, and operation of the first mobile offshore drilling rigs capable of drilling in 100- and 150-foot water depths."

Upon graduation from Baylor University as a business major in 1949, Ralph took a job with Glasscock. He said, "They needed somebody to help in the production department so they said, 'You go out on the rig and learn the ropes, and then we'll have something for you.' They thought that since I was college educated, I could do anything."

He got the full roughneck education.

He told of one incident involving Lonnie Glasscock, Jr.

"We were on the tenth floor. They tell the story about this roughneck that came in. Mr. White was his accountant and bookkeeper and all that. He made out the checks. Mr. White would get to the office about 6:30 or 7. This roughneck got fired, and he came in at about 7 o'clock and said he wanted his check.

"Mr. White said he'd have to have a pink slip.

"He said, 'I want my check now. I got fired. This is how much I worked, and if you don't [pay me], I'm gonna throw your butt out the window.' So they made him out a check. When Lonnie, Jr., came, Mr. White told him the story. He said he didn't like that kind of business and he needed some kind of protection, and Lonnie said, 'Mr. White, you're just going to have to wear a parachute.'"

Ralph Storm performed a variety of work with Glasscock. "The first work I did was digging a manhole to put an anchor for a pumping unit over in Taft for the Glasscock Company," he said. "Finally I got to drive a truck hauling pipe to rig sites. Then I'd stay there and test the well, run the pipe. I got the full facade of the oil business.

"They even had me out leasing and curing leases. They always said the best clearing of a title was the drilling of a dry hole. When you hit a good well, everybody owned it."

After he was transferred to Dallas in the reorganization plan, Jimmie Storm resigned from the Dallas firm in 1962. In 1959, "with a whole lot of help from Jim," Ralph had organized Storm Drilling Company, which bought three land drilling rigs. In April 1965 Storm Drilling Company Rig No. 1 drilled the second-deepest hole in Texas and the deepest drilled by an independent contractor up to that time when the Phillips Petroleum Company No A-1 Nueces Land and Livestock Company reached a total depth of 24,220 feet, where it was plugged and abandoned. The Jurassic wildcat in McMullen County fell short of another record set by Phillips Petroleum Company in 1959, when the No. EE-1 University wildcat in Pecos County went to a depth of 25,340 feet.

Jimmie formed Storm Marine Drilling Company in 1964. *The Corpus Christi Caller-Times* reported that "Storm Marine will have no interest in drilling on land or in the shallow waters of the bay. Its efforts will be concentrated on offshore drilling operations."

James C. 'Jimmie' Storm

Jimmie Storm rose from roustabout to vice president of a successful drilling firm, then resigned, took his money, borrowed a bundle, and bet it all on a new venture.

Jimmie Storm (at right) in the field. The other men are unidentified.

During the 1960s Jimmie Storm designed, built, and patented slant-hole drilling rigs that would spud at an angle and thereby increase the lateral area from which offshore platforms could obtain production from a shallow field.

In June 1964 the Bethlehem Steel Corporation's Beaumont shipyard delivered Stormdrill I, the first of four mobile oil-drilling platforms that the company would acquire. The fourth, Stormdrill IV, was commissioned on February 4, 1966. All four operated in the Gulf of Mexico under long-term leases with major oil companies.

The Storm operation was extremely successful until the late 1960s, when activities in the Gulf came to a standstill. On July 1, 1968, it was announced that Dearborn Computer Company of Chicago had purchased Storm Drilling Company and Storm Marine Drilling Company for $9 million. At this point Ralph Storm retired at age 39, but Jimmie went on to form the Marine Drilling Company, which constructed, owned, and operated seventeen mobile offshore jack-up drilling rigs between 1969 and 1980.

In 1975 Jimmie said he had sold his interests on three occasions, "but never by my own choice.... I was probably the only man in the world foolish enough to invest my entire estate, then go into debt in an effort to re-establish myself in the offshore drilling industry."

Jimmie Storm was a great influence in the development of the Port of Corpus Christi. He became a member of the Port Commission of the Port of Corpus Christi Authority in 1973 and served as chairman from January 14, 1983, until his death after a sudden heart attack on June 24, 1991. In 2003 he was named to the Offshore Energy Center's Industry Pioneers Hall of Fame.

In a 1991 tribute to Storm, Harry G. Plomarity, the Port's executive director, said, "Mr. Storm's death is an enormous loss for both the port and for South Texas. His strong and able leadership was instrumental in shaping the Port of Corpus Christi into the sixth largest port in the United States."

The Port of Corpus Christi *Channels* for the fall of 1991 said, "Mr. Storm is mourned by all those who had the opportunity to know him ... a compassionate, generous leader, a great man."[3]

Another oilman and civic leader well known for compassion and generosity was Paul R. Haas, president of Corpus Christi Oil and Gas Company. Through a long and distinguished career, his knowledge of business, figures, and money and his knowledge of and friendship with those who had money made him extremely successful in finding financing for ventures that most oilmen would be unable to tackle.

His longtime associate and partner, Laurence McNeil, had this description of him:

Storm Drilling Company activities on the barge dock adjacent to the ship channel entering the Port of Corpus Christi

Paul R. Haas

'People in the oil and gas business were constructive. They did something or found something that, except for them, would not have occurred.'
—Paul R. Haas

"That guy is a financial genius. I mean that guy can sell nickel pencils for five dollars if you give him a chance to work at it. Financially he's just out of this world."

When Haas was starting out, however, he had no idea he would end up in the oil business. He began at age nineteen as an office boy with the Arthur Andersen Accounting firm in New York City, where he became head of the department handling Securities and Exchange Commission matters. In 1937 he was transferred to Houston. He traveled over the area as a senior accountant doing all kinds of audits, including those of public utilities, banks, and oil and gas companies.

The more he associated with the people in the oil and gas business, the more he found that "they were constructive," he said. "They did something or found something that, except for them, would not have occurred. Accountants, or auditing, is looking at the past and seeing what has been recorded, then doing some estimating for the future, but as far as accomplishing something physical, it's not that kind of business."

In 1941 Robert T. Wilson offered him an auditor's position in La Gloria Corporation, a gas-gathering company Wilson had established in 1940. Haas had turned down several offers to join cycling companies but joined La Gloria immediately, as it was the type of job he was looking for. It was the largest cycling business in the country.

Such companies had been created recently in the gas fields to recover liquids from natural gas and return the stripped gas through input wells to the formations to keep the pressures up. Wilson was La Gloria president, a role model and inspiration for the officers and employees of the company, and, just as importantly, for the entire city of Corpus Christi.

La Gloria's top officers included John F. Lynch, Ted Scibienski, Haas, Emmet Wilson, who was not related to Robert, and others. Their involvement in civic and charitable affairs was complete. Robert Wilson's influence extended down to the company department heads and others who were encouraged to work for the common good of the city.

In 1946 the Corpus Christi city government imploded in controversies and corruption. Mayor Roy A. Self stepped down and massive resignations occurred.

Wilson was elected mayor on a ticket that promised to clean up corruption among officials and employees. He did just that in less than a year before his unexpected death in 1947 at age 51. He had paved miles of roads and installed sidewalks, and he was preparing to lead the city into a manager-council form of government.

Haas was an accountant but became vice president and treasurer. He said he "learned a lot from Mr. Wilson, who had great business ability and who was also civic-minded. In my opinion he would be able to do just about anything he wanted to in the business world because of his

Paul Haas felt that oil-field workers like (from left) Joe Wright, Mack Rusk, and Buster Bishop performed a constructive service.

extreme ability. We became fast friends, and I was able to acquire a great deal of business knowledge from him, as well as, I hope, a great deal of empathy for other people."

Wilson treated his employees well and instituted a profit-sharing plan. Haas would also be generous to his workers. After Wilson's death La Gloria Corporation was liquidated and the assets acquired by La Gloria Oil and Gas Company, owned by Lynch, Scibienski and Haas. Haas explained that the transfer came because Robert T. Wilson's brother, Charles E. Wilson—president of General Motors Corporation and chairman of the board and owner of much La Gloria stock—was Secretary of Defense. Because La Gloria manufactured isobutane, an important component of 100-octane gasoline used by the military, Charles Wilson had to give up both posts to avoid the appearance of a conflict of interest. Ownership passed to the three through a series of complicated financial moves, including loans from banks and insurance companies.[4]

McNeil joined La Gloria by a different route. "I grew up in Oklahoma where my grandfather and father owned a hardware store in Cushing during the oil boom of the 1920s," he said. "When I was seven or eight years old, I would see wagons loaded with pipe coming from the train station bogged down in mud. The roads were dirt, and the wagons were pulled by teams of horses. I think that the romance got into me then."

Actually he more or less flew his way into the business. He took civilian pilot training and got his license while attending school at Texas A&I College at Kingsville just before World War II. He got many more

Laurence 'Larry' McNeil

Laurence 'Larry' McNeil worked in La Gloria plant on weekends, attended college, and flew the company plane when it was necessary.

flying hours in the Army Air Corps, piloting transport planes over the Hump through the treacherous mountains of Burma.

After the war he returned to A&I to study natural gas engineering. He had a part-time job testing gasoline and natural gas samples in La Gloria laboratory. With ninety dollars a month from the GI Bill and powering his old car with gasoline collected from the samples, he was prepared to go to school.

When La Gloria President John Lynch offered him on-the-job training at two hundred dollars a month, McNeil said he'd rather go to college and get his degree. Lynch learned he was a veteran pilot, one of very few in South Texas qualified to fly by instruments. The mountains had taught him well. As it turned out, he worked in the plant on weekends, attended college, and flew the company plane when it was necessary.

"I was a busy son of a gun," he said.

It helped that Lynch was a good friend of Dr. Frank Dotterweich, professor and founder of the Natural Gas Engineering School at A&I.

"When Mr. Lynch wanted me to fly, he'd call Doc and say, 'I need Little Mac'—that's what they called me—'I need Little Mac to fly for me,'" he said.

"Doc would say, 'Okay.' Then he would give me assignments for a week. And he marked me present during the week."

His job at La Gloria was assured, but changes were on the way. He became an engineer and worked with the drilling rigs. He was promoted to chief engineer and in 1956 became manager of production.[5]

Chester Wheless, who would later form his own gas company, had worked for La Gloria since its earliest days—using a pick and shovel and hauling concrete in wheelbarrows, constructing the recycling plant where he would later work.

He recalled that one of the three original owners of La Gloria was a "very astute entrepreneur who operated very efficiently," he said. "One of his basic principles—after being away from the office for some time chasing deals, papers would stack up on his desk. He would push the accumulation of mail from one end of the desk off the other end into the trash basket. He'd say, 'If it's anything important, I'll get a duplicate of it.'"

Wheless's very first work was in La Gloria Field, where there were twenty-five to thirty different reservoirs "stacked like pancakes one on top of another, except, of course, where there was shale or some separation between the reservoirs. It was discovered by Magnolia Oil Company and eventually covered 5,000 acres."

Wheless, who had a petroleum-engineering degree from the University of Texas, resigned from La Gloria in 1955 and used his knowledge of recycling to form Santa Rosas Gas Company to strip gas,

Dr. Frank H. Dotterweich

'Doc' was real pioneer

Dr. Frank H. "Doc" Dotterweich pioneered the first Natural Gas Engineering Department in the nation at Texas A&I College at Kingsville in 1937. He retired as dean in 1971 but continued to teach and work on technical projects until the year before his death in 1990.

He was popular with his students and highly regarded in the industry and academic circles. He earned a doctorate in chemical engineering from Johns Hopkins University and served in the Navy in World War II.

During winter commencement in 1989, he received Texas A&I University's first Presidential Citiation. For his "Outstanding accomplishments in gas processing research and technology and for excellence in engineering education," Gas Processors Association gave him the prestigious Donald L. Katz Award.

In 1928 he was a member of the United States Olympic lacrosse team in Amsterdam, the only time the sport was played in the games. The team won first place, but its gold medals were never delivered.

While he was in Europe, he took time to study the design and operation of gas plants there.

<http://www.engineer.tamuk.edu.Homepage/InfoDotterweich.htm>

Horse-drawn oil-field wagons loaded with pipe inspired young Laurence McNeil's interest in the industry.

In 1957 La Gloria Oil and Gas Company made an exchange of stock and merged with Texas Eastern Transmission Corporation of Houston.

pressurize it, and deliver it to pipeline companies after he saw a need for the process in the Fort Stockton Field in Pecos County.

"It was a matter of taking a good idea from South Texas and putting it to use in another location," he said.

He sold the "very successful" operation seven years later.

Recounting his experience in putting deals together, he remembered a crusty West Texas rancher who refused to talk about leasing his land. He visited the ranch several times. The rancher, who "was tough as nails," would talk but not about leases.

"It occurred to me that he lived way out in the country, had no telephone, no television. I told him, 'Henry, I have a deal I think you're going to like. You're always out taking care of your ranches, and your wife is way out there all by herself. For that lease, let me have a TV installed while you're out.'

"Well, that caught his attention. He agreed and I got the lease. He wasn't interested in money, but he liked the TV. Barter worked better than money."[6]

In 1957 La Gloria Oil and Gas Company made an exchange of stock and merged with Texas Eastern Transmission Corporation of Houston. Haas and Lynch became vice presidents of Texas Eastern. Haas stayed for two years. The only advantage in working in a large corporation, he found, was "you do not get as much blame for the bad decisions." Because he did not want to live in Houston, he returned to Corpus Christi to "engage in my own ventures."

McNeil wasn't happy either. Only occasionally did he get out in the field. Mostly he sat among piles of invoices. And everything, no matter how small, needed approval. One problem he cited occurred over a kitchen vent-a-hood. He visited a company house in Freer, where it was "hot as hell." The wife of an employee said everything was fine except that the kitchen got so hot.

He told her, "Why don't you have a vent-a-hood put in?"

She did, and the company was faced with buying two hundred more for other employees. McNeil said, "I got in all kinds of trouble about giving a vent-a-hood to somebody without clearing it. I could spend up to five thousand dollars without anyone's approval, but I put a vent-a-hood in, and that really caused a stink."

He questioned the logic of the large company's inventory system. In their small company, "if we have a well drilling and we are going to need casing, once we made up our minds to 'case' at two o'clock in the morning, I picked up the telephone and ordered the casing and it was on its way."

At Texas Eastern, if a requisitioned item was in stock at the Houston office, the office would send it out rather than authorize a local purchase. If it was not in stock, the purchasing agent would take bids before purchasing it. McNeil compared the costs in the two methods, showing that it was bad business to hold up the rigs to wait for casing. Rather than authorize local purchasing, however, the company set up warehouse stocks in Corpus Christi, Falfurrias, Houston, and Lafayette, Louisiana.

"The stock in each one of these places must have been worth $100,000," McNeil said. " . . . I know large companies must have checks and balances for various reasons. This is the reason I prefer a small company."

Emmet C. Wilson had joined La Gloria as an accountant when he got out of the Navy in 1946. He had remained with the company through the merger and had become a vice president in charge of acquisitions with Texas Eastern.

In November 1959 the group of Gorman, Gierhart, and Howe asked John Crutchfield to evaluate their Prado Field, near Hebbronville, "with the idea in mind that they were ready to sell out." Sun Oil Company had originally held the 1943 lease, on La Parra pasture of the East Ranch. Of three wells Sun drilled, one hit gas and the other two missed. Sun gave up, but after the lease went back to the landowner, Gorman, Gierhart, and Howe took it over. Joe Gierhart, who had worked for twenty years as a geologist for Humble, was looking for the Prado sand but actually found a sand known as the Loma Novia, which was a great producer. He drilled seventy-four wells in the Prado Field.

Larry McNeil's bosses at Texas Eastern weren't happy when he approved a Vent-A-Hood for a family living in hot company houses in an oil-field camp like this one.

'My zeal and enthusiasm for the Prado property had become so high that my first reaction was "let's get a group together and buy it ourselves."'
—John Crutchfield

With the sellers asking $20 million to $21million for the property, Crutchfield tried to sell it to a major oil company, but his efforts were unsuccessful. Haas, Emmet Wilson, and McNeil suggested to Texas Eastern that it buy the property. Company officials were not interested.

"By this time," Crutchfield said, "my zeal and enthusiasm for the Prado property had become so high that my first reaction was 'let's get a group together and buy it ourselves.'"

He approached Haas, a good friend, who had recently left Texas Eastern and was seeking another opportunity. The two of them worked on the project and decided that $16 million was the maximum they could pay. Through long complicated negotiations, Haas managed the purchase, and Prado Oil Company was born, with Haas, McNeil, Wilson, and Crutchfield as partners.

In three years the partnership drilled 191 wells and brought the drilling cost from thirty thousand dollars down to nine thousand dollars a well and eleven thousand dollars for a dual producer. Production, under tight control, rose from 270 to 7,000 barrels a day, McNeil reported. Prado, purchased in 1960, produced for them until 1965, when they sold it to Standard Oil Corporation of Ohio.

McNeil and Haas flew to Chicago to close the sale.

"Mr. Haas took a safety pin and pinned that $34-million cashier's check inside his pocket," McNeil said. "Knowing Mr. Haas, he wanted to get that check in the bank so it would start paying interest. The first thing in the morning, we deposited it in First State Bank. It was almost as much as the bank's total assets."[7]

McNeil told a story that illustrates the company's generosity with its employees. There was an employee who kept the gauge records in Perryton, Texas, although math was not his strong suit. He would tell his wife what each gauge was, and she would write it on the books. In spite of this weakness, he was an excellent employee. He would check the production when the weather was bad, and, instead of hiring people to paint the tank batteries, he would say, "That gives me something to do. I'll just paint my own tank batteries."

"When we sold Prado," McNeil said, "we told him, 'We've sold out, but SOHO will want you to work for them. We're going to give you fifty thousand dollars.' This was in about 1964, and fifty thousand dollars was a pretty good slug of money."

On his first day on the job with SOHO, a company man came out to accompany him.

"[The former gauger] puts the key in the ignition," McNeil said. "The guy turns it off and says, 'We don't start the truck until we have the safety belt fastened.' They backed out. The first well was a few hundred feet across the road. As they stopped, [the former gauger] unstrapped his seat belt, and the new manager said, '. . . we don't unbuckle until we turn off the ignition.'

"The guy wouldn't get out and open the gate, so [the former gauger] gets out, opens the gate, gets back in the car. He has to be reminded to buckle up. Drives through the gate. Turns off the ignition. Closes the gate. Reminded again to buckle up before turning on the key. They drive up to the well. Same thing. First well. He opens the tank and the pressure blows.

"'. . . we don't open a tank like that. You have to punch the button down here so it will vent. We just don't do it that way.'

"[The former gauger] had the car keys in his hand. He dropped them through the hatch into the tank filled with oil and said, 'I quit.' He walked to the house and told his wife to pack up, 'We're leaving.'"

With the fifty thousand dollars he bought a peanut farm west of Pleasanton.

"At least fifteen years later," McNeil said, "I was playing golf with a guy who said he was from Pleasanton, president of the bank. I said, 'There is a gentleman up in your area that used to work for me that you may know.'"

McNeil told him the man's name.

The man replied that he did, that a man of that name was chairman of his board of directors.

"I said, 'That's not the [man] I know. He can't even write his name, leave alone be chairman of the board of a bank.'

The banker said, yes, that's the man, all right. "'He's the chairman and owns the bank, and his boys are all in the bank, too.'

"I said, 'His wife had to do all the figuring and everything.'

"He said that was the man. I asked where he got his money.

"He said, 'The peanut farm he bought happened to be right in the middle of that oil play around Pleasanton, and he had something like 960 acres right in the middle of it. I can personally tell you [the former gauger] is worth over $20 million.'"[8]

The organization and methods the partners developed with Prado served as a blueprint for their offshore El Gordo project, which was even larger. Because offshore operations require a lot of money and a lot of expertise, few independents had been involved in them. Haas took chances in his long career, but they were so carefully planned that they could not be considered gambles. Houston Lighting and Power Company and Soloman Brothers of New York were two of the prime investors in the venture that became El Gordo.

Furthermore, Haas, McNeil, and Emmet Wilson, chief officers of the company they formed in 1968 as Corpus Christi Oil and Gas Company, assembled as many top-rated experts in many specialties as any other independent who ever operated on the Texas Coast. H. W. "Bill" Volk joined the team early on, as vice president in charge of exploration, geology and geophysics, and land-lease transactions.

Volk's interest in geology had started at a very young age in his father's stone quarry, where he worked summers.

In three years the Prado partnership drilled 191 wells and brought the drilling cost from thirty thousand dollars down to nine thousand dollars a well and eleven thousand dollars for a dual producer.

Bill Volk

Bill Volk was establishing himself as an independent when Paul Haas asked if he wanted to 'go after a few elephants.'

"I saw all these fossils and I noticed bedding planes and I noticed changes in coloration and it just totally fascinated me," he said. "I went to the University of Minnesota, and I majored in geology, of course, because that's what I wanted to do."

Tidewater Oil Company sent him to Corpus Christi in 1946 after he served thirty-seven months in the Navy. He was with Northern Pump Company for thirteen years, serving as geologist, landman, and operator, but in 1965 Northern Pump was sold and his office closed just as he and his wife were building a new house.

"There I was, without a job, with two little boys and a house I couldn't afford. That was the low point of my career," Volk said.

He was establishing himself as an independent when Paul Haas asked if he wanted to "go after a few elephants."

Haas wasn't talking about wild game. He was referring to discoveries that they might find drilling offshore in the Gulf of Mexico. And waiting for them offshore was El Gordo.

When the drillers first saw the log on the well ten miles off the Texas coast in the northwest Gulf of Mexico, this was their reaction:

"We're running this thing again. This couldn't be right," drilling engineer Larry Rodolph told Clyde Stout.

The production superintendent gasped. Only once before had he even seen such an indication on a log. It showed a rich natural gas sand 242 feet thick, a once-in-a-lifetime discovery. A show of 50 feet is exceptionally good.

"No, the log is right," Stout said. "I might not live long enough to see something like this again."

Charles Menut, geologist and offshore manager who ran the log, said, "I broke into a cold sweat when I saw it."

Stout remembered, "I've got to call McNeil. He's getting ready to fly up to Houston."

But how to tell him? The radio-telephone was patched in so all the oil installations on the Texas coast could hear his message on their loud speakers. And this well—located at a spot they had called "485"—was a "tite well." No information was being leaked about it. He had to tell Larry McNeil back in Corpus Christi the good news, but he had to tell it in code. Otherwise he would alert other companies, who would be anxious to buy up some of the unleased tracts in the vicinity.

Interest in that offshore area had slackened after Shell and Humble Oil companies had drilled dry holes 1400 feet from the location. Had they drilled 500 feet deeper, they would have hit the bonanza.

As McNeil told it, "Clyde called and said he needed to see me very much as he had something very important to talk about. I asked him if I was smiling (good news) or frowning (bad news). Clyde said, 'You are laughing.'

"I said, 'Really?'

"He said, 'How high are you standing?'"

The previous day, in the Corpus Christi Oil and Gas Company office high in the 600 Building on the bluff overlooking Corpus Christi Bay, a low-lying fog bank hung over downtown. Stout had asked how much visibility Mac needed to take off in his plane.

"I am standing here 250 feet above the ground," McNeil had replied. "It takes two hundred feet and half a mile visibility to fly. If I can see the ground from my window, I can fly because I am standing about 250 feet above the ground."

"I know how high I'm standing," McNeil answered and rushed to inform partners Paul Haas and Bill Volk that Clyde had called a cryptic message that that they had 250 feet of pay.

McNeil immediately flew to Port Lavaca to see the log. It wasn't 250 feet of pay, but it was 242 feet of good sand.

"I got goose bumps," he said, "because I had never seen a log this good."

Later Menut and Volk were casting about for a name for their discovery. Corpus Christi Oil and Gas had successfully opened several fields, with most named for sites along the coast. Freeport Field was named for the town, Cowtrap for a lake. Two bench marks on the shore were Mata, "dead" in Spanish, and Gordo, "fat."

"One of us said, 'Gordo' and the other said 'El,'" Menut said. "We both envisioned the cartoon character, Gordo, 'the fat one.'"

El Gordo had a special significance for Paul Haas, president of the company.

"When El Gordo was discovered, there was an unbelievable set of circumstances, including what happened in my own situation," he said. "I was having lunch with [oilman] Jack Flaitz at the Eldorado Country Club near Indian Wells, California."

Haas and his wife spent several weeks every year at a cottage they owned there but remained in contact with Corpus Christi. He was aware that a log was being run on what would become the first El Gordo well.

"During lunch I received a telephone call from Larry McNeil," he said. "He gave me the news of the indicated 200 feet of sand section in the well. I found it difficult to believe."

When he returned to the table, he was in such a state of shock that everyone asked if he was ill.

"This was one of the greatest moments in the history of my life in the oil and gas exploration business," he said. "It is a feeling that is very difficult to find in many businesses. The exploration business is romantic in a sense, it is a gamble, and it affords you the feeling that you have added something to the wealth of the world that might never have been found without you."[9]

The discovery was a result and a payoff for the effort Haas, McNeil, and Emmet Wilson had put forth in assembling the C. C. Oil and Gas

(Above) El Gordo, with 242 feet of rich natural-gas sand (Top) Helicopters ferried men and supplies to the rigs.

team of experts. It was Bill Volk who assembled oil and gas lease blocks and recommended the site to explore for untapped reserves.

"I was getting along . . . until Paul Haas came to me and asked if I wanted to join a company he was forming, Corpus Christi Oil and Gas," Volk said. "Of course I wanted to. Also, I had great admiration for Paul Haas."

It was fortuitous for both Volk and C.C. Oil and Gas that Haas had approached him with the offer to join the firm. Volk brought in Dr. John Marr, a pioneer in the "bright spot" technique of locating natural gas, as the consulting geophysicist.

"Bright spotting" was a new method of determining the prospect of gas in an area. In a paper he wrote about a geophysical survey in Louisiana, Dr. Marr explained that when the sound of the shock wave going through the earth hits shale, it moves very fast. When it hits sand, it moves slower, and if the sand has gas in it, the speed is even slower, the action showing up as a bright spot on their equipment.

Volk took on Charlie Menut of Houston as the lead geologist on the team, and Joseph G. Putman III was hired as chief offshore geophysicist. As the company's financial dealings had become more and more complex, Malcolm Alexander, an accountant experienced in handling audits and tax and insurance problems in the oil industry, was hired as vice president in charge of finance.

Then, in 1973, came the leasing of some 60,000 acres in Gulf waters for "a total of something in the order of $3 million." The offshore leases were good for five years and would expire unless wells were drilled.

Offshore rigs were scarce, McNeil said, and they desperately needed rigs to meet their drilling schedule.

"You've got to understand my motto has always been 'There ain't no can't,'" he said.

At this point Jimmie Storm, a friend who owned Marine Drilling, came to the rescue. He worked out an arrangement with Conoco for CCOG to use drilling rig Stormdrill V for six months out of the year.

"Jimmie was a tremendous, tremendous help to get started," McNeil said.

The first offshore well they drilled was the 722S No. 1 off Calhoun County, which resulted in the discovery of the Sherman Field.

"We had trouble getting gas from the Sherman Field," McNeil said. "It had to go fourteen miles across Matagorda Island. The Air Force did not like the idea of putting a gas line across their bombing range. It took six months to get that straightened out. Then there were the whooping cranes. I could write a book on the subject of whooping cranes. We had to move the lines to get out of the whooping cranes' way. Then we could only do the work from March until October, so we had to delay production for a year. I figure it cost us $50,000 per crane, and there were fifty of them on the island."

In *El Gordo . . . El Magnifico* Volk explained their progression:

An artist's rendering of El Gordo

"Then we went north to offshore Brazoria County and discovered what is known as the Freeport Field. From there, we went to offshore Matagorda County and discovered Cowtrap Field. We also discovered a field called Playa about this same time. The depths of the pay sands in these fields varied from 4200 feet to 8200 feet. All of this activity carried us through 1975, and on January 13, 1976, we spudded the El Gordo Well—520 L, NW/4 # 6."

It is not often independent operators can compete with the big guys in buying vast ocean leases and getting financing for a very expensive venture. It takes confidence and more than a little bit of daring to go where the biggest and best-equipped companies had gone and failed. It was this type of confidence that led Volk to recommend that they drill at the location of the El Gordo discovery, a thousand feet from the well Humble had abandoned.

"We did it because we thought they hadn't gone quite deep enough. They almost did, but they stopped when they got to pressure," Volk said. "Pressure is a change from the normal pressure gradient to a steep pressure gradient which requires a lot of different things, like setting intermediate casing so you can build up your mud weight so you can control the pressures from that depth."

He said that if mud weight is not increased, it is not possible to drill through overpressured zones.

"They didn't go deep enough to do that. So we did," he said.

He said the major oil companies at the time had more or less neglected the Texas part of the Gulf because "they felt that the fields that were being discovered there were not quite big enough to satisfy their needs and parameters. This created a great opportunity for us because many independents did not want to go out there because the expenses and hazards are greatly different from onshore.

Glasscock Rig No. 5, a bay rig, was an antecedent of the rigs used later in the deeper Gulf of Mexico discoveries.

"We decided we would take that chance and bought a huge block of acreage in 1973. And we drilled on our best prospect, and it turned out to be our best field," he said.

"What we were doing was taking the reserves that the majors knew about, but they weren't big enough for them to make money off of it," McNeil said. "We were able to make money by cutting costs."

"The majors were after the real elephants then," Putman said. "Nothing else interested them."

Volk considered the discovery date of El Gordo—February 20, 1976—the high point in his career. Describing El Gordo, he said, "It's Miocene .... 220 feet of pay in one sand, gas pay ... is very unusual. It is still producing. It was a very big thing for the company. We went on and developed other areas in the Texas Gulf Coast and then went into South Louisiana. We had good success there but it was nothing like El Gordo.

"I had very good people working for me, and the management was excellent, and Jay [Endicott] was a good right hand man, to say the very least," he said. "It's all been very meaningful and very satisfying for me, and I'm grateful for all it has done."

In 1985 he and Endicott resigned from the company. Volk opened his own office. He was still doing oil and gas work in 2004.

That first well in El Gordo Field assured the success of Corpus Christi Oil and Gas Company. Later No. 16 came close to destroying it.

When Larry McNeil was flying the Hump in World War II, the planes flew between and around the mountains because they could not fly over them. On one occasion he was exhausted, and the co-pilot took over the controls. McNeil was in a deep sleep when suddenly he sat upright and discovered his reliever had dozed off and the plane was heading for a mountain. Even so, he called his experience with Well 16 in El Gordo Field "the closest call I have ever experienced, any where, any time."

Trouble with the well hit when pressures suddenly built to an alarming level and would not bleed off. The men found that the casing had broken and there was imminent danger of the reservoir blowing.

"We realized that all of our pipe strings were ruptured, and we knew we had big trouble," he said. "The whole platform could be lost, and three wells would go."

There was no fire, but one was possible. "We called Coots and Boots, which is a fire-fighting firm. Both men had formerly worked for Red Adair. Coots Matthews was a big old country boy but very, very astute in all his decisions and very methodical."

They closed a safety valve but did not know if they could pump heavy mud in to kill the well. Pressure was more than 6,000 pounds. The Otis Company, which manufactured the safety valve, thought they could pump the mud in. Halliburton trucks were barged out. Mac and Coots stayed by the well, and others were ordered back.

"I could see that Coots was nervous because he was chewing tobacco and there was a flat place down below us. It looked like it was raining tobacco juice," Mac said, "because he was spitting over the side of the platform about every minute or two. When we got on the platform, I said, 'Coots, you know we don't even have a rope to get off this.'

"'Goddamit,' Coots replied, 'if this thing goes, you aren't gonna have time to look for a rope.'"

McNeil told him they were going to get off their precarious perch. Then he noticed he had neglected to put on a life jacket or hard hat.

"If we jump off, we're going to get hurt and hurt bad," Coots said. "I don't know how bad, but if we jump five stories off this thing, we're sure as hell going to get hurt."

They pumped mud all day and all night. Finally they managed to repair the casing, pull the tubing, and get things under control. They never found a trace of all the mud they pumped into the well.

A Bill Pugh net lifts workers onto a rig.

A special Sunday dinner

In the early days of Corpus Christi Oil and Gas Company's offshore drilling, Haas, McNeil and Wilson invited all the employees, their wives, girlfriends, or boyfriends out to the rig for Sunday dinner, Clyde Stout recalled.

They were transported to the rig on a crew boat, but nobody told the ladies they would be having a ride on a Billy Pugh net, which lifted them by crane far above to the well deck.

Passengers stood on the narrow ledge on the outer edge of a circular disk and held on to rope netting.

"It was a white-knuckle ride that would more than equal any ride at the State Fair," Stout said.

Some of the ladies—and likely some of the gentlemen, if they would admit it—dreaded the return trip to the boat, but nobody ended up in the drink.

El Gordo–El Magnifico, 108-109

'We ran according to the rules. We were just in the right spot at the right time to get it done.'
—Larry McNeil

"That was the nearest thing to a genuine catastrophe," McNeil said. He estimated a blowout would have cost a minimum of $50 million. Eventually the company had about twelve different offshore platforms, at that time unmanned. Some of the unorthodox methods they used as cost-cutters could not have been used later. For example, they used solar power to transmit the flow rates back to shore.

"We could radio out there and shut those platforms off by remote control," McNeil said. "The federal government inspected everything and passed on it."

They also had equipment that would take out all the oil or distillate in the water produced from the well and run the water through straw pipe over the side. Later it became illegal to dump any water overboard or to operate unmanned offshore wells.

"Some of the platforms we had are still in operation, and they had to hang living quarters on the side of them," McNeil said. "We ran according to the rules. We were just in the right spot at the right time to get it done."[10]

Another expert in his field who joined the firm was James Herring, who was hired after he retired from the Railroad Commission.

"Paul and Mac could make a decision and not be wishy-washy," he said.

They were working offshore, and after all his experience with the state, Herring had a chance to deal with the federal bureaucracy. Among the many reports to be filed was one on archaeology on a spot 700 feet under water because "at one time during the ice age when the ocean level was low, that spot may have been occupied."

In 1990 the partners sold El Gordo Field to Pacific Gas and Electric Company, ending one of the greatest chapters in South Texas exploration. However, they continued to operate C. C. Oil and Gas and remained active in other endeavors.

In addition to his position as chairman of the board and president of CCOG, Haas continued to busy himself with a vast number of business, civic, and charitable activities. In 1957-58 he served as president of the Corpus Christi Independent School Board of Trustees at a time when school desegregation was a major issue.

"I probably spent an inordinate amount of time in the early days on the local school board," he said, "because I was president during the difficult period of the amalgamation of the black and white situation. We had about the same percentages as Little Rock, but we did a better job of controlling it. We got through it nicely."

As recognition for his work in education, the Corpus Christi district chose Paul R. Haas as the name of a middle school opened in 1968.

"That's been a very gratifying situation," Haas said. "We attend the graduation each year, and we give a book for each kid each year."

Every year he and his wife, Mary, also provided scholarships for ten of the school's students.

"We enjoy helping," he said. "A favorite part of my activity has been philanthropic. I've served that field in a number of situations as well as the education field . . . . I felt everyone has a responsibility to do that sort of thing. And if you can make the time to do it, you should spend the time."

In 1980 Haas was termed one of the top twenty most powerful men in Texas, although few people knew it. *Texas Business* noted, "Haas' power is wielded in behind the scenes fashion and often through operatives. He keeps a low profile and spends part of his time at his home in Palm Springs."

McNeil also was active in the field of education, serving on the board of directors of Texas A&I University, later Texas A&M University-Kingsville, and on the board of the university's Ex-Students' Association. Laurence McNeil died August 1, 2004, in Corpus Christi.

In John David Scott's *El Gordo . . . El Magnifico*, McNeil had reflected on his experiences:

"I don't know what the future holds for sure, but I hope there are not too many more whooping cranes and No. 16 wells in store for McNeil. I predict there are some big things and some exciting things to come, but if not, I have had my share of luck and thank God, it has, for the most part, all been good!"[11]

The Corpus Christi middle school named for Paul R. Haas

Paul R. Turnbull

Frank P. Zoch

Paul R. Turnbull and Frank P. Zoch, Jr., had drilled primarily in the Southwest area until they joined as partners with William P. Clements and went worldwide.

While Corpus Christi Oil and Gas was one of the first independents to successfully go offshore, drilling in the Gulf of Mexico, another Corpus Christi company would play a role in operations that eventually circled the globe.

Turnbull and Zoch Drilling Company was a modest five-rig drilling company when it started in Corpus Christi in 1955, but from its roots grew the biggest drilling operation in the world.

Paul R. Turnbull had been an independent for two years before he hooked up with Frank P. Zoch, Jr. Prior to that he had been Southwest Division engineer for Humble Oil and Refining Company for twenty years and served La Gloria Corporation as manager of drilling and production.

Turnbull and Zoch drilled primarily in the Southwest area until they joined as partners with William P. Clements and went worldwide.

Clements, who was later to gain fame as the first Republican governor of Texas after Reconstruction, won his spurs in the oil business in South Texas.

Clements played high school football for the Highland Park Scotties in Dallas, where he gained a reputation as a hard charging blocking back and defensive end. He became the first Highland Park player ever named to the All-State team and was editor of the high school yearbook. He planned to enter a military academy, but his father was left penniless by the Depression and he had to forego college. Not only would he wait for higher education, he would have to work to help support his family.

He joined a geophysical crew at Sinton and learned to operate the torsion balance and seismograph equipment and to survey, read instruments, and plot the data. The work took him as far as Victoria and McAllen. He also played on the Sinton baseball team.

After his father found a job, he went to SMU for a year, but he and Coach Matty Bell did not see eye to eye. He transferred to the University of Texas but soon tired of college and lost interest in football. Graduate engineers were paid $110 a month, and he knew he could make $180 a month in the oil field.

That was his pay for eighteen months with Trinity Drilling Company, where he learned every phase of drilling an oil well, mostly in Southwest Texas. In his next job, with Oil Well Supply, he rose to be manager of the San Antonio office.

At a friend's urging he joined a venture in which the two received financial support to buy two rigs and start a drilling company. Their backer was Toddie Lee Wynne, partner of Clint Murchison. January 1, 1947, was the start of Southeastern Drilling Company, which was to become SEDCO, Inc. in 1968.

It wasn't long before Southeastern rigs were at work offshore and in Louisiana, West Texas, East Texas, New Mexico, Mississippi, and the Gulf of Mexico. Another was at work off Trinidad.

In the meantime Turnbull learned that Stanvac, a company owned jointly by Standard Oil of New Jersey and Socony Vacuum, was looking for a contractor for a program in India and East and West Pakistan. As Turnbull and Zoch was not big enough to tackle such a project, he called Clements.

They won the contract and signed a partnership on a 50-50 basis under the name Southeastern Asia Drilling Company. Soon they had four rigs at work and won another large contract to drill in Iran. By 1956 the firm had rigs at work in Pakistan, two in Iran, one in Turkey, British Somaliland, and the Persian Gulf.

William P. Clements

William P. 'Bill' Clements, who was later to gain fame as the first Republican governor of Texas after Reconstruction, won his spurs in the oil business in South Texas.

Like these members of a 1937 crew, Bill Clements did surveying work in South Texas early in his career.

A deepwater platform under tow to its offshore destination

Turnbull learned of another huge job: a venture to drill 1000 wells in Argentina in 1958 as Southeastern Drilling Company of Argentina. They got the job for some $50 million, at the time the largest drilling contract ever made.

By 1972 SEDCO had land rigs and three offshore rigs in the Near East. It had a total of thirteen offshore rigs of various types in the Persian Gulf, Arabian Gulf, Borneo, Western Africa, the North Sea, Australia, Eastern Canada, and Qatar, as well as eighteen land rigs in Algeria, Iran, Abu Dhabi, and Oman.

On New Year's Day 1972 Paul Turnbull died of a heart attack, and the partnership of SEDCO and Turnbull and Zoch was terminated. Zoch died a year later. On September 14, 1984, SEDCO was sold to

Schlumberger. At that time Forex, the company's drilling division, had 67 land rigs and 18 offshore rigs.

In 1999 Transocean Offshore, Inc. and Sedco Forex Holdings Limited merged, forming the world's largest offshore drilling company. It grew larger still on January 31, 2001, when it acquired R&B Falcon Corporation. All operated under the name Transocean, which had 150 mobile offshore drilling units.

In 1973 Bill Clements had taken on governmental responsibilities when President Richard Nixon appointed him to be Deputy Secretary of Defense. He was elected governor of Texas in 1978 and again in 1986.

Ironically, Clements's biggest headache as governor came from a blowout on one of his company's offshore rigs, the Ixtoc 1, in the Bay of Campeche on June 3, 1979. It was drilling for Pemex, the Mexican national petroleum company.

The runaway well dumped thousands of barrels of crude into the Gulf of Mexico, alarming residents all along the coast and creating a great embarrassment for a governor so greatly involved in the energy business. It did little good to point out that the rig was leased to the Mexican company that was operating the well.

It was summer, and the spill spelled disaster for the upcoming tourist season. At first it appeared the huge oil slick would miss the Texas Coast, but two weeks later beaches from Port Aransas to Boca Chica at the Rio Grande were black with sludge. Eventually the tides changed, carrying the oil away from shore.

Red Adair's well fire-fighting crew brought the burning well under control on March 23, 1980, almost ten months after it blew out.

Clements's political opponents made hay out of the disaster. When the wrecked rig was towed out into the Gulf and sunk in deep water, Texas Attorney General Mark White, Clements's bitter enemy, claimed the action was taken to hide evidence of the cause of the blowout.

Sedco eventually paid more than $4 million to settle damage claims filed against it.[12]

To some at the time, the spill was reminiscent of the early days of the oil industry, when it was like a troublesome child who did not know what it was doing. It allowed crude oil to cover the countryside and send salt water and oil down creeks and rivers. Industry representatives opened valves to create gushers to impress investors, cut down forests, and left a terrible, dangerous mess in earthen pits and over the despoiled countryside when they moved on.

So much oil flowed over that land that some entrepreneurs found it profitable to salvage oil from the creeks and sell it for boiler fuel on rigs. Gradually, however, the state stepped in with stringent regulations, and companies assumed responsibility for protecting the environment in the interest of public relations.

In 1999 Transocean Offshore, Inc. and Sedco Forex Holdings Limited merged, forming the world's largest offshore drilling company.

A worker prepares to dock an Esso tanker so it can take on a load from the Humble storage facilities at Harbor Island.

Then there occurred an event that the *AAPG Explorer*—in a special century edition published by the American Association of Petroleum Geologists—termed "a tectonic shift in the world's attitude toward industry in general and the oil industry in particular." The heading for the article read "Another Day That Lives in Infamy."

It referred to the afternoon of January 29, 1969, when a well was capped on a Union Oil Company platform six miles west of Summerland, Calif., as it was beginning to blow out. Pressure built up, causing five surface ruptures in an unexpected east-west fault on the ocean floor, releasing a huge flow of oil and gas.

Crews struggled for eleven days to stem the flow. In the meantime thirty-five miles of coastline was inundated with a coating of tar, covering beaches on Santa Cruz, Santa Rosa, and San Miguel islands.

For days the world saw pictures of dead birds and oil being cleaned from others. A group called GOO —Get Oil Out—was founded. Californians were urged to boycott Union Oil and burn their Union credit cards. Some 100,000 signed petitions to ban offshore drilling.

Another black mark against the oil industry occurred twenty years later, on March 23, 1989, when the tanker Exxon Valdez strayed from the shipping lanes and ran aground on Bligh Reef in Alaska's Prince William Sound. This event was caused by a serious misconnection among the ship's crew. Conceivably, it could have happened anywhere in the world. Unfortunately it occurred at a spot where it caused the maximum damage to the environment.

Again there were countless pictures of the cleanup and the cleaning of birds and animals, and all the rancor from Santa Barbara was re-ignited, reinforcing opposition to Alaska drilling and solidifying support for no-drill zones off the Atlantic and Pacific coasts.

The Santa Barbara oil spill in itself was not a big event, but its repercussions were explosive, giving rise to the modern environmental movement and galvanizing millions to the cause. One result was the formation of Earth Day in April 1970. Later the Environmental Protection Agency was established, and environmental statutes passed.

Environmental concerns are not antithetical to oil interests, but extremism in the movement seriously damaged its credibility with the industry. Most oil men would probably agree with geophysicist Randy Bissell, who said in 2004, "I am all for the environment, but I am not an environmentalist."[13]

The situation in the Gulf of Mexico exemplifies his position. Despite the Ixtoc blowout, most of the drilling in the Gulf has been safe and efficient. Instead of being detrimental, drilling has proved beneficial to the ecology, as drilling and production platforms have provided protection and breeding grounds for marine life off the Texas coast.

Quite a trek for the turtles

Tom Davidson, a geologist who worked near Refugio for Bishop Oil Company, cited examples of the type of regulation oilmen found objectionable.

"We were on the O'Connor property, west of Highway 77, which, as the crow flies, is fifty or sixty miles from the Gulf of Mexico. We got this notice from Washington. They wanted us to do a study because they were concerned that our activity might interfere with the Kemp Ridley sea turtle. They were asking us to hire a marine biologist to do a study to make sure there were no sea turtles there.

"Those turtles would have to travel a long ways to cross Highway 77, which could really be dangerous for them . . . . We didn't actually have to do the study but it took a few days to contact this person and telephone him."

Then there was the gauger who found a bird nest on a well head in Redfish Bay.

"He called the Railroad Commission, and they had to hire a licensed nest remover to come out and remove the nest," Davidson said. " I never knew there was such a thing as a licensed nest remover."

Tom Davidson interview

Refinery Row

From the earliest days of the settlement of Corpus Christi, deep water had been the dream, but it was a long time coming. Hall's Bayou was little more than a muddy ditch dividing Corpus Christi from North Beach, which was outside the city limits and removed from the niceties and morals of the city. Later North Beach became a fine residential area before it was wiped out by the Great Storm of 1919.

It was this catastrophe that gave birth to the Port of Corpus Christi, which opened in 1926. Almost immediately warehouses and storage tanks were installed. The discovery of vast area oil and gas fields in the 1930s and the harbor, with its ready access to the Gulf of Mexico and adequate water and natural gas supply, made Corpus Christi a natural location for refineries. Dozens of small plants provided markets. Over time they were sold, merged, and grew along the channel.

In 2004 analysts were pointing out that no new plants had been built in recent years. However, existing plants had grown to the extent they equaled the capacity of several new refineries. Corpus Christi joined the complex on the upper Texas Coast to constitute the major fuel supplier of the United States.

The alkaline unit at the Suntide Refinery in the 1950s. Suntide became a part of Koch Refining in 1981.

The Port of Corpus Christi was already inadequate in the 1930s as plants sprang up along the harbor. The Bascule Bridge became a bottleneck and was replaced in 1959.

(Above) The Benavides Plant of Duval Gasoline Company in October 1938 (Top) Steam belches from an early-day South Texas gas plant.

Port stimulated refinery growth

Early wildcatters had a rough enough time getting equipment over crude dirt roads that more often than not were cattle trails, then cutting through South Texas cactus and mesquite thickets to the well site. If they persisted and made a well, they still had a problem—how to sell the oil. There were few pipelines, and big operators with access to markets sometimes forced them to sell out.

To solve the marketing problem as fields sprouted all over the area, spider webs of pipelines spread throughout the coastal plain, led by Houston Oil Company, and little "teakettle" refineries sprang up in nearly every new oil field. In 1937 there were thirty-seven refineries and nine natural gasoline plants in the area. The largest by far was the Humble Oil and Refining Company plant at Ingleside, with a daily crude processing capacity of 20,000 barrels a day. Smallest was San Antonio at Southton, which handled only 100 barrels a day. Amsco Refining at Corpus Christi refined 10,200 barrels, and Alto Refining Company at Clarkwood processed 250 barrels a day.

In 1938 there were four refineries in San Antonio—Humble, The Texas Company, Phoenix, and Carle. At one time there were refineries at Thrall, Mirando City, Benavides, Edinburg, McAllen, Pettus, Taylor, Harlingen, Sullivan City, and Port Lavaca. Although a few, like Magnolia at Luling, survived, most faded.

It was Corpus Christi—with its new port and access to deep water, ample natural gas and water, and connections with new and prolific oil fields—that offered the most promise to the refiners, and it was there that the industry took firm hold.

In that same year there were eight refineries in Corpus Christi: Amsco Refining Corporation at Port and Summers; Barnsdall Refining on Nueces Bay Boulevard; Corpus Christi Refining Corporation on Poth Lane; Pontiac Refinery on Nueces Bay Boulevard; Taylor Petroleum Company on Poth Lane; Terminal Refining Corporation, one-fourth mile north of Poth Lane; Southwestern Oil and Refining Company, at Port and Summers; and the Texas Company, at the turning basin.[1]

It was Corpus Christi—with its new port and access to deep water, ample natural gas and water, and connections with new and prolific oil fields—that offered the most promise to the refiners, and it was there that the industry took firm hold.

Construction in progress at the Corpus Christi Refining Corporation in 1936. Tankage was being added to serve the refinery and to constitute a terminal for deepwater shipping from a dock to be built near the Southern Alkali Corporation plant, which is visible in the left background.

In 1942 American Smelting and Refining Company and the Sinclair Refining Company were established, as was the Celanese Corporation of America chemical plant at Bishop.

Sixty years later—as a result of a series of mergers and acquisitions—the port area of Corpus Christi was one of the most productive modern refining and chemical complexes in the nation.

One of those twenty-first century refineries, CITGO, could trace its origins to 1920, when the Cities Service Company was founded, and to 1937, when Saul Singer, a banker who used his banking connections to raise funds for the project, established Pontiac Refining Company. Legend has it that he named the refinery in honor of a trusty old Pontiac automobile that had carried the family around the country.

Under the direction of Saul's son, Edwin Singer, the refinery flourished. It merged with Gulf Oil Corporation in 1960, and Singer, widely known as a philanthropist and patron of the arts, became a director of Gulf Oil. In 1967 Celanese Corporation of America, parent company of Champlin Petroleum Company, bought the stock of the former Pontiac Refinery from Gulf. In 1970 Union Pacific Corporation bought Champlin. In 1976 the facility refined 125,000 barrels of crude per day.[2]

Eleven years later Union Pacific formed a partnership with Petroleos de Venezuela, which was created in 1976 when Andres Perez nationalized his country's oil production. In the meantime, in 1965, Cities Service changed its marketing brand name to CITGO, and in 1982 Cities Service merged with Occidental Petroleum. In 1983 Southland Corporation acquired CITGO. In 1986 the Venezuelans bought half interest in a huge Louisiana refinery owned by CITGO. Venezuela bought fifty percent of the Champlin Refinery in 1987 and a year later announced the purchase of the remaining half when Cities Service sold out to the Venezuelan firm.[3] In 2003 the Corpus Christi refinery was actually three plants on two tracts, the East Plant and the West Plant, totaling 564 acres.

Another giant on refinery row in 2003 was Flint Hills Resources, whose antecedents included Suntide Refining Company and Southwestern Oil and Refining Company.

Carl Newlin, who went to work for Suntide before the plant was built and retired after forty years, gave a brief history. Originally the refinery was to be a joint venture between Sunray DX Oil Company and Tidewater, hence the name Suntide. However, before the plans could be finalized, both parties withdrew.

Floyd Martin, executive vice president of Sunray-Midcontinent, liked the idea of building a refinery on the Texas Coast, so he left Sunray, went to New York, and borrowed $17 million to arrange the financing himself. The plant went on stream in 1953. By 1957 Suntide had reached No. 404 on the Fortune 500 industries. In 1968 Suntide became a subsidiary of Sun. The name remained Suntide Refining Company when the company became Sun Petroleum Products Company, but the

In the fall of 1937 crude oil was loaded onto tankers to be shipped from the Port of Corpus Christi to refineries in the East. Before the year 2000 the tankers were bringing foreign oil in to the city's refineries.

company name reverted to Suntide Refining Company shortly before it became a part of Koch Refining in 1981.

Southwestern Oil and Refining Company had been organized in 1936 by S. S. Seltzer, Sr., and S. S. Seltzer, Jr., using some of the equipment from the previous Barnsdall Refinery. In 1950 equipment from AMSCO was incorporated into the plant. Kerr-McGee purchased Southwestern on February 25, 1974, and on August 8, 1995, Koch purchased the Kerr-McGee Southwestern Refinery as its East Plant.[4]

In 2002 Flint Hills Resources was formed as the independent refining arm of Koch Industries, and the two refineries became Flint Hills East and Flint Hills West.

The story of the other major Corpus Christi refiner was also the story of William "Bill" Greehey, who presided at the funeral of one energy company and celebrated the birth of another. Five years later it seemed almost certain that he would conduct another funeral for the newborn.

The baby in question was Valero Energy Corporation, created by the Texas Railroad Commission in 1980 in the order that dissolved Oscar Wyatt's LoVaca Gas Gathering Company. Greehey, who had been working with the nearly bankrupt subsidiary of Wyatt's Coastal, was named Valero CEO.

(Opposite) During Suntide Refinery construction a LUMMUS designed crude unit with vacuum tower is exposed while workmen prepare to give it a dress of aluminum cladding. Vapor lines and condensing lines would come later, in 1953. (Below) An aerial view of the Suntide facility in its early days

A Millenium Moment for Valero

Valero kept its employees happy. On New Year's Eve 1999 there was panic that Y2K—the arrival of the year 2000—would cause drastic damage to computers originally programmed to accept only two-digit dates. Employees were kept on duty watching for computer glitches and other possible problems. Several weeks later every Valero worker on duty that night received a silver clock and a personal letter of thanks from CEO William "Bill" Greehey. It was a small gesture but one that workers remembered.

Fortune Archives, http://www.fortune.com/fortune/2000/05/29/ten.html

After five years it appeared that the fates had conspired against him. Miners shut down the mines in Britain, OPEC cut back exports, and the price of crude skyrocketed. As a new diversification program, Valero had invested heavily in Saber Refinery, a small high tech plant in Corpus Christi. It lost $53 million the first year, and stockholders were getting panicky.

The Saber installation was no ordinary refinery. It boasted all the environmental safeguards mandated by the federal government in the Clean Air Act, and it had room for expansion. Many refineries across the country, unable to afford the expense of modernization to produce clean-burning gasoline, closed. That was in 1985.

In 1997 Valero left its natural-gas roots in a merger with PG&E Corporation and kept its refining and marketing operations as a new independent company with the Valero name.

Then the sun came out. The miners went back to work, OPEC increased the world supply, and the refinery was making money. The dark clouds went away.

Many readers saw in newspapers in May 1996 that Bill Greehey had announced his retirement. He was stepping down June 30. He was going to laze around his ranch and travel some. That didn't last long. Valero needed him as much as he needed Valero. With Greehey at the helm as chairman and CEO, Valero continued to experience remarkable growth.

In 1997 the company left its natural-gas roots in a merger with PG&E Corporation but kept its refining and marketing operations as a new independent company with the Valero name. In 2001 it merged with Ultramar Diamond Shamrock Corporation, making it one of the top three refiners in the nation and No. 50 among Fortune 500 companies.

Valero employees were encouraged to participate in community affairs, and the company did likewise. From No. 93 in 2000, Valero rose to No. 23 in 2005 on Fortune magazine's annual rankings of the nation's "100 Best Companies to Work For."

However, Greehey's Valero was not the first San Antonio-based refinery to make the Fortune 500 list. That honor fell to Tesoro, which was founded in 1964 by Dr. Robert V. West, Jr. West had been chief petroleum engineer and production superintendent of Slick-Moorman Oil Company, president of Slick Secondary Recovery Corporation, vice president of Texstar Corporation, and president of Texstar Petroleum Company, which became Tesoro.

Based on its fiscal 1973 operations, Tesoro was named to the list in June 1974. In 2004 Tesoro's holdings included six refineries with a capacity of 560,000 barrels per day and included retail operations on the West Coast and in Alaska and Hawaii.[6]

Another chapter in the saga of Oscar Wyatt's Coastal Corporation began with the twenty-first century—the $22.6-billion merger of Coastal with El Paso Energy Company. It was a marriage, announced with great fanfare, that was soon in danger of unraveling.

El Paso's average monthly stock price plummeted from a high of $66.04 in December 2000 to $4.80 in February 2003. The world's largest pipeline company had major problems with its trading practices. Wyatt owned 4.6 million shares of El Paso stock. Voicing concern about the value of El Paso stock owned by Coastal employees, he joined others in an attempt to unseat the company leadership, but the vote fell short. With the Coastal States refinery in Corpus Christi under the control of El Paso, which was reeling under a loss of $447 million in 2001, the company sold the 115,000 barrels-per-day facility to Valero.[7]

In 2003 Valero had fourteen refineries in the United States and Canada, with a flow-through capacity of two million barrels per day, including the 115,000-barrel-per-day Corpus Christi Coastal plant,

A tank farm across from the area that became the harbor of the Port of Corpus Christi illustrates the role of refineries in the history of the city.

the former Saber Refinery, also at Corpus Christi, and a Three Rivers refinery acquired in the Diamond Shamrock merger.

Coming full circle, the Corpus Christi Coastal plant had originated as Corpus Christi Refining, the harbor's first tenant, which had been bought by Howell Refinery in 1957. In 1972 Howell became a partner with Quintana, and in 1981 Howell sold the refinery to Quintana. It closed down in 1984 but on August 30 of that year was revived by Coastal States, which had also taken over the old Sinclair Refining Company refinery in 1962. In 2000 the Coastal plant was merged into El Paso Natural Gas, from whom Valero acquired it the next year. At the end of 2003, total production at the two Corpus Christi plants was 340,000 barrels a day.

On March 5, 2004, El Paso sold its 315,000 barrel-per-day refinery in Aruba, also a former Coastal property, to Valero for $465 million.

The Caribbean facility gave Valero a total refining capacity of 2,441,000 barrels of refined products a day, the largest of any independent refining operation in the United States and representing 12 percent of the nation's production.[8]

In April 2005 Valero announced the acquisition of Connecticut-based Premcor, one of the nation's largest refiners, for $8 billion. The deal made Valero the largest refinery in North America, with nineteen refineries and a refining capacity of 3.3 million barrels per day and moved the company to fifteenth on the Fortune 500 list.[5]

And Oscar Wyatt could not be counted out. In a June 13, 2004, interview with the *San Antonio Express-News*, he complained that energy companies were being run by MBAs from Harvard and Yale who believe that "stockholders aren't entitled to profits, only management is entitled to profits," and that they haven't taken genuine risks. With them "it's all hocus-pocus."

In the fall of 2004 he was busy with a new start called NuCoastal and with a group of investors was negotiating to buy the huge Enron natural gas pipeline system for something over two billion dollars.[9]

A Changing World

Young men who came home from World War II and earned degrees in geology and engineering were radically different from the pioneer geologists, who trudged the outdoors, studying rocks, sighting hills and ridges, searching for seepages, sniffing for gasses, and tasting cores on well sites for a sign of oil.

The oldtimers found a lot of oil because it was there. Their successors did better with seismic, and science made them more efficient in their jobs. Computers and other high-tech equipment gave them far more information about underground formations than the pioneers had thought possible.

The progression continued at an ever-accelerating rate until a new breed could go exploring without leaving their offices. They probed deeper and more efficiently with heat-resisting instruments, improved fracing (pronounced "fracking") compounds, and new drilling techniques.

Taking a new look at what their fathers or even grandfathers worked over, they were optimistic about the future. There was no limit to how knowledgeable they would be or how sophisticated their tools would become.

A truck and crane of Bill Findley's Alice FESCO conduct slickline operations.

Ingleside's Kiewit Offshore Service prepares BP's huge Thunder Horse drilling platform for its deepwater location in the Gulf of Mexico.

A Grey Wolf crew zeroes in on high pressure gas at a drilling site near Driscoll.

Future promises challenges, hope

In the spring of 2003, Bill Carl said, "The oil business isn't oil anymore."

This statement was highly significant, in more than one sense. At that point Carl was speaking of the financial structure of the energy companies.

"What's happening," he said, "... is that ... the guys that are running many of these companies are financial types. They end up as CEO of a company they just inherited ... that was doing exploration, and they alter the program. It's business. It changes the atmosphere. It used to be our people were spending their own money and raising money from individuals. They were entrepreneurs, and their word was their bond.

"Enron is an example of arrogance and greed," he said, speaking of the Houston energy giant that crashed in a wave of scandal in December 2001. "The oil business was once really a fraternity and a wonderful way of life, and it's changed terribly. They are all in the stock market, and they have all suffered. The oil and gas business has the worst PR in the country. They spend a lot of time and money on it, but they can never seem to get a handle on it.

"There has never been a merger philosophically between the majors and the independents. The majors make their money from chemicals, transportation, and selling gasoline. Independents make their money selling crude oil and natural gas. They are not really in the same business."

However, there was an additional truth in Carl's statement that the oil business wasn't oil. By 2003 it was becoming clear that the future of the South Texas petroleum industry was not so much in oil as in natural gas.

Ten years earlier Carl had said it was not cost efficient to look for natural gas. "There is an oil/gas BTU equivalency of six to one—600 MCF of gas to a barrel of oil ...," he said in 1993. "If oil is selling for twenty dollars, gas should be selling for $3.16. ... And it is not. It is selling for more like $2.16 right now. ... The drilling for natural gas—and that is my end of the business—is absolutely minimal. But I do

By 2003 it was becoming clear that the future of the South Texas petroleum industry was not so much in oil as in natural gas.

'There is plenty of gas in the world, but the cheap gas is gone.'
—Dan A. Hughes

think that natural gas is the fuel of the future, and it is the future of the independent. Just geologically if you drill on the Gulf Coast . . . and you find new reserves, I would guess eighty percent of them would be gas as opposed to oil. We're not drilling for it right now. But I think it has got to happen. . . ."

Events in the next decade proved Carl right. The rising demand for and price of natural gas were serving as incentives to exploration and deep drilling throughout South Texas and in the deep waters of the Gulf of Mexico.

By 2002 Beeville's Dan Hughes felt that gas wells were being overproduced to meet the higher demands, using up two-thirds of the reserves in one year.

"U. S. reserves and those in the Gulf will not last at the present rate of production, and prices will continue to rise," Hughes said. "There is plenty of gas in the world, but the cheap gas is gone. Liquefying gas in other countries and bringing it to this country takes a lot of money to do, and it will take several years to set up. A million cubic feet of natural gas isn't very much in a gas well.

"If you liquefy a million cubic feet of gas," Hughes said, "you will get 10,000 gallons of liquid. If you liquefy a lot of gas, you're going to have a lot of ships to carry it. . . .It must be kept cold. They are importing some gas this way now. The biggest reserves are in Northern Siberia. We are talking about hundreds of trillions cubic feet of gas. Next is Iran, then South America and offshore from Trinidad. There are also great reserves in the Gulf of Mexico."

Carl agreed. "There is a big, big discovery out in the Gulf of Mexico, 9000 feet of water," he said. "The wells are drilled from the surface of the water, 29,000 feet. The field is called Crazy Horse."

Carl said he was told that the gas would be piped to the surface to floating barges and liquefied.[1]

In 2003 Larry McNeil pointed out that natural gas was powering about sixty percent of the country's electricity and that eight dollars an MCF would be a minimum natural gas price.

"That means that electric bills are going to be high," he said. "I think gas is going to get so important—so scarce—that the federal government is eventually going to step in and do something to regulate it. Our problem is that the people of the United States don't really realize what our energy problem is. They don't understand why these prices are so high."

Carl agreed that the price of natural gas was high, but "money you pay for gas is not all going back to the people who take it from the ground. It's going to traders and pipelines and things in between that are raking in all the money," he said.

"I have to laugh when they say we are going to war over oil," Carl said, as war with Iraq loomed in the spring of 2003. "We can always buy the oil. It's not going any place and they have to sell it to us. But gas is another matter. There is plenty of gas around the world, but it is going to be expensive getting it here.

"Some people believe we can use hydrogen cells," he said. "They think you can pull the hydrogen from the air and put it in a cell and run your car. You make hydrogen from water and electricity. And you make the electricity by firing natural gas. There's that problem again."

Randy Bissell, a young Corpus Christi geophysicist, saw promise in the future of the area natural gas industry but said it would come at a cost. He predicted that a cartel of natural gas-producing nations that controlled the supply would form something akin to OPEC and, from their own perspective, decide what the right price of natural gas ought to be.

In a 2004 interview Bissell, senior geoscientist for a successful independent oil company involved in drilling operations on the Kenedy Ranch, noted that half as many rigs were operating in the United States in 2004 as compared to 1985.

"We are seeing over and over again that there is a reduced opportunity in the sheer numbers of prospects that are being generated," he said. He predicted that there would be a depletion of natural gas in the United States and that the challenge would come from liquified natural gas (LNG) .

Bissell, who graduated from the University of Southern Mississippi and Oklahoma State University, credited excellent training programs offered by Exxon with making him an expert in several specialties.

"I am an earth specialist," he said. "I would recommend young geologists to spend some time with a major integrated company. There are many experiences you do not get with an independent."

He had worked for Exxon Corporation in drilling off the coast of Africa, where wells with tremendous potential in both gas and oil were discovered. The oil couldn't be produced because there was no way to dispose of the tremendous amount of gas.

"There are hundreds of trillions of cubic feet of gas in Africa and Southeast Asia—enough to sustain the world supply for decades," he said, pointing out that gas plants were being built to liquefy those reserves and that all sorts of investments were being made in technologies related to the tankers, liquefaction, storage, and other aspects.

"Once the infrastructure is built and paid off, the price will come down a bit," he said. "The price is tremendously sensitive. That's one of the dangers of LNG, if you look at what OPEC countries have done with oil. They will periodically flood the market with oil in order to reduce activity levels around the world so they can raise the price and not have any competition. I think the United States is in danger of being in the same situation with natural gas."[2]

By 2004 the importation of liquefied natural gas had achieved its predicted importance, as both demand and price continued to rise

'Some people think you can pull the hydrogen from the air and put it in a cell and run your car. You make hydrogen from water and electricity. And you make the electricity by firing natural gas. There's that problem again.'

—Bill Carl

LNG companies were interested in the Corpus Christi area because it had a deepwater channel, significant local gas demands, nearby pipelines for both local and national delivery, and large tracts for plant sites.

to new heights. As Bissell predicted, foreign natural gas producers were poised to make an invasion similar to the one of the latter years of the twentieth century, when foreign oil had flooded the country as domestic production waned. That combination of price and demand made it feasible to construct large plants near vast overseas natural gas fields and to acquire expensive fleets of liquefied natural gas tankers. LNG people said the price of the gas would eventually drop from highs of more than six dollars a thousand cubic feet. However, they said, it would not drop too much because of the extra expense of processing and transmission.

George Tanner, president of Mestena, Inc., and Mestena Operating Company, agreed. "I think LNG will serve to keep the production and prices level," he said.

However, he believed that LNG importation would not have much effect on the price of natural gas because of the situation in Texas and the Gulf of Mexico. "The Gulf of Mexico depleted 37 percent in one year, and the State of Texas depleted 28 percent in one year," he said.

Even so, he looked to the future with optimism. The reserves in Brooks and Jim Hogg counties amounted to millions of barrels of oil and billions of cubic feet of gas in four geological trends.

"Every year when we have our reserve studies done, they just keep extending life to the fields. I would say that the average life in the reserve is about eighteen years," Tanner said.

Tanner said the future was in the "deep stuff." He had identified "at least seven Wilcox structures which are extremely large and unfortunately extremely deep." He was using 3-D to get a clear definition of the seven structures so he could put together a group to drill it. "The lease covers a lot of acres," he said.

Chester Wheless was also optimistic about the area's future. "I think there's a lot more gas to be found in South Texas," he said.

In the summer of 2004, seven LNG receiving terminals were planned along the Texas Coast from Sabine Pass to Brownsville, three of them on the northern shores of Corpus Christi Bay.

William D. "Bill" English, vice president of development for Cheniere LNG, Inc., which was locating a plant adjoining the Reynolds Sherwin plant at Ingleside, said, "Either demand is going to get crunched down by industries closing up, shutting down shop, and moving away, or we're going to find new gas supplies. There aren't any other solutions."

He cited the abundance of natural gas around the world—in the Middle East, Africa, the Caribbean, Venezuela, and other areas. In the summer of 2004 there were 125 existing LNG liquefaction plants in twelve countries, forty-two LNG receiving terminals in eleven countries, and 113 existing LNG storage tanks in the United States.

LNG tankers had a capacity of three billion cubic feet, "about the capacity of one of our three tanks," English said.

He said the Cheniere plant would process 2.5 billion cubic feet of gas a day, an Oxidental Petroleum plant, 1.1 billion BCF, and an Exxon-Mobil, 2 BCF a day. Cheniere Energy planned additional plants at Brownsville and Freeport. Exxon-Mobil also planned facilities at Sabine Pass and Sempra Energy one at Port Arthur.

Contrary to popular belief, English said, LNG is not stored under pressure. It is liquefied by lowering the temperature to minus 259 degrees Fahrenheit. The vapor-to-liquid ratio is 600-1.

"We are an open access firm," English said. "We build the terminal and make it available to people around the world. We just make it available. We store the gas, warm it up, and send it on its way in whichever pipelines they want us to put it into.

"People have objected, saying, 'What if terrorists came and blew it up or flew an airplane into one of the ships or tanks and there would be a great explosion and blow up the town?'"

That couldn't happen, he said. "If you took a match to it, it would not explode. You could have a fire, but not an explosion. Nothing dramatic. That is true of the ships. Now, gasoline would blow up. In the worst case scenario, a tank is ruptured and spilled, and then catches fire. Then you would have a big fire. The tank is 1,600 feet from the property line. It has never happened in twenty years."

English explained why the LNG companies were interested in the Corpus Christi area. It had a deepwater channel, significant local gas demands, nearby pipelines for both local and national delivery, and large

Artist's rendition of the proposed Cheniere LNG plant at Ingleside

An International Services, Inc., crew mixes cement.

tracts for plant sites. The companies were also looking "for a community that is familiar with oil and gas industry development," English said.[3]

In keeping with his description, the city was ready to continue the role it had developed decades earlier as a center for industrial development. Although the pace of technological advances accelerated exponentially during the 1990s and beyond, some of them had had their beginnings in those early years in the city and surrounding areas of South Texas.

As production began to taper off after the heavy demands of World War II, the overworked wells required stimulation. One company, Texas Acidizers, began acidizing oil and gas wells producing from sand formations. A new fracing procedure had entered the picture as early as 1963, the same year the *Oil & Gas Journal* listed the Luling Oil Field as one of the "giant" fields of the nation.

Then came the drastic slowdown of the 1980s. Weldon Oliver, president of International Services, Inc., of Luling, said most of the "stimulant" firms went out of business—only the big survived the selloffs and mergers. At one time there were a number of major service companies. Then there were three majors and nearly all of the small independents. By 2003 there were hardly any of them left.

"I guess I am one of the best equipped small cementers in South Texas," Oliver said. He picked up a core at the Luling Area Oil Museum to demonstrate

A pumping crew from Weldon Oliver's International Services, Inc., of Luling. The firm used the latest fracing methods to increase production.

fracturing in a well. "This is Austin Chalk, which runs throughout Texas. You can see this line on the map. It runs all the way to Laredo.

"On some of it you can get away with using a little acid, but most of it is such a hard, dense formation that it has to be fractured. What we do is with those big high-pressure, high-horsepower pumps, we pump down the casing and actually spread the formation apart, fracturing it. Then we start mixing sand with fluid and pumping it in there to prop it open. That makes the well produce better," he said.

He demonstrated how the untreated chalk was almost inpenetrable before treatment. After a frac job, oil readily seeped through it.

"Four years ago they would come in flowing 600 to 800 barrels a day. Now the same wells or an offset well will probably yield 300 a day. In say sixty days it will quit flowing, and they will have to put it on a pump. Still, for a 3000-4000-foot well, 200 to 300 barrels a day is not bad at today's prices. . . ."

Another description of fracing came from W. E. "Bill" Findley of Alice, owner of a diversified oil field services company.

"Fracing means that you inject in a gel which is a heavy substance like jelly," Findley said.

"It will flow. If you put this gel in a hopper with a special sand, that solution is pumped into the well through perforations. The bottom hole pressure is, say, 5,000 pounds. We might frac it up to 10,000 pounds and the sand opens the production sand and the flow is increased."

Then the sand that is not held by the formation is removed so it will not cut pipe and regulators.

"Flowback, or cleaning the sand, is a large portion of what we do," Findley said.[4]

Findley, who had a petroleum engineering degree from the University of Texas, had had no idea where Alice, Texas, was until an Oklahoma company sent him there as a sales representative selling oilfield equipment in 1947.

Planning a career in sales engineering, he went to work for BS&B Corporation, an equipment manufacturer. It didn't take long for him to be the biggest thing to happen to the small Texas city.

"After two years I noticed there was a need for a gas-well production-testing company, so I started my company in December of '49," he said. "I started my company with one truck, and the first two years of starvation started."

He and Garman Kimmell of Oklahoma City designed the truck and built a portable laboratory—a split-stream lab truck—"designed to determine optimum separation conditions for gas-condensate wells at the well site."

He named his company Findley Engineering Service Company, which became FESCO. With Findley still at the helm, FESCO grew into a company with more than 450 employees, 350 vehicles of all kinds, and 12 district offices, with headquarters in Alice.

"The first year in business I did $8,500 worth of work, and the bank note was $5,000. I starved to death for a year and a half. The

second year was $24,000 and the third year was $48,000. Then I hired some help," he said.

In 2004 the firm had four laboratories—one in Alice, one in Ozona, one in Bryan, and one in Lafayette, Louisiana—and was doing about $50 million in business a year.

Findley recalled how markets for natural gas really developed when Tennessee Gas opened its line to the East in 1945. "They were paying fifteen cents a thousand, which was great. Before that it was two or three cents," he said. "Now the price is between five and six dollars. So that meant a lot of those shut-in gas wells had an outlet. As they tied in, our volume increased for testing those wells.

"As you know," he said, "South Texas—from four or five counties around Laredo—produces probably over 60 percent of all the gas in Texas. So we are in gas country."

(Above) W. E. 'Bill' Findley, Jr., waves from the van of the company he started in 1949. (Left) The Alice headquarters of FESCO, Findley's diversified oilfield services company, in 2004

'Alice is doing well in oil well services. As a result the average per capita income is greater than that of Corpus Christi, Kingsville, or other surrounding towns. We are the largest oil-well service center south of Houston.'

—W. E. 'Bill' Findley

FESCO primarily tested production of oil and gas wells with portable separator heaters when a well was completed.

"We test it for oil, gas, and water. We get samples of each and bring them into our laboratory for testing. We do not do electric logging or cementing or perforating, but we do about everything else—wireline work, down-hole work, high-pressure lubricating, bottom-hole pressure measurements, and flowbacks."

He explained that flowback means the well might produce ten million cubic feet of gas a day if it is a good well. If they want more, they frac it and increase the flow, sometimes doubling it.

Some had said that excessive production in a gas well damaged the gas-sand formations.

"In the old days you would seldom 'pull' a well over one, two or three million cubic feet because the sand thickness of the formation might be 10, 12, or 20 feet," Findley said. "Below most gas strata is a water strata. If you pull the well too hard, you'll bring the water up. And sometimes it's 100 percent water, and you've killed the reservoir.

"Nowadays, the thickness of [pay sands is] 80 to 100 feet.... For some of them you don't have the problem of pulling water up if you pull the well hard. So they do it, and these wells will flow at 200 degrees—hot, hot, hot, besides having high pressure. The highest we've encountered down in this area is 12 to 13,000 pounds. Very high. We have better technology for the deep wells for both heat and pressure. So today, business is deeper, but our production is tremendous, with the increase in prices, so the major companies are doing well," he said.

The deepest well his company had worked on was 23,000 feet.

Alice didn't have much production within a 10-to-30-mile radius. "We had ten or twelve major companies operating in Alice [that] moved out years ago," he said.

In the slump of the 1980s, his company revenue was cut in half. "It was sad," he said. But in 2004, "Alice is doing well in oil-well services. As a result the average per capita income is greater than that of Corpus Christi, Kingsville, or other surrounding towns. We are the largest oil-well service center south of Houston," he said, noting Alice's position as the "Hub City."[5]

Some second- and third-generation oil-family members also took advantage of the new technology.

Sitting in his office overlooking the Corpus Christi Bayfront in the spring of 2004, Floyd Nix was a personification of the history of the South Texas oil industry. Some oilmen could remember the past as pioneers. Others worked in the present with a firm eye on the future. Floyd Nix could do both.

Early in his career Nix was roughneck, derrick man, and driller on some of the first Humble Oil and Refining Company rigs drilling on the King Ranch.

"I got out of the service [after World War II] and wanted to be a natural gas engineer," he said. But he had a family and had to drop out of Texas A&I College "because we couldn't live on $75 a month the government was paying."

He was looking for steady work. "I was like an old dog that found a bone when I went to work for Humble. They were good to work for," he said.

He knew what he was getting into because his family history was closely aligned with Humble.

"My dad went to work for Humble in 1927," he said. "That was just ten years after Humble was organized in 1917. His first job was on a wooden rig six miles south of Kingsville."

The well came in with a potential of 80 million cubic feet of gas in the Kingsville Field. It was a significant discovery because three wells half a mile to the north were completed as oil discoveries from the same sand at 2,250 feet. News of this activity drew attention in other parts of the state and caused exploration to begin in earnest along the Texas Gulf Coast.

"My dad was a roustabout, roughneck, and sometimes a driller, but he actually preferred the job of firing the boilers," he said. "We had a farm about twelve miles southwest of Kingsville next to the King Ranch. When we went to town, we carried a five-gallon can and bought

Floyd Nix's father had worked for Humble for ten years when this picture was taken in 1937. When Floyd got his Humble job, he said, he 'was like an old dog that found a bone.'

As drilling superintendent for an independent, Floyd Nix monitors his wells by computer from an office above Corpus Christi's Shoreline Drive. In a career that spanned more than fifty years, his working conditions had seen radical changes.

kerosene," he said. "We used it for cooking and heating. For some reason we always used a potato to cap off the spout.

"When I was a kid and heard of a blowout, he'd say, 'There's a well on fire. Let's go look at it.' We'd get in the old Model A and take off," Nix said.

In later years the elder Nix was given lighter work as a pumper and gauger until his retirement in 1967. He died in 1970.

Of his own start Floyd said, "I lived in the [Humble] Stratton Camp for quite a while. Your utilities were $3 a month for gas, water, and lights. Course, we weren't making but about $1.25 an hour, but we got overtime and riding time. They kept steady crews. You worked for Humble, you're going to be there the rest of your life. The crews stayed together as a team. We drilled wells on the Kenedy Ranch, the East Ranch, the King Ranch. We had leases on all of them."

He reflected on the changes he had seen. "When I started, we had standard derricks. Rig builders bolted steel pieces together. They'd build two at a time so that when you finished one well, they would have the derrick ready for another one.

"There is still a lot of work around the wells, but machinery makes it a lot easier," he said. "You always use a shovel around a drilling rig but not as much as you used to. The driller today sits in a covered area with all the controls, and he can read [from a computer] everything that's going on right there in front of him. He controls all that. You've still got to have people. If he sees something wrong, they have speakers around the rig so they can keep in contact with everybody."

He retired as drilling superintendent for Exxon in 1986 after forty-one years of service. He had good retirement pay, medical and dental insurance, and other benefits, but "after six years of watching television and mowing the yard, I jumped at the chance when they offered me a job. That was in 1992."

He worked on a contract basis as drilling superintendent for an independent operator.

"They don't even take anything out of my check," he smiled. "I even pay my own taxes."

Though Nix was old enough to talk about the old days, he was in every way active in 2004. He scanned a computer screen showing everything happening on a drilling rig a hundred miles away—depths, mud weight, pressures—everything he would need if he were on the derrick floor. As he talked, his cell phone rang, and he reported that a slight misalignment of the casing on a well had been corrected.

He talked of other developments, including the coming of liquefied petroleum.

"On the King Ranch when we hit gas, we just capped them over. There was no demand for gas then," he said. "Now they're talking about tankers carrying five billion cubic feet of liquefied natural gas—more than a good well will produce in its lifetime."

He noted other developments, including the use of casing as drill pipe. Conoco had been using the procedure for some time in the Laredo area. "We haven't reached a point to where it would be advantageous to us to do it," he said.

At seventy-six he was as much at home with the new technologies in a high rise office as he was when he was running a rig out in the brushlands.[6]

Some did not welcome the new technology with such enthusiasm. Jon Spradley said, "You talk computers. I'm a dinosaur. All I know is how to turn one on. I can recover my e-mail, and that's about it."

Even so, Spradley pointed out the importance of 3-D technology. "As far as independents go, unless you have it, you don't have anything to sell anymore," he said. "I would go to Houston with a deal, and if I don't have 3-D, there's a big bell that goes off at Sugar Land to alert somebody downtown that this guy doesn't have 3-D. I just turn around and come back."

Spradley said that he was putting deals together with Richard Phillips and a young man named Arthur Porter. Porter was their lead man. "We try to come up with subsurface ideas and sell it or acquire some

Jon Spradley

'As far as independents go, unless you have 3-D, you don't have anything to sell anymore. I would go to Houston with a deal, and if I don't have 3-D, there's a big bell that goes off at Sugar Land to alert somebody downtown that this guy doesn't have 3-D.'

—Jon Spradley

Super carries the day

Ted A. Snelling Jr., drilling superintendent for Grey Wolf Drilling Company, had seen a few blowouts in his time.

"We had this blowout in the Huisache Field up near Goliad," he said.

"We were on a Flournoy Drilling Company rig that caught fire when a pipe ruptured. We could see it was going to go. Me and the tool pusher, Darryl DeForrest, turned to run. Darryl went and got in his truck."

Snelling jumped into his truck, and the two pulled away, trying to get off the location.

"The fireball overtook us and sucked all the oxygen out of the air. It killed the engines," Snelling said. "After the flames left, I opened the door and started running. I had run about a hundred yards when somebody said, 'If you would just slow down and put me down, I believe I can probably walk now.'

"It was Darryl. I didn't even know I had picked him up."

Telephone interview with Ted Snelling, November 3, 2004

seismic. . . .This 3-D seismic that they use now has had monumental successes and some monumental failures," he said.

Spradley explained 3-D as "a 360-degree picture the way they lay the lines out. So your prospect at 15,000 feet, you have 30,000 feet of coverage. What they do is set off little earthquakes and take the reflections back up and measure them. Then they reduce it to a CD-rom, and you've got a base map that shows where the different lines are. You can move that mouse around, and you can take a slice of the earth's crust, where it shows a cross section and shows you the faulting. They have it now to where they have 4-D," which shows the changes over time.

"They have all sorts of tweaking ability now on all that data."

Spradley said that independents were hurt when the major companies were the only ones that could afford 3-D. "They would go in and take the option on acreage, and then they'd do their shooting, which was about a year turnaround. Then they'd make their leases. What it did for two or three years, it would take that big block of acreage out of play. . . . So it was not a boon to us."

He expressed concern over the change in prices. "We used to drill a well turnkey—a 6,000-foot test operator's turnkey figure would be $3.50 a foot. . . .When I was working for Jake L. Hamon, we used to figure that a well that didn't require protection pipe, we could drill it for $6 a foot with logs and everything. . . .If you had an 8,000-foot test, you were going to spend $50,000 on it. If we had to set protection pipe, we used a number of $12 a foot, which doubled the cost. If you drilled it in the water without pipe, it was going to be $25 a foot; and if it was with pipe, it was about $45 a foot, 'cause you had to haul the pipe and all that stuff out there. . . .

"At today's prices, I think probably $25 a foot is not unusual. So a 10,000-foot well without protection pipe is going to cost you half a million with the seismic that goes along with it and everything. You're going to have close to a million dollars easy, plus lease costs, etc. You have to pay $25,000 for a tap into a pipeline.

"In the old days we used to say it wasn't much of a deal unless you could make 25 to 1 on your money. That was back when gas was $2.50 and oil was $5 a barrel. Jake Hamon said if you didn't get 15 to 1 on your money, the odds of your making a well vs drilling a dry hole, you only make one out of nine and one out of five would get your money back.

"With 3-D and the other science it isn't that risky anymore. Still, it is much more expensive to drill a well. Arthur and I have a little ol' deal over in San Pat County north of Gregory. It's going to be a 9500-foot test and I think it's half a million bucks. We've got partners.

"Back in the old days, we used to get a 32nd override and a ham sandwich."[7]

Brian O'Brien not only emphasized the importance of 3-D but also evoked memories of pioneer drillers in discussing his Maverick County discovery.

In addition to oil, the discovery well also produced lots of fresh water. "When I drill for oil and hit a lot of good fresh water, I have made producing wells, filtered it, and turned it over to the landowner," he said. "You can't afford to drill 7,500 feet for water."

O'Brien had a ranch in Maverick County and noted that it was 2004 before two lakes in the Rio Grande were getting back to the conservation level of 1991.

"There will come a time when water might become more valuable than oil," he said.[16]

His statement seemed prescient, as the Brazos River Authority, based in Waco, was negotiating in 2004 with T. Boone Pickens's Mesa Water, Inc., which controlled the rights to billions of gallons of groundwater in the Texas Panhandle.[8]

When it came to the philosophy of conducting the business, the Killams of Laredo had possibly the best of both worlds. While Radcliffe Killam said in 2001 that he tended to run his company conservatively, his son David seemed to have inherited some of the risk-taking genes of his grandfather, O. W. Killam.

Radcliffe had learned the business working for his father in the refinery and as a roughneck. "He paid me the same salary everyone else was getting, not a bit more," he said. Like his father, he got a degree in law but never practiced. His degree was from Harvard.[18]

David also grew up working in the oil field. He went to college, he said, not believing he would be in the business because "they kept telling me how bad it was. Later in college I changed my mind and fell in love with it. I was always drawn to geology. It's exciting to know that you can do something nobody else has ever done. It's an immediate business. You know right away if you've been successful or not when you drill a well. It's interesting because of the history of the area."

He exhibited a willingness to turn from strictly shallow drilling to explore new deep horizons and seek natural gas.

"I don't look at anything shallow. I feel it's already been done," he said. "We're looking at things nobody has had a chance to look at. We do that with 3-D and other means. There's still romance here."

David defended his grandfather's use of doodlebugs, saying he used them as sort of a sixth sense and as an incentive to go out and drill a well. Having a grandfather as a legend "sort of puts it on my shoulder. I still work with a lot of things he accumulated. It's an exciting viewpoint of the tradition in it.

"My father says it was more fun in the early days. I think it's still a lot of fun. My grandfather could go out and get a lease that covered 15

A prediction from the past

The importance of natural gas in the 2000s came as no surprise. It was 1942 when Railroad Commission Chairman Ernest O. Thompson predicted a rosy future for natural gas in Texas.

"It is well within the realm of possibility that gas may bring more income to Texas than oil in the very near future," he told The Associated Press.

"Technology is developing so many uses for gas that it is going to be more valuable," he said. He pointed out that once gasoline was poured into streams because refiners considered it a useless byproduct in their desire to obtain kerosene. Natural gas was undergoing a similar transition.

"Gas flares are becoming scarcer in all fields . . . and it may not be long before it is not permitted to go into the air at all," he predicted.

Construction of the interstate pipelines that opened huge markets for natural gas soon proved him right.

"Natural Gas May Be MoreValuable Than Texas' Oil," *Corpus Christi Caller-Times*, Januaray 21, 1942.

to 20,000 acres. Now you have to work like crazy for a year to pick up a lease of 640 acres."[9]

More than thirty years after he returned to San Antonio as a consulting geologist, Wilford Stapp found reworking records rewarding. In an October 2002 interview he said, "I'm working on one 11,000 to 14,000-foot well where salt water is highly charged with gas. There's been a study of salt water saturated with gas at high pressure. It is passed through a separator, and you recover the gas and put the salt water back in the shallow sand. That's what they are about to do. The geo pressure is about twice as much pressure as you would normally find.

"There's another project I'm working on down in South Texas—17,000-foot abandoned wells where the pressure is so great it is dangerous to pull the casing out," he said. "So they walk off and leave everything except the tubing.

"I think you can go back in there and perforate those saltwater sands. The pressures are about 10,000 to 12,000 pounds and the temperature is typically 400 degrees.

"You aren't getting any water out of there. You're getting steam. You can pass that through turbines and create electricity. You recover the saturated gas, which is quite a lot. The water is not all that salty. It can be used for irrigation."[10]

Some of the veteran geologists did not remain active in their later years, but they all applauded the technological advances. In 1998 O. G. McClain confessed that he was not familiar with modern techniques. He said, "I can understand them when they talk geology, but when they start talking geophysics, they lose me. It's marvelous what can be done. I'm devoted to it."

Richard Wilshusen agreed. "There has been a lot of progress in just the past year," he said in 2004.[11]

Among the experts in the changing technology the veterans cited was Jake Venable. You could say that Venable had a hard time playing it straight or that he would give a slanted view of the oil business. You'd be right, because of his expertise in directional drilling for companies that dealt with the very deep, very hot, and very high-pressure wells being drilled in South Texas in 2004.

Although Venable was not a college man with a degree in geology or engineering, he supplied the tools and worked with some of the most successful hi-tech operations in the region. He was on the cutting edge of a new wave of drilling instrumentation that led exploration to new depths in new projects and screened through old fields for new or overlooked formations. His specialty removed the guesswork from drilling, taking the driller to a remarkably accurate destination thousands of feet below, even though the target may have been hundreds of feet from the drill site.

In the 1990s directional drilling companies began to utilize multilateral drilling, which involved drilling from a single surface well bore two or more laterals (well bores) that extended at different depths and/or in different directions into the reservoir or reservoirs.

Jake got his start in the industry as a high school student working for his uncle, Jack Leonard Shelton, who invented the Key Seat wiper, which cleans ledges or bridges that occur during drilling in the well bore.

"He was a fishing-tool operator with Homco when he started," Jake said. "I would come down and sit on the rigs in the summer and help him. He used to get a big thrill out of sending me up on the floor and telling those old drillers what to do and watching their reaction. I was about seventeen."

He returned to school in Abilene. After he graduated, his uncle, who by that time had the directional drilling company Shelton Oil Service, asked for his help.

"The work was around Carancahua Bay," Jake said. "I was watching two rigs and my brother was watching two, and we split one. I was doing the directional work, but if they were short, I would roughneck. I actually broke out in directional in 1958."

Jake had his own company, Venable Directional Drilling Consultants, Inc.,(VDDC), for 13 1/2 years.

Directional drilling had been used since earlier days, for such purposes as relieving blowouts or reaching locations that were otherwise inaccessible, but like other aspects of oil-patch technology, it had undergone radical advances. In the 1990s directional drilling companies began to utilize multilateral drilling, which involved drilling from a single surface well bore two or more laterals (well bores) that extended at different depths and/or in different directions into the reservoir or reservoirs.

"I sold a bunch of horizontal jobs when it first started in the nineties," Venable said. "I was using Smith International Tools, but they didn't have the hands to do horizontal. I said I had never done one, but Ben Soileau with Smith said there's nothing to it. So I went out on the job."

After successfully completing that job, Venable became an expert horizontal hand. He had his own company, then hired out to Smith. After Halliburton bought out Smith, he worked for Halliburton until he took early retirement to help his brother, Marshall Venable, who had a stroke. At that time Marshall was working on several directional wells in Zapata. After about two years of this work, Jake went back to Halliburton as a directional hand until he himself had a five-way heart bypass. After recovery he came back as a salesman and later directional coordinator with Strata Directional Technologies for about three years. He then joined Multi-Shot as the regional drilling manager.

"I was happy there, but PathFinder made me an offer they couldn't match," he said. After working for Pathfinder for more than a year, he decided to go back to work for Strata. As South Texas regional manager he was able to closely monitor the directional work done by his hands and three sons, James, Justin, and Jason Venable.

Jake Venable points to a bottom landed joint stinger (mule shoe) for an MWD probe used in guidance while drilling a directional or straight hole.

Directional drilling with MWD (measurement while drilling) and LWD (logging while drilling) and multishot capabilities was on the cutting edge of technology in the early 2000s. However, directional drilling had been used since much earlier days. (Right) In March 1937 workers were setting up this rig to drill a directional well to kill a blowout north of Texas Highway 44. (Below right) A bent housing motor used for orientation for twenty-first century directional wells. All this equipment aided in finding a particular productive area.

He didn't go out in the field any more "unless they have trouble I can't handle on the phone." He described one major change for the personnel in the field.

"When I first started, you took a case of whiskey in your trunk. At the location you gave the tool pusher a bottle and one to each one of the drillers," he said. "Now you can't even have it in your car, Or even beer cans in your car."

And drilling crews returning from a week's respite had to be aware that drug testing was a possibility at any time.

Venable described elements in "MWD" and "LWD"—measurement while drilling and logging while drilling. He referred to PZIG—pay zone inclination gamma, PZS—pay zone steering, magnetic survey systems, gyro systems, directional motors, and other advanced technology.

"The geologist will say, 'This is the sweet point right here. I want you to land within a two-foot square of that sweet point,'" he said. "In the past we've been anywhere from 57 to 65 feet above the bit with our MWD information. PZS using LWD attaches directly above the drill bit to provide real-time inclination and gamma ray measurements."

The system, extremely accurate, removes the guesswork. "It eliminates a whole bunch of dry holes," he said.

Many of the directional wells extend 2,500 feet or more from the surface drilling site.

"With on-site satellite information, the geologist can sit at his desk in his office or at home and watch the drilling on his computer," he said. "This satellite transmission is available with LWD or MWD services."

Scientific progress was rapid in 2004, and he expected the improvements for directional to continue. As he displayed racks of directional tools with various motors, he said testing equipment for servicing the deep, hot wells, which had been able to withstand only 350 degrees, was expected to survive 450 degrees or more.[12]

Directional drilling, with the ability to obtain production from a large area with a single surface well bore, also offered the possibility of allaying some environmental concerns. Venable drilled in the Aransas Wildlife Refuge, with little or no impact to the environment. Bill Miller also drilled at the refuge, where preservation efforts succeeded in increasing the number of near-extinct whooping cranes from 15 or so to around 200.

MWD—measurement while drilling—technology enabled crew members to monitor progress by computer and satellite anywhere from an air-conditioned portable building at well site to their home or office.

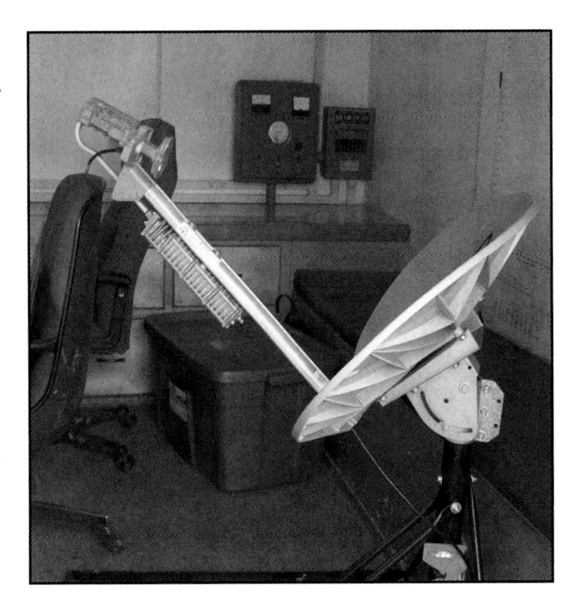

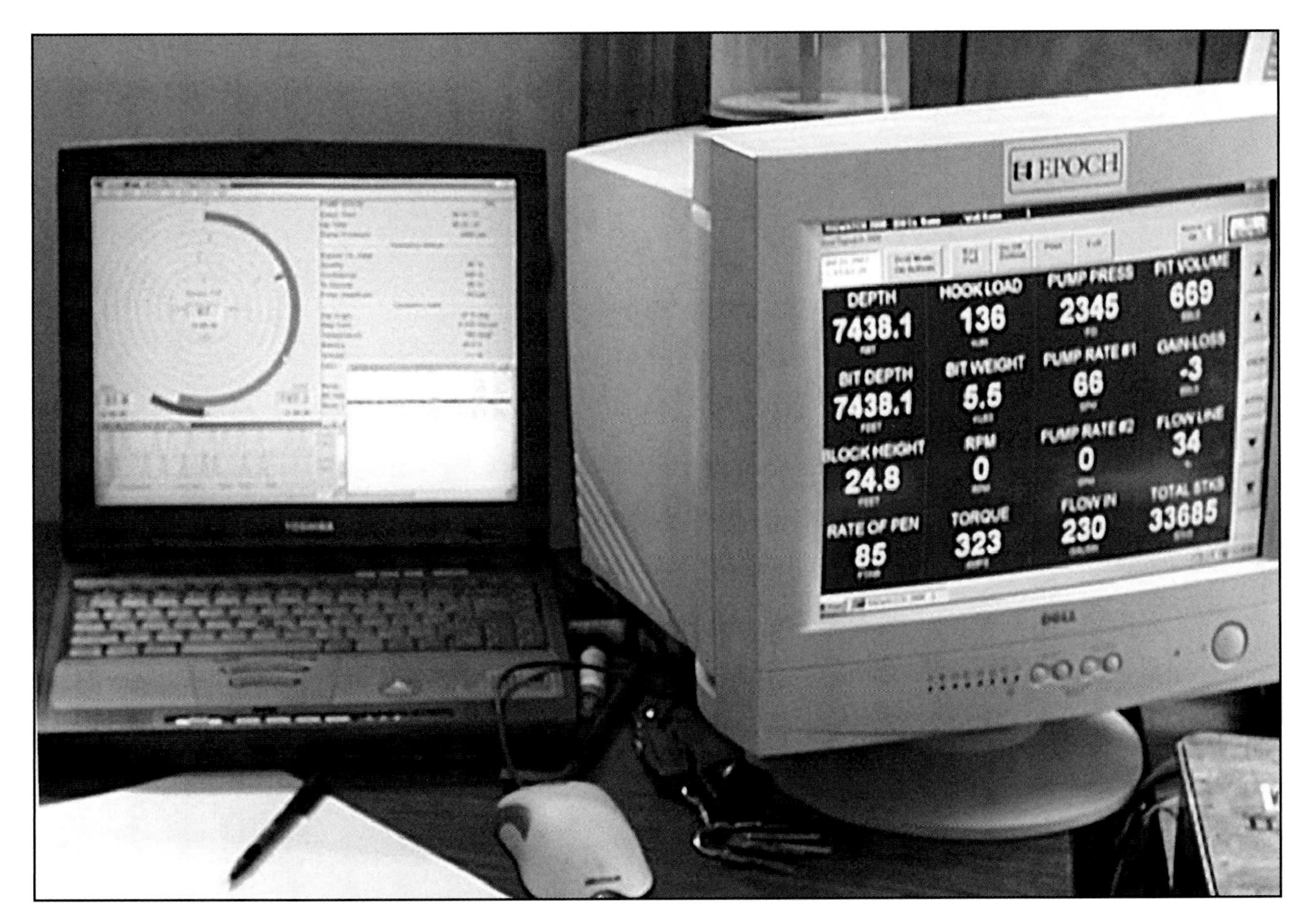

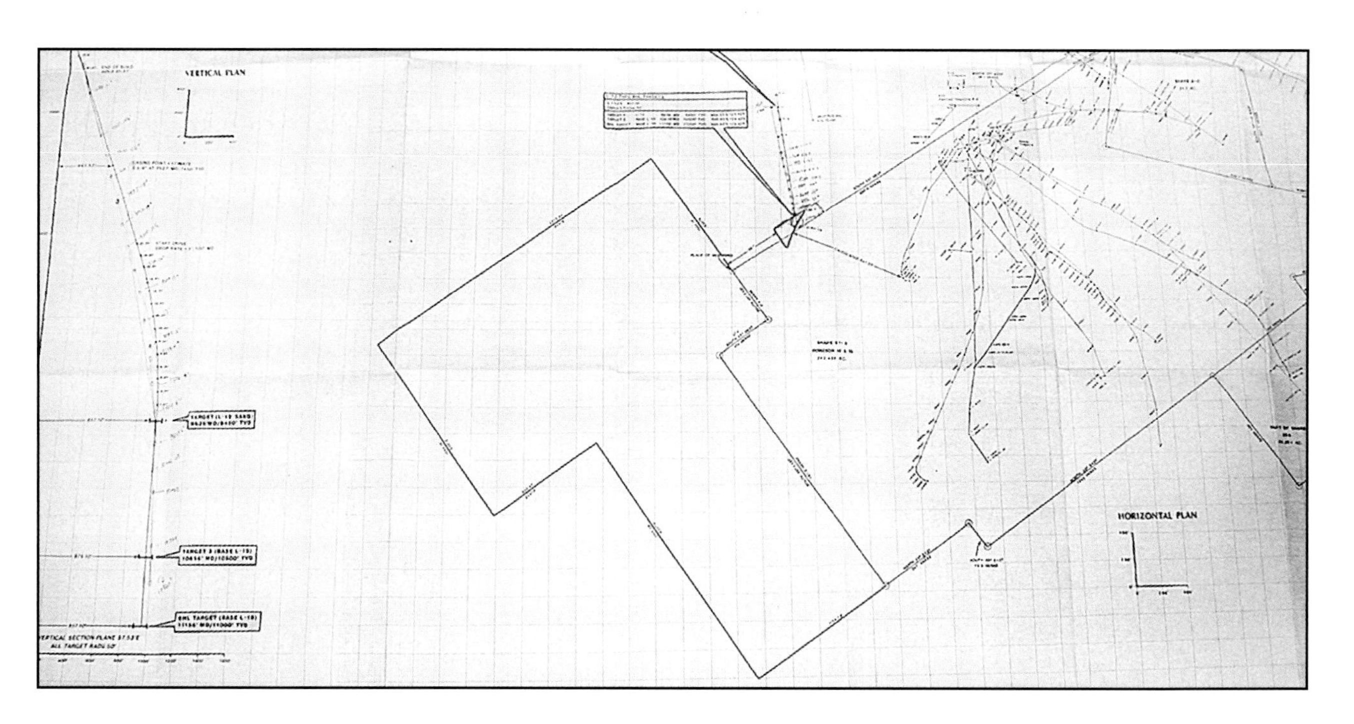

A 'spider map' illustrates how different well bores can extend from a single surface well hole.

Miller, who had entered the industry working for Halliburton in 1950 and later became an independent consultant, emphasized the need for more exploration but expressed the hope that such preservation efforts would continue.

"We are going to have to develop our Alaskan reserves, offshore, and in state and federal game preserves. We could do in this country as we have in the Aransas game preserve," he said. "We shut down drilling during the whooping crane season, during their migrations. Continental Oil Company started drilling there in 1950 and did a super job of helping fish and game people in developing resources. There is no telling how much Continental had invested in these cranes. It is my wish that this type of development can continue and the oil and gas industry through their efforts can stay and develop the game preserves that we have here in this country."

In a 2003 interview he said, "The frightening thing I see about the industry is the fact that we had 4,800 to 5,000 rigs drilling in North America. Now it is around 1,000 and continuing to decline."[13]

Jerry Clark and Dan Pedrotti also expressed some concerns.

The old guard is gone, Clark said, and his era was about to end, too. He worried about what was to follow.

"Bright kids find the jobs, but those who are hard workers with drive can't get in the front door," he said. "You have a whole class of people who have never been associated with or worked for a major oil company. That's where we learn."

The bright ones could do wonders with computers, but he believed they should have concepts they could visualize.

"It is really an artistic talent," he said, "for imagination cannot be taught. And small independents can't compete with large firms that can drill 16,000 to 20,000 feet. But it does leave a window of opportunity for shallower pays.

"We are completing wells now that are in reservoirs that twenty years ago you wouldn't have looked at," he said. "With improved fracing and completion technology we are making those rattier formations quite lucrative.

"In this business I don't know what luck is. Do you make your luck? Timing and imagination help," he said. "Some of the biggest oil finders in history never went to school in their lives. I'll bet H. L. Hunt never got out of high school."

"In the seventies our generation were the top people," Pedrotti said. "We knew what was going on. We understood the geology and exploration business as it was in the seventies. We basically missed the ... Rowling era with all the 2-D seismic. They were the masters of the eighties. So were Bill Maxwell and that group at Harkins. Then came the nineties. It was time to move to the next phase, which was 3-D. It took five years to get ourselves in the position where we could do 3-D shoots, which we did on the Cage Ranch, and as a result of that, we doubled the production on Cage Ranch to fifty billion cubic feet of gas in that field. That enabled our technical staff to move forward into the '90s.

"Now I am concerned about the 2000s. What is going to be the next step forward? Any young men interested in exploration need to get very well versed in seismic and computer technology. All the easy stuff has been found. It will take real good people...," Pedrotti said.[14]

One young man of the type Pedrotti had in mind was Tom Davidson, who was president of the Corpus Christi Geological Society in 2000-2001 before moving to San Antonio in 2003.

When Davidson graduated from the University of Texas in 1986, the oil business was in the middle of a major slump. Oil was $9 a barrel, and people were getting laid off. Banks were folding and real estate was in trouble.

Davidson said, "I had a choice between McDonald's and Sears or taking a job with my dad, who was an independent geologist in Corpus Christi."

He chose to work with his father, Phil Davidson, who had worked for Texaco, Cosden Oil, American Petrofina, Fina, and finally Samedan before going independent in 1980.

Tom was an independent who often worked for Bishop Oil Company, which worked the Tom O'Connor Field, the largest producing field in Texas. He said that basically the work in the twenty-first century was the same as that of earlier days—that prospecting required hard work, study, and utilization of new technology.

'We are completing wells now that are in reservoirs that twenty years ago you wouldn't have looked at. With improved fracing and completion technology we are making those rattier formations quite lucrative.'

—Jerry Clark

'I believe that in my lifetime there will be some movement in the Corpus Christi area for wave power. And there is the wind. It's always there. Wind, wave, sun. They're all going to be big.'

—Bill Findley

"As for the old timers, they were pretty smart individuals. They didn't leave much behind," he said. "There are a lot of people who have 3-D seismic who go back in and look for leftover scraps that people previously missed. What they are finding is that they didn't miss much."

Some aspects of government regulation were major concerns for Davidson.

People don't pay much attention to the oil business until prices are high, he said. "They are angry, and they put pressure on the government. It doesn't make any difference if they are Republican or Democrat and they pass bad legislation. When we need the legislation, when prices are low, 95 percent are happy, and congressmen don't listen to the five percent."[15]

Oilmen of different generations were united in their concern over government and environmental regulations.

Bill Carl said, "In most of the prospective reserves, we can't get permits to drill them because of environment restrictions. . . . You can't drill off California, and you can't drill off the East Coast, you can't drill in the Federal lands in the Rockies and in Alaska. It just pretty much limits natural gas, and it is going to be a problem. You can't get the politicians to realize that."

Larry McNeil expressed concerns about environmentalists who opposed drilling in some of the richest oil reserves, such as the Rocky Mountain area, off the coast of California, and in Alaska north of Prudhoe. He said no one knew how much oil and gas was in Siberia, but there were "probably 10 to 15 billion barrels of oil" in Alaska, and "it's all part of the same thing.

"They talk about a 1,300,000 acres up there . . . and if you develop the whole oil reserve, you would be talking about somewhere around 10,000 acres."

He also recommended the use of nuclear power.

"In France they standardized their nuclear plants so the most workable system prevails. It is cheaper and more efficient than having different companies using different designs, different valves. The most efficient would lead to efficiency and lower costs."

Bill Findley had another prediction about the future of energy. He pointed out that developing nations, including China, will have to have more oil and gas, and he saw more use of solar power, already in use in some equipment.

"I believe that in my lifetime," the 82-year-old said, "there will be some movement in the Corpus Christi area for wave power. And there is the wind. It's always there. Wind, wave, sun. They're all going to be big."

Continued on page 338

Best laid plans don't always stay on track

"We lost the rig," the voice on the telephone said.

"I didn't know you had spudded yet," the drilling company man answered.

"You don't understand," the voice said. "The rig is destroyed."

Silence.

"It got hit by a train."

A Corpus Christi Drilling truck driven by Bobby James Chambless and carrying a 120,000-pound rig hit high center on the track at the U. S. Highway 77 entrance to the O'Connor Ranch near Refugio on April 28, 1994. Chambless saw he couldn't get out of the cab in time, so he desperately tried to back up. At the last moment the wheels caught and moved a few feet. He was thrown clear but survived.

The train didn't fare so well, either. Twenty-two cars derailed and a mile and a half of track was damaged. The collision dumped milo, plywood, animal fat tallow, and Union Pacific machinery into a pile of wreckage. Although some of the train's cars carried flammable liquids, they were far enough back to remain on the tracks, so there was no danger of explosion.

Chambless and three trainmen were airlifted to hospitals for treatment.

None of the drama showed on the Bishop Petroleum log. It read

"4-26-94 Will move in dozer to dig reserve pit and scraped location today.

4-27-94. MI dozer scraped and leveled location. Dug water pit.

4-28-94. Dug reserve pit. DRLG Rig hit by train. Will move in Rig #5 today."

(Top) Tom Davidson in his office in Boerne, Texas, with a news clipping about the incident

In the summer of 2002, a Grey Wolf drilling rig utilizes the latest technology at a site near Driscoll, Texas.

Richard Wilshusen felt that exploration was a good field for those already in it but said he would not advise young people to choose it as a career.

"I wouldn't advise a kid to go into the field and spend their whole life in it," he said. "I had a brother who felt the future was in foreign oil, and he did very well in it. If they're interested in the really big stuff, foreign would be the place to go."[16]

However, some considered a bright future for the area possible. Randy Bissell found reasons to be pessimistic and reasons to be optimistic about the future of South Texas.

The pessimist looks at the glass as 95 percent empty, he said. Optimists say "what a huge basin and there's so much hydrocarbons here. Just a fraction of it being left is still plenty for me."

He remained optimistic about opportunities on all the major ranches of the area.

"The challenge for South Texas," he said, "is that we are going to have to become leaner and meaner and compete with offshore supplies of natural gas in an area where opportunity is decreasing. So we are going to have to become more reliant on new techniques and technologies to help us do our job better.

"We can talk about going deeper, but a lot of opportunities will exist within and between known fields," he said. "The seismic technologies have advanced our ability to do our job better in the last ten years, and there will be refinements to them that will allow us to even see and interpret the rocks more clearly."

Deep wells were nothing new to South Texas. They had been drilled since the 1960s.

"What's made the deep more commercial is the stimulation technology—the ability to get gas and condensate out of tighter rocks and how very large fracture stimulation jobs allow recovery of reserves that would have been noncommercial in the past," he said. The combination of technology and price support "has allowed us to get reserves that would otherwise not have been produced."

This new stimulation technology had wells producing as much as 50 million cubic feet of gas a day. The cost of drilling the wells might run as high as $10 million, but the price of gas made for a very short payout.

The downside was that the production decline was often rapid.

"It might make 50 million cubic feet for the first few days of production. A month later it's making half that and two months later a quarter of that," he said. "It will still be making a profit but significantly lower than the initial rate."

As for natural gas reserves that were new at that time, he said, "We are exploring the very deepest part of the Gulf of Mexico. We've kind of reached the depth there. For political and environmental reasons we have excluded certain areas of the country from drilling. We have these moratoriums on the East Coast and West Coast."

The challenge, Bissell said, was to find opportunities in a very mature basin area.

"We are blessed in South Texas with having an extremely prolific hydrocarbon province," he said. "There is an extremely complicated relationship between the structures and the sands. In that complexity, there is a tremendous opportunity for those of us with courage to drill wells and find oil and gas."

Brian O'Brien agreed with this assessment, disputing those who thought that all South Texas oil had been found.

"There are lots of areas to be worked over," he said. "Seismic has improved. There are some deep basins that have hardly been touched yet. With technology and money, oil will be found."[17]

Listings indicate a deep future

The futuristic trend of extremely deep exploratory drilling wass exhibited in listings in the *American Oil & Gas Reporter* in 2003 and 2004 showing drilling contractors and proposed depths of projects in various South Texas counties.

The charts indicated thirty-two wells with predicted depths of 20,000 or more feet. Of those, seven were projected to 25,000 feet and two to 35,000 feet. Of the latter EMSCO was listed as contractor in a Nueces County test and National Drilling Company in a Kenedy County project.

The 25,000-foot sites were in Duval, DeWitt, Webb, Victoria, Hidalgo, Jim Hogg, Lavaca, Live Oak, Bee, and McMullen counties.

Epilog

For more than a century, it has been a long and glorious ride for those who would challenge the earth to give up her tightly held treasures. Through ambition, curiosity, perseverance, resourcefulness, creativity, and common sense, they introduced America to the Machine Age and beyond. They celebrated booms that gave birth to boisterous new towns and endured depressions that stacked rigs of bankrupt operators. They made fortunes and lost them in the blink of an eye. Those who survived left a heritage as colorful as that of the American cowboy.

At first petroleum was a powerful stimulant, leading the country on a binge of wasteful overproduction. The market served to curb the excesses. Government controls were welcomed at first, then cursed as production was cut back, sending small operators into bankruptcy. Taxes as high as 90 percent took the incentives from the search.

Foreign oil was a villain until petroleum reserves began to drop and overseas oil had to be tolerated.

As the twenty-first century began, there were signs of despair and signs of hope. Oil was getting scarce. Reserves were low. Natural gas was replacing oil as the fuel of Texas, but finding it called for deeper and deeper drilling at greater and greater cost.

During the glory years the government and schools of Texas were carried by taxes on the oil industry. As this is written, revenue has dwindled, and the state finds itself in a dilemma. It is promising to avoid new taxes but has yet to find a replacement for the rich source it has relied on.

But as the heyday of the industry dims, the people can enjoy the fruits of the labors of successful oilmen who have endowed libraries, museums, hospitals, schools and universities, charitable foundations, and dozens of other public facilities.

Even if the industry fades, the legacy of the riches it has provided will continue to lift the quality of life for Texans.

Charter Members

With travel to San Antonio for meetings curtailed by gasoline rationing and other considerations during World War II, Corpus Christi members of the South Texas Geological Society formed their own society in 1942. Charter members pictured in this photo taken on March 6, 1967, the group's twenty-fifth anniversary were (front row, seated, left to right) Reagan Tucker, Guy B. Gierhart, Edwin A. Taegel, H. N. Seevers, Dale L. Benson, W. L. Day, Frith C. Owens; (back row standing) C. I. 'Al' Jennings, Henry McCallum, C. E. Buck, Dr. W. Armstrong Price, W. A. Maley, Gaston Parrish, C. J. Cunningham, James D. Burke, Henry Cardwell, and Ira H. Stein.

Corpus Christi Geological Society History Committee

Ray Govett, Chairman
Alan Costello
Tom Davidson
Jay Endicott
Brent Hopkins
Joe McCullough
Dennis Moore
Fermin Munoz
Dan Neuberger
Dan Pedrotti
Tubby Weaver
Sebastian Wiedmann

The South Texas Geological Society

In 1929 Charles H. Row, later a Sun Oil Company official, was president of the South Texas Geological Society. Other early members of the society were R. F. Schoolfield and Kenneth D. Owen. The second president was Douglas R. "Doc" Semmes.

Other early presidents were Hershel H. Cooper, Ed W. Owen, a well-known consultant; L. F. McCollum, 1933, later president of Continental Oil Company, and Arch Maley, later an executive with Humble Oil and Refining.

Frank Zoch was first president of the Corpus Christi Petroleum Club. One of his chores was hosting a more-than-lively national convention of oil scouts.

A History of the Corpus Christi Geological Society

Geologists in the area joined together to advance the science of geology. They met in San Antonio and did not adopt a name until the San Antonio Geological Society was organized March 24, 1929. It first hosted the annual convention of the American Association of Petroleum Geologists (AAPG) in 1931.

Later the name was changed to the South Texas Geological Society, and meetings were held in both Corpus Christi and San Antonio. Geologists from Beeville, Pleasanton, Laredo, and other South Texas cities were part of this Society.

Past Officers
South Texas Geological Society

Year	President	Vice President	Secretary-Treasurer
1929	Chas. H. Row	R. F. Schoolfield	
1930	Douglas R. Semmes	Herschel H. Cooper	Kenneth D. Owen
1931	Herschel H. Cooper	L. W. McNaughton	Ed W. Owen
1932	Ed W. Owen	Olin G. Bell	Chas. A. Stewart
1933	F. McCollum	Fred P. Shayes	Julian Q. Myers
1934	T. J. Galbraith	A. E. Gatzendaner	T. R. Banks
1935	Joseph M. Dawson	Philip S. Schoeneck	Adolph Dovre
1936	Adolph Dovre	Wm. G. Kane	Harry H. Nowlan
1937	Harry H. Nowlan	W. A. Maley	Stuart Mossom
1938	W. A. Maley	W. W. McDonald	C. C. Miller
1939	Willis Storm	Dale L. Benson	Ira Brinkerhoff & R. N. Kolm
1940	Fred P. Shayes	Gentry Kidd	R. N. Kolm
1941	Gentry Kidd	G. B. Gierhart	W. W. Hammond

In 1942 World War II had started, gasoline and tires were rationed, and it became difficult for geologists to travel between Corpus Christi and San Antonio. Geologists from the Corpus Christi area decided to start their own society. The Corpus Christi Geological Society started in 1942 with forty-two charter members. The late W. Armstrong Price was elected the first president. The South Texas Geological Society continues in San Antonio. W. Carlton Weaver may have been the last surviving charter member of both the South Texas and the Corpus Christi Geological Societies.

Past Officers
Corpus Christi Geological Society

Year	President	Vice President	Secretary-Treasurer
1942	W. Armstrong Price	L. B. Herring	C. I. Jennings
1943	Frith C. Owens	E. D. Pressler	J. Bruce Scrafford
1944	Ira H. Stein	Henry D. McCallum	Elsie B. Chalupnik
1945	R. D. Hendrickson	W. E. Greenman	O. G. McClain
1946	W. E. Greenman	Dale L. Benson	O. G. McClain
1947	Dale L. Benson	R. D. Hendrickson	H. C. Cooke
1948	Henry D. McCallum	Norman D. Thomas	James D. Burke
1949	James D. Burke	W. H. Wallace, Jr.	Bill Volk

In 1950 the secretary-treasurer position was divided. World War II had ended, and geologists who had served in the military had returned to their civilian jobs. Many veterans who had attended college on the GI Bill were graduating and entering the profession. The oil industry enjoyed a brief postwar boom that ended in 1951, and membership in the society grew. Weekly meetings were held during this period. The only known surviving charter member in 2005 is O. G. McClain, who now lives in Houston.

Year President	Vice President	Secretary	Treasurer
1950 O. G. McClain	W. M. Chadwick	J. C. Meacham	Richard N. Spencer
1951 W. M. Chadwick	J. C. Meacham	Richard N. Spencer	Murphy L. Walker
1952 W. H. Wallace	Thomas D. Barber	Jean H. Story	William B. Oliver
1953 William B. Oliver	W. Gus Bealmear	Joseph W. Lea	C. Wayne Holcomb

The position of vice president was divided in 1954. The first vice president was responsible for recruiting committee chairmen for the jobs needed to carry out objectives of the society, the second vice president obtained speakers for the meetings. Meetings were changed from weekly to monthly in the 1950s, except during summer months, when no meetings were held.

Year	President	First Vice President	Second Vice President	Secretary	Treasurer
1954	Joseph W. Lea	Wm. W. Henry	P. F. Oetking	Martha Bybee	F. Clarkson
1955	Wm. W. Henry	Holland Mondy	Wilford Stapp	Mary Payne	Charles Baker
1956	Joe O'Brien	Bill Volk	Alan Lohse	Louise Clarkson	James Ramsay
1957	Ralph Beeker	James Ramsay	Robert Morton	F. Clarkson	R. Wilshusen
1958	Robert Owen	Wm. Ledbetter	John Currie	J. E. Gordon	Bernard Dietz
1959	Wm. Ledbetter	Wm. D. Dobbins	Charles Baker	Art Tschoepe	F. I. Brooner
1960	Charles Baker	Dale Olson	Joe McCullough	Mary Magaw	Fred Thompson
1961	Joe McCullough	Jim Dennis	Ted Cook	Wm. W. Hoag	Joe Sockwell
1962	Jim Dennis	Joe Sockwell	Richard Hicks	Dan Pedrotti	Eleanor Hoover
1963	Joe Sockwell	Dan Pedrotti	Dave Sheridan	Don Rothschild	Byron Dyer
1964	Cleo Buck	Byron Dyer	Don Clutterbuck	Jim Miller	Jack Stanton
1965	Byron Dyer	Jim Miller	Don Boyd	Herschel Walker	Fred Thompson
1966	Don Boyd	Fred Thompson	Paul Strunk	Jim Rutland	George Fegan
1967	Paul Strunk	Jim Rutland	John Schultz	Don Haynes	Herman Loeb
1968	John Schultz	Don Haynes	C. R. Burnette	Les Giddens	Earl Melton
1969	Don Haynes	Earl Melton	Jack Birchum	Ed Heath	Don Kling
1970	Earl Melton	W. R. Payne	Jack Sulik	G. Heinzelman	Willis Oakes
1971	Jack Sulik	Don Kling	Byron Moore	Jim Collins	John Oliver
1972	Don Kling	Jim Collins	R. P. Marshall	Don Kieth	Mark Wolbrink
1973	Jim Collins	W. D. Dobbins	Wm. A. Atlee	Mark Wolbrink	Leroy Manka

Year	President	First Vice President	Second Vice President	Secretary	Treasurer
1974	Wm A. Atlee	Mark Hodgson	Jay Endicott	Bill Floyd	Bob Giltner
1975	Bob Giltner	Bill Floyd	Ralf Andrews	George Fegan	John Hyndman
1976	Ralf Andrews	Bob Travis	George Fegan	R. W. Maxwell	Carroll Pitzer
1977	George Fegan	Carroll Pitzer	R. W. Maxwell	Helen Klein	Cliff McTee
1978	Carroll Pitzer	Cliff McTee	Bill Heintz	Ronnie Thomas	David Brock
1979	Cliff McTee	R. W. Maxwell	Leroy Manka	Jim Brewster	Ronnie Thomas
1980	R. W. Maxwell	Leroy Manka	David Becker	Curtis Mayo	Owen Hopkins
1981	Leroy Manka	Bill Heinz	Curtis Mayo	Craig Anderson	Robert Manson
1982	Ray Govett	Craig Anderson	Bob Espeseth	Bob Rice	D. Chisholm
1983	Wm. D. Dobbins	Curtis Mayo	Charles Wisdom	John Drake	Sherie Harding
1984	Curtis Mayo	G. Heinzelmann	John Drake	Duncan Chisholm	Julie Moore
1985	G. Heinzelmann	John Drake	Duncan Chisholm	Sebastian Wiedmann	Tom Henderson

To help maintain continuity of leadership in the society, the position of first vice president was changed to president elect in 1986. The duties of president elect included finding committee chairmen. Membership in the society had grown to 475 members in 1967 and 880 members in 1981.

Year	President	President Elect	Vice President	Secretary	Treasurer
1986	John Drake	Duncan Chisholm	Tom Henderson	Julie Moore	Patrick Nye
1987	D. Chisholm	Tom Henderson	Patrick Nye	Bill Walker	Bruce Brown
1988	Tom Henderson	Patrick Nye	Bill Walker	Carroll Pyle	H. Marshall
1989	Patrick Nye	Carroll Pyle	Rick Railsback	Charles Franck	Wayne Croft
1990	Carroll Pyle	Rick Railsback	Wayne Croft	Jon Herber	Owen Hopkins
1991	Rick Railsback	Wayne Croft	Greg Dennis	Sebastian Wiedmann	Frank Cornish
1992	Wayne Croft	Sebastian Wiedmann	Carlos Maggio	Jim Claughton	Larry Billingsley
1993	Seb. Wiedmann	Jim Claughton	Jim Collins	Gloria Sprague	Larry Billingsley
1994	Jim Claughton	Larry Billingsley	Gloria Sprague	Maggie Moorehouse	Wayne Sandefur
1995	Larry Billingsley	Gloria Sprague	Herman Vacca	Tommy Dubois	Bob Rice
1996	Gloria Sprague	Herman Vacca	Tommy Dubois	Virginia Henderson	Bob Rice
1997	Tommy Dubois	Charles Franck	Scott Wruck	John Carnes	Matt Franey
1998	Charles Franck	Scott Wruck	Bill Heintz	John Carnes	Matt Franey
1999	Scott Wruck	Alan Costello	David Hatridge	John Carnes	Tom Jones
2000	Alan Costello	Tom Davidson	Dan Neuberger	John Carnes	Tom Jones
2001	Tom Davidson	Dan Neuberger	Fermin Munoz	Lou Lambiotte	Jay McGovern
2002	Dan Neuberger	Fermin Munoz	Brent Hopkins	Lou Lambiotte	Bob Rice
2003	Fermin Munoz	Brent Hopkins	Jay Heidecker	Erika Brown	Dennis Moore
2004	Brent Hopkins	Dennis Moore	Erika Brown	Cori Lambert	Duncan Chisholm

These are the men and women who held offices in the Corpus Christi Geological Society, but they are not the only ones who contributed to its success. Each year there are ten to twenty committees, which do much of the work for the society. Often, but not always, these committee chairmen take one of the offices listed above. Committee chairmen work with officers to make the society function.

Caroline Duffield was a long time aide to the Corpus Christi Geological Society. She started working for the society years ago and continued until her death. She printed and mailed the bulletin and other mailings. She helped maintain membership rolls and mailing lists. She printed guidebooks and other society publications. Her dedication to the Corpus Christi Geological Society was unequaled.

Another individual that younger geologists will never have the privilege of knowing was Jeff Jeffries. Jeff was a founder and part owner of Nixon Blue Print Company. Early in Nixon's operations, Jeff saw a need for someone to maintain electric logs from South Texas. Nixon Blue Print Company maintained copies of the logs, and Jeff was always eager to help in identifying a log that might have a different name or something else that made it difficult to find. He had a phenomenal memory when it came to South Texas electric logs. Both Caroline and Jeff were given special recognition by the society.

The objective of the Corpus Christi Geological Society has always been to promote the science of geology, particularly as it relates to the geology of South Texas. To accomplish this objective, the society has undertaken a variety of projects. Since its inception, the society has met to discuss geology. Speakers at these meetings have been both members and guests. Members, and others, have contributed to publications to aid fellow geologists to better understand South Texas geology. The bulletin has grown from a newsletter type publication to a magazine with technical articles, news of members, and activities of the society. Field trips are sponsored by the society to help members interpret what they may suspect in the subsurface. Often guidebooks are published for these field trips which may help other geologists. Type logs of certain fields are published to help in establishing a consistency of terminology. Field studies and an Rwa catalog have been published. Cross sections showing general geology of the Gulf Coast are available from the society. The society regularly offers continuing education courses for all geologists. As new ideas and concepts are developed, the society tries to present them to the membership.

The Gulf Coast Association of Geological Societies was formed in 1951 to aid the local societies in their objectives of furthering the knowledge of geology. The first meeting of the GCAGS was in New Orleans, Louisiana, and the second was in Corpus Christi. W. M. Chaddick, Jr., who was president of the Corpus Christi Geological Society in 1951, was the second president of the GCAGS. That meeting drew over 1,000 geologists and their spouses to Corpus Christi, with 800 of the attendees from outside of the city. The GCAGS affiliated with AAPG in 1967 and is now the Gulf Coast Section of AAPG. Meetings of GCAGS are rotated among the member societies. Corpus Christi held the convention again in 1958 with E. A. Lohse as president and R. C. Wilshusen as convention chairman. The convention returned in 1964 with Earl Knott as president and Byron Dyer as convention chairman. Don Boyd was president when it returned in 1972, and Jerry Sides was convention chairman. Paul Strunk was president of the 1981 convention and Wilson Humphrey chairman. The 1989 convention had Bill Payne president and Gerry Heinzelmann chairman. The most recent GCAGS convention held in Corpus Christi was the 1998 meeting with Bob Travis as president and Gloria Sprague as convention chairman. Paul Strunk and Don Boyd were named honorary members of GCAGS in recognition of the support they had given to that organization. GCAGS awarded Gloria Sprague the Distinguished Service Award in 1999 in recognition of her work for the organization.

At the 2000 meeting of GCAGS, held in Houston, the Don R. Boyd Medal for Excellence in Gulf Coast Geology was created to honor Don Boyd. The Boyd Medal is to be awarded for (1) excellence in research, (2) professional leadership, and (3) oil and gas exploration. It is not necessary that the award be made each year, but when it is awarded, the Boyd Medal is to be the highest award given by GCAGS. This honor given to Don Boyd is unique for GCAGS. Don passed away December 20, 2000, and the first Don Boyd Medal for Excellence in Gulf Coast Geology was given to his widow, Patricia, at a meeting of the Society on April 18, 2001. Don Boyd served as secretary of AAPG, on the Advisory Council, and on other committees. He was twice nominated for president of AAPG. Don was made an honorary member of AAPG at the 1989 convention.

Paul Strunk has served as treasurer of AAPG, on the Advisory Council, and on other committees. Paul was made an honorary member of AAPG, April 11, 1999, at the convention in San Antonio. Fred Dix, who retired as Executive Director of AAPG, was a member of the Corpus Christi Geological Society,

Joe Dawson, a geologist who lived and worked in Corpus Christi for many years, received a posthumous award. The Corpus Christi Independent School District named an elementary school the Joe Dawson Elementary School in his honor. This is a unique honor for a Corpus Christi geologist. Joe was not only a long time geologist but also a hero during the Normandy Beach landings in World War II.

The Corpus Christi Geological Society is represented in the AAPG House of Delegates. The number of delegates is based on society membership and is currently two, Gloria Sprague and Brent Hopkins.

Members of the society have held national or state offices in both the American Institute of Professional Geologists (AIPG) and the Society of Independent Earth Scientists (SIPES). The late Fred Thompson was an officer in the society and served several positions in the national SIPES organization, including president. The late Dick Peterson was a long time member of CCGS and served as president of the SIPES Foundation. Robert Owen was vice president, natural resources, for SIPES. Jack Sulik, Paul Strunk, and Charles Lundberg are, or have been, directors of the national SIPES organization. Curtis Mayo, a former member of the Society, was president of the Texas section of AIPG, and Paul Strunk has also served as president of the Texas Section of AIPG.

The Corpus Christi Geological Society has bestowed honorary membership on nine of its members. Dale Benson was an early president of the society and represented Corpus Christi when it was part of the South Texas Geological Society. Carroll "C.C." Miller was an officer in the South Texas Geological Society and remained active in the local society for many years. W. Armstrong Price was the first president of the Corpus Christi Geological Society and remained active in it until shortly before his death in 1987 at the age of 98. Frith Owen was the second president of the society. Cleo Buck was a president of the society and remained active until shortly before his death. O. G. McClain was a charter member of the Corpus Christi Geological Society, held several offices, and is still a member, although he now lives in Houston. He recently contributed an article to the bulletin. Don Boyd was an honorary member until his death. Paul Strunk and Owen Hopkins are honorary members still active in the society.

In 1991 the society started honoring members with a designation of "Distinguished Geologists." Joe Uri, W. Carlton Weaver, Bill Colson, Joe McCullough, Dick White, Bill Volk, Jay Endicott, Frank Cornish, John Vreeland, Lawrence Hoover, Louie Sebring, Jr.,and Bill Carl have been honored by the society with that title. Of this group, Joe Uri, W. Carlton Weaver, Bill Colson, Dick White, John Vreeland, and Louie Sebring, Jr., have died.

The society has several avenues for serving the community and the membership outside of meetings and courses. Speakers are provided to schools and other organizations requesting information about geology or the oil industry. Members work with the Corpus Christi Museum of Science and History and local universities and colleges. A booth is sponsored at the annual Bayfest, which provides an avenue for the younger segment of the population to become familiar with geology. The society sponsors a blood drive twice a year, which provides for not only the membership but also the community. Scholarships are made available to local college students, and many in the society provide part-time employment for some of these students. Geologists assist Boy Scouts in obtaining merit badges in geology and science.

Social functions, athletic events, and other activities are held each year for members, spouses, and guests. Many of these programs are sponsored with the society auxiliary and other groups, such as the landmen and engineers.

Any history is never complete. It is not unusual for an officer to be elected and someone else serve the term because of transfers. Many geologists have contributed to the continuing success of the Corpus Christi Geological Society, and not all are mentioned in this brief history.

Ray Govett
November 2004

Donors to the Corpus Christi Geological Society History Fund

Platinum

Corpus Christi Exploration Company Foundation

Gold

EOG Resources, Inc.

Silver

Killam Oil Co., Ltd.
Manti Resources, Inc./Futureus Foundation
Mestena Operating, Ltd.
Mr. & Mrs. Dan Pedrotti
Sipes of Corpus Christi

Titanium

Ed Rachal Foundation
B & T Robertson Foundation
Schlumberger Oilfield Services
Coastal Bend Geophysical Society
Phil Davidson
Fulton Family Fund
Headington Oil Company
O.G. McClain
Mr. & Mrs. Joe McCullough
Jack Sulik
The Rachael & Ben Vaughan Foundation

Tungsten

American Shoreline
Bob Rice
Mark Minahan
Newell Fisher

Tungsten

Jerry Clark
Mr. & Mrs. Randall Bissell
Mre. & Mrs. Alan Costello
Jerry Dewbre
The James R. Dougherty, Jr., Foundation
Dougherty Foundation
Chris Douglas
Les Giddens
Hiwood Exploration
Neu Oil & Gas, LLC.
Volk Oil and Gas Company
W. Carlton Weaver
Charles Winn

Nickel

Frank Cornish
Tom Jones
Bruce Nacci
Joe Mueller
Mr. & Mrs. Jeff Cobbs
Core Laboratories, Inc.
Tom Davidson
Tommy DuBois
Jay and Frances Endicott
Curtis Mayo
Paul Mueller
Jay & Amy Heidecker
Owen Hopkins & Susan Hutchinson
Dan Neuberger
Mr. & Mrs. Richard E. Paige
Eduardo Riddle
Gregg Robertson
Mr. & Mrs. Scott Rutherford
Shoreline Gas, Inc.
John C. Worley
Charles Brocato
Steve Davidson
John Griesbach

Nickel

Brent Hopkins
Doug Sartoris
Paul Strunk
Dennis Moore
Robert Graham
Nora Govett
Bruce Heun
Louis & McConnell Lambiotte
Chuck Lundberg
Bill Maxwell
Tom Medary
Mark Miller
Gloria Sprague
Tom Swinbank
Mr. & Mrs. Sebastian Wiedmann

Copper

Ralph Andrews
Jack Barcklow
Joe Baria
Jim Brewster
Hank Ellsworth
George Fegan
Charles Franck
Borden Jenkins
Craig Mullenax
Fermin Munoz
Robert Steinberg
Rick Tuley
James & Jeneva Bryant
Joe Butts
Charles Forney
Lawrence Hoover
Dale Phipps
Carroll Pitzer
Huck Walker
Bryan Bishop
David Miller

Copper

Elizabeth Chapman
Bruce Servant
Jim Frazier
Cori Lambert
Michael Bertness
Barbara Beynon
John Burnette
Chuck Collins
Bruce Fields
Carlos Maggio
E. S. Middour
James Morren
Mitchell Nielsen
Patrick Nye
Winston Sexton
Raymond Welder
Cameron Gates
Jim Gordon
John Goss
Pete Graham
James Gresham, Jr.
Walter Light
Brian O'Brien
Tom Fett
Bill Parmley
John Channas
Hewitt Fox
Matt Franey
Jeff Osborn
Richard Parker
John Sangree
Jon Spradley
William Harrison
Val Neshyba
Ron Nickle
William Smith
Jack Smitherman

On behalf of the Corpus Christi Geological Society, I thank all who contributed to the completion of this book. Without the support of the Society, and financial contributions from members and others, the book could not have been written. The full cooperation the Society during research and writing of the book ensured its success, and all should be proud of their efforts.

The History Committee of the Corpus Christi Geological Society debated the best way to finance this book and other projects for a year or more. It was finally decided that the task would merit support from members of the Society and others in the community, so that route was taken. The Geological Society provided funds to create a promotional video, and articles were published in the Bulletin of the Society to initiate interest in the project. Paul Haas and Larry McNeil provided funds to encourage members and the community to support the effort, and the response was gratifying.

Alan Costello was president of the Society when the history project began. Tubby Weaver, Bill Volk, Jay Endicott, Joe McCullough, and Sebastian Wiedmann joined Ray Govett on the History Committee. Due to the amount of money involved in creating the final product, it was requested that the President and President Elect of the Society be part of the Committee. Bill Volk resigned from the Committee but made significant contributions prior to quitting.

This book represents the first phase of the History Committee project. Recording the history on videos is continuing. The Committee determined that the book would be finished before the video, and finishing the book became the priority. Some of those interviewed were recorded on videos, and this will be incorporated into that phase of the history project.

In order to raise the funds, it was necessary to do it through a vehicle that would allow contributors a tax deduction. Rather than start a new tax exempt corporation within the Society, it was decided to use the Coastal Bend Community Foundation as the agency to help the Society handle funds raised.

Tom Davidson replaced Alan Costello as president of the Society and initiated the funding. Members of the Society immediately started to contribute. Dan Pedrotti joined the Committee during the tenure of Tom. Dan Neuberger succeeded Tom, continued raising funds, and signed a contract with the Walravens to do the writing. Sufficient funds had been raised to ensure that the book would be written and printed before the contract was signed.

Fermin Munoz replaced Dan Neuberger, and during his tenure much of the book was written by the Walravens. Brent Hopkins succeeded Fermin and signed the contract to have the book printed. Dennis Moore will replace Brent and have the responsibility of getting the book to all who wish to read it, as well as continuing with the video portion of the history.

Now you know why Ray Govett agreed to be Chairman of the History Committee—others do all of the work. This history project represents the largest financial commitment the Corpus Christi Geological Society has undertaken and hopefully will be financially successful. All profits from sales of the book and videos may only be used for educational purchases. The Society has a scholarship fund which grants aid to geology students at the Universities and Community Colleges in South Texas. Funds for these scholarships have come from members of the Society, and hopefully, the History Committee will soon be able to contribute to the scholarship fund. The Society supports local schools in their science programs by furnishing speakers and exhibits at all levels.Money from this project could be used for that program. Members of the Society acquaint students with geology by sponsoring a booth at the annual Bayfest, as well as aiding in Boy Scout geology programs. These are some of the educational endeavors of the Society. Virtually everything the Society does is intended to educate either members of the Society or the community about geology or the oil industry.

Tubby Weaver was a tremendous contributor to the success of the project. He personally knew some of the older members of the profession and arranged interviews with them. Tubby loved to talk about the oil and gas business and could do so for hours at a time. There were many like him in the industry who died before their stories could be recorded. Perhaps this book will inspire the recording of additional industry stories to be used in a future history of the oil and gas industry.

Ray Govett, Chairman
History Committee of the Corpus Christi Geological Society

New Field Discoveries Districts 1, 2, and 4

This is a list of new field discoveries from the beginning through 2004. Minor fields are not listed. It is intended that the list show how hydrocarbons were discovered in South Texas and how the area developed. With production calculated through 2004, fields that have produced more than one hundred million barrels are noted. Some of these have produced much more than the 100 million barrels. Gas was converted to barrels at six MCF of gas equals one barrel of oil. some fields were combined into other fields and the original name dropped. In most casesthe old name is not on the list.

The list was compiled from the International Oil and Gas Development book published by the International Oil Scouts Association and data provided by the Texas Railroad Commission. As a general rule, when the Scout data did not agree with the Railroad Commission data, the Scout data was used for determining the date of discovery for older fields. Generally only one name is used on the list. For example, the Railroad Commission lists 190 reservoirs for the Seeligson Field. All of these reservoirs are considered as one field in this work. They may be a north, south, east, or west attached to the name of some fields, and most of these appendages are not listed.

Field	Date	County	Depth	District
Piedras Pintas	Early	Duval	180	4
Calliham	Nov. 1902	McMullen	848	1
Conoco Driscoll	1908	Duval	500	4
Reiser	1909	Webb	380	4
Somerset	1911	Atascosa-Bexar	1254	1
Gas Ridge	1912	Bexar	250	1
Charco Redondo	Jan. 1913	Zapata	147	4
Kingsville	Jan. 1913	Kleberg	2940	4
Jennings	March 1914	Zapata	500	4
Thrall	Feb. 1915	Williamson	900	1
Refugio, Old	Sept. 1920	Refugio	3650	2
Edna	Jan. 1921	Jackson	3850	2
Minerva-Rockdale	Feb. 1921	Milam	1700	1
Mirando Valley	April 1921	Zapata	1425	4
Carolina	Nov. 1921	Webb	1270	4
Mirando City	Dec. 1921	Webb	1510	4
Southton	May 1922	Bexar	600	1
Aviators	May 1922	Webb	1501	4
White Point	Early 100 million barrel field	San Patricio	1600	4
Luling-Branyon	Aug. 1922 100 million barrel field	Caldwell	2100	1

Taylor-Ina	1922	Medina	290	1
Mt. Lucas	Jan. 1923	Live Oak	3500	2
Saxet	Jan. 1923	Nueces	2300	4
Noack	Oct. 1923	Williamson	900	1
Ina Adams	Nov. 1923	Medina	925	1
Henne-Winch-Ferris	June 1924	Jim Hogg	1944	4
Cole Bruni	July 1924	Duval-Webb	1600	4
Lytton Springs	March 1925	Caldwell	1161	1
Los Olmos	July 1925	Starr	250	4
Adams	April 1926	Medina	900	1
Randado	April 1926	Jim Hogg	1200	4
Kohler	June 1926	Duval	1773	4
Alworth	Nov. 1926	Jim Hogg	1000	4
Jacob	1926	Live Oak-McMullen	1050	1
Cuellar	Feb. 1927	Zapata	1332	4
Cole, West	April 1927	Webb	2333	4
Escobas	May 1927	Zapata	533	4
Dale-McBride	Aug. 1927	Caldwell	1920	1
Roma	Oct. 1927	Starr	182	4
Albercas	1927	Webb-Jim Hogg	250	4
Salt Flat	May 1928	Caldwell	2250	1
Larremore	June 1928	Caldwell	1310	1
Palanga Dome	June 1928	Duval	442	4
Government Wells, N	Aug. 1928	Duval	1568	4
Eckert	Sept. 1928	Bexar	690	1
Yoast	Nov. 1928	Bastrop	1415	1
Agua Dulce	Dec. 1928 100 million barrel field	Nueces	2025	4
Oakville	1928	Live Oak	230	2
Chicon Lake	June 1929	Medina	395	1
Buchanan	July 1929	Caldwell	950	1
Darst Creek	July 1929 100 million barrel field	Guadalupe	2240	1
Manford	Aug. 1929	Guadalupe	2290	1
Martinez	Aug. 1929	Zapata-Jim Hogg	1860	4
Palo Blanco	Sept. 1929	Brooks	2460	4
Pettus	Oct. 1929	Bee	3600	2
Lamar	Oct. 1929	Aransas	6900	4
Sandia	Nov. 1929	Jim Wells	2900	4
Normanna	Dec. 1929	Bee	1920	2
Hordes Creek	1929	Goliad	4149	2
Chapman-Abbott	Jan. 1930	Williamson	1820	1
O'Hern	Jan. 1930	Webb-Duval	2760	4
Riverside	Feb. 1930	Nueces	5037	4
Saxet	Feb. 1930	Nueces	4060	4
Cosden	May 1930	Bee	3659	2
McFaddin	June 1930	Victoria-Refugio	3742	2
Chittum	July 1930	Maverick	5000	1
Holzmark	Sept. 1930	Bee	3200	2
Dunlap	1930	Caldwell-Guadalupe	2123	1
Schimmel-Batts	May 1931	Bastrop	1450	1
Refugio-Fox	May 1931	Refugio	3400	2
Von Ormy	June 1931	Bexar	600	1

Dec. 1931–June 1935

O' Conner-McFaddin	Dec. 1931	Refugio	3100	2
Mt. Lucas	1931	Live Oak	3500	2
Refugio, New	1931	Refugio	1900	2
Keeran	Jan. 1932	Victoria	5600	2
Bentonville	April 1932	Jim Wells-Nueces	5770	4
Sarnosa	April 1932	Duval	2390	4
Richard King	May 1932	Nueces	3860	4
Rio Grande City	May 1932	Starr	1348	4
Laurel	June 1932	Webb	1775	4
Tuleta	July 1932	Bee	3925	2
Hagist Ranch	Aug. 1932	Duval	2110	4
Bateman	Sept. 1932	Bastrop	2160	1
Moca	Dec. 1932	Webb	844	4
Carroll	1932	Bastrop	2300	1
Jacob	1932	McMullen-Live Oak	1050	1
Peters	Jan. 1933	Duval	2092	4
Hilbig	Feb. 1933	Bastrop	2350	1
Premont	April 1933	Jim Wells	3240	4
Greta	May 1933	Refugio	3500	2
	100 million barrel field			
Barbacoas	May 1933	Starr	2440	4
Cedar Creek	June 1933	Bastrop	1678	1
Wray	June 1933	Zapata	348	4
Coletto Creek	July 1933	Victoria	2776	2
Mathews	Sept. 1933	Williamson	850	1
Hoffman	Sept. 1933	Duval	3372	4
Eagle Hill	Oct. 1933	Duval	1406	4
Guerra	1933	Starr	2047	4
Pearsall	Feb. 1934	Frio	5350	1
	100 million barrel field			
Caesar	May 1934	Bee	3055	2
Dirks	May 1934	Bee	3850	2
Sinton	May 1934	San Patricio	5400	4
Tom O'Conner	June 1934	Refugio	4392	2
	100 million barrel field			
Bruni	June 1934	Webb	3400	4
Lopeno	July 1934	Zapata	2000	4
McNeill	Aug. 1934	Live Oak	4350	2
Colmena	Aug. 1934	Duval	1501	4
Comitas	Aug. 1934	Zapata	814	4
Sam Fordyce	Oct. 1934	Hidalgo	2650	4
Labbe	Nov. 1934	Duval-McMullen	2912	4
Loma Novia	Dec. 1934	Duval	2486	4
Port Lavaca	1934	Calhoun	6120	2
Angelita	1934	San Patricio	5367	4
Plymouth	April 1935	San Patricio	5479	4
	100 million barrel field			
Placedo	April 1935	Victoria	6003	2
Ray	April 1935	Bee	3915	2
Loma Alta	May 1935	McMullen	2210	1
Seven Sisters	May 1935	Duval	2186	4
Mauritz	June 1935	Jackson	5615	2
Lopez	June 1935	Webb-Duval	2126	4

Baldwin	July 1935	Nueces	3875	4
Plummer	Aug. 1935	Bee	3050	2
San Salvador	Aug. 1935	Hidalgo	6658	4
Kimbro	Sept. 1935	Travis	660	1
Charamousca	Sept. 1935	Duval	1450	4
Piedre Lumbre	Sept. 1935	Duval	1346	4
Staples	Oct. 1935	Caldwell-Guadalupe	695	1
Appling	Nov. 1935	Calhoun	7641	2
Clara Driscoll	Nov. 1935	Nueces	3773	4
Corpus Christi	Nov. 1935	Nueces	4070	4
Rancho Solo	Nov. 1935	Duval	2535	4
Taft	Dec. 1935	San Patricio	4872	4
	100 million barrel field			
Mineral	1935	Bee	7820	2
Rutledge	1935	Bee	3270	2
Loma Vista	Feb. 1936	Duval	2914	4
Alta Verde	March 1936	Brooks	914	4
Heyser	May 1936	Victoria-Calhoun	5475	2
Alta Mesa	June 1936	Brooks	1200	4
Aransas Pass	July 1936	Aransas	6535	4
El Mesquite	July 1936	Duval	2850	4
Flour Bluff	July 1936	Nueces	6650	4
Colorado	Aug. 1936	Jim Hogg	2829	4
Weser	Oct. 1936	Goliad	3655	2
La Blanca	Oct. 1936	Hidalgo	6650	4
Sullivan	1936	Nueces	5765	4
Telferner	Jan. 1937	Victoria	2500	2
London	Jan. 1937	Nueces	4698	4
Lundell	Jan. 1937	Duval	1347	4
Ricaby	Feb. 1937	Starr	1200	4
Sweden	Feb. 1937	Duval	5840	4
Chapman Ranch	March 1937	Nueces	6450	4
El Tanque	March 1937	Starr	1706	4
Ganado	April 1937	Jackson	5080	2
Benavides	April 1937	Duval	3495	4
Killam	April 1937	Webb	1920	4
Midway	April 1937	San Patricio	5227	4
Rancho Solo	April 1937	Duval	1849	4
McFaddin, North	May 1937	Victoria	6211	2
Stratton	May 1937	Nueces, Kleberg-Jim Wells	6291	4
	100 million barrel field			
Ezzell	June 1937	McMullen-Live Oak	1450	1
Luby	June 1937	Nueces	4298	4
Burnell, South	July 1937	Karnes-Bee	3675	2
Oilton	Aug. 1937	Webb	1885	4
Las Animas	Oct. 1937	Jim Hogg	1778	4
Ramirena	Jan. 1938	Live Oak	5073	2
Turkey Creek	Jan. 1938	Nueces	5380	4
Weslaco, South	Jan. 1938	Hidalgo	8604	4
Normanna, North	Feb. 1938	Bee	4250	2
Alfred	Feb. 1938	Jim Wells	3223	4
Rincon	Feb. 1938	Starr	4175	4

April 1938-Oct. 1939

	100 million barrel field			
Spiller	April 1938	Caldwell-Guadalupe	2100	2
Alice	April 1938	Jim Wells	4800	4
Tom Graham	April 1938	Jim Wells	5380	4
Rhode	May 1938	McMullen	1750	1
Casa Blanca	May 1938	Duval	1005	4
Fitzsimmons	May 1938	Duval	4300	4
Fairfield	June 1938	Bexar	1175	1
Riddle	June 1938	Bastrop	1705	1
Tesoro	June 1938	Duval	5054	4
La Reforma	June 1938	Starr-Hidalgo	5917	4
Munson	Aug. 1938	McMullen	1290	1
Walnut Creek	Aug. 1938	Caldwell	1230	1
La Rosa	Aug. 1938	Refugio	5350	2
West Ranch	Aug. 1938	Jackson	5100	2
	100 million barrel field			
Longhorn	Aug. 1938	Duval	4008	4
Kelsey	Sept. 1938	Jim Hogg-Brooks	4690	4
	100 million barrel field			
Sun	Sept. 1938	Starr	4818	4
Tulsita	Oct. 1938	Bee	3632	2
McAllen	Oct. 1938	Hidalgo	5970	4
Francitas	Nov. 1938	Jackson	7384	2
Bird Island	Nov. 1938	Kleberg	7205	4
Cedro Hill	Nov. 1938	Duval	900	4
Seeligson	Nov. 1938	Jim Wells	3200	4
	100 million barrel field			
Campana	1938	McMullen	2492	1
Mission River	1938	Refugio	4600	2
Sarco Creek	1938	Goliad	5020	2
Melon Creek	Feb. 1939	Refugio	5850	2
White Creek	Feb. 1939	Live Oak	1280	2
Lavernia	March 1939	Guadalupe	1250	1
Gandy	March 1939	Nueces	5400	4
Sullivan City	May 1939	Hidalgo-Starr	3394	4
Cordele	April 1939	Jackson	2592	2
Magnolia City	April 1939	Jim Wells-Kleberg	5440	4
Southland	April 1939	Duval	5284	4
Adami	May 1939	Webb	980	4
Ben Bolt	May 1939	Jim Wells	3978	4
Camada	May 1939	Jim Wells	5627	4
Minnie Bock	May 1939	Nueces	5410	4
Lake Pasture	June 1939	Refugio	5188	2
Chiltipin	June 1939	Duval	4760	4
Mathis, East	June 1939	San Patricio	5265	4
Reynolds	June 1939	Jim Wells	5102	4
Volpe	June 1939	Webb	2449	4
Texana	Aug. 1939	Jackson	5053	2
Robstown	Aug. 1939	Nueces	5548	4
Wade City	Aug. 1939	Jim Wells	4780	4
La Gloria	Sept. 1939	Brooks-Jim Wells	6560	4
Odem	Sept. 1939	San Patricio	6995	4
Fannin	Oct. 1939	Goliad	5053	2

Holland	Nov. 1939	Starr	2959	4
Heard Ranch	Dec. 1939	Bee	4700	2
Powderhorn	Dec. 1939	Calhoun	4620	2
Victoria	Dec. 1939	Victoria	2606	2
Tarancahuas	Dec. 1939	Duval	2010	4
Bonnie View	1939	Refugio	4190	2
Clark Muil	1939	Jim Wells	5304	4
Fagan	Feb. 1940	Refugio	5261	2
Thomaston	April 1940	DeWitt	7855	2
Glenn	April 1940	Webb-Zapata	2125	4
Bridwell	May 1940	Duval	4321	4
Orange Grove	May 1940	Jim Wells	5025	4
Glen	April 1940	Webb-Zapata	2123	4
Lolita	May 1940	Jackson	5254	2
Orange Grove	May 1940	Jim Wells	5025	4
Boyle	June 1940	Starr	3149	4
Cole	June 1940	Duval	1746	4
Gallagher	June 1940	Jim Wells	5190	4
Manila	June 1940	Jim Hogg	2542	4
Jay Welder	July 1940	Calhoun	5634	2
Baffin Bay	July 1940	Kleberg	7452	4
Margo	July 1940	Starr	2124	4
Patel	July 1940	Jim Hogg	3120	4
Tenney Creek	Aug. 1940	Caldwell	2335	1
Shield	Aug. 1940	Nueces	6600	4
Washburn Ranch	Sept. 1940	LaSalle	4822	1
Yzaguirre	Sept. 1940	Starr	4612	4
Elroy, East	Oct. 1940	Travis	665	1
Henshaw	Oct. 1940	Jim Wells	5128	4
Nichols	Oct. 1940	Hidalgo	3480	4
Willamar	Nov. 1940	Willacy	7574	4
	100 million barrel field			
Kreis	Dec. 1940	Duval	3195	4
Carrizo	Jan. 1941	Dimmit	2204	1
Coloma Creek	Jan. 1941	Calhoun	5887	2
Maurbro	Feb. 1941	Jackson	5200	2
Vienna	Feb. 1941	Lavaca	8440	2
Jones	Feb. 1941	Jim Hogg	4184	4
Genevieve	March 1941	Bee	4169	2
Agua Prieta	April 1941	Duval	4708	4
Haldeman	April 1941	Jim Wells	3710	4
Woodsboro	May 1941	Refugio	5876	2
Chaparosa	May 1941	Jim Hogg	2855	4
St. Charles	June 1941	Aransas	9308	4
LaWard	Sept. 1941	Jackson	5087	2
Petronilla	Sept. 1941	Nueces	6940	4
Brushy Creek	Oct. 1941	DeWitt	7710	2
Holbein	Oct. 1941	Jim Hogg	2780	4
Yturria	Nov. 1941	Starr	4193	4
Provident City	Dec. 1941	Lavaca	8540	2
Lentz	1941	Bastrop	2230	1
Mayo	Jan. 1942	Jackson	5067	2
McLean	Jan. 1942	Webb	3218	4

Jan. 1942-1943

Neuhaus	Jan. 1942	Jim Hogg	3435	4
Medio Creek	Feb. 1942	Bee	4855	2
Penitas	Feb. 1942	Hidalgo	5923	4
Rooke	Feb. 1942	Refugio-San Patricio	7132	4
Sejita	Feb. 1942	Duval	5784	4
Garcia	March 1942	Starr	3673	4
Koopman	March 1942	Jim Wells	4113	4
Yorktown	April 1942	DeWitt	7184	2
Alamo	April 1942	Hidalgo	5505	4
Harmon	May 1942	Jackson	5360	2
Frost	May 1942	Starr	4184	4
Green Branch	June 1942	McMullen	5740	1
Poth	June 1942	Wilson	4091	2
Quinto Creek	June 1942	Jim Wells	4860	4
Sterling	July 1942	Jackson	4820	2
Stewart	July 1942	Jackson	4991	2
Imogene	Sept. 1942	Atascosa	7560	1
Nordheim	Sept. 1942	DeWitt-Karnes	8991	2
Dinn	Sept. 1942	Duval	2479	4
Cadiz	Oct. 1942	Bee	3731	2
Caesar, South	Oct. 1942	Bee	6458	2
Cadena	Oct. 1942	Duval	3460	4
East	Oct. 1942	Jim Hogg	4702	4
Tijerina-Canales-Blucher	Nov. 1942 100 million barrel field	Jim Wells-Kleberg	6186	4
Collier	Nov. 1942	Jackson	2169	2
Tynan	Dec. 1942	Bee	4195	2
Dulup	Dec. 1942	Webb	1803	4
Hobson	Feb. 1943	Karnes	3954	2
Strake	Feb. 1943	Duval	3142	4
Wilson	Feb. 1943	Jim Wells	7345	4
Hondo Creek	March 1943	Karnes	6557	2
Violet	March 1943	Nueces	7195	4
Weil	March 1943	Jim Hogg	5037	4
Kittie	April 1943	Live Oak	1432	2
Lockhart	April 1943	Starr	1528	4
Huff	May 1943	Refugio	4992	2
Slick	May 1943	Goliad-DeWitt	7503	2
Blanconia	June 1943	Bee	4944	2
Porter	June 1943	Karnes	3969	2
Ross	June 1943	Starr	2845	4
Yaeger	June 1943	Jim Hogg	3732	4
Goebel	July 1943	Live Oak	7025	2
Cameron	Aug. 1943	Starr-	4120	4
Coquat	Sept. 1943	Live Oak	7542	2
Armstrong	Sept. 1943	Jim Hogg	3218	4
Robinson	Sept. 1943	Duval	4850	4
Little Kentucky	Oct. 1943	Jackson	5896	2
Thomas Lockhart	Nov. 1943	Duval	4671	4
Runge	Dec., 1943	Karnes	6595	2
Terrell Point	Dec. 1943	Go!iad	4650	2
Scott & Harper	Dec. 1943	Brooks	6875	4
Washburn, South	1943	LaSalle	6032	1

Clara	1943	Duval	3768	4
Chapa	Jan. 1944	Live Oak	8166	2
LaSalle	Jan. 1944	Jackson	5062	2
Brownlee	Jan. 1944	Jim Wells	4950	4
Dougherty	Jan. 1944	Nueces	5550	4
Verando	Jan. 1944	Jim Hogg	2869	4
Warms!ey	Feb. 1944	DeWitt	7085	2
Yougeen	Feb. 1944	Bee	3741	2
Retamia	Feb. 1944	Webb	4946	4
Gregory	March 1944	San Patricio	8496	4
Rosita	March 1944	Duval	3904	4
Sellers	March 1944	Jim Wells	4938	4
Weesatche	May 1944	Goliad	7980	2
Cabeza Creek, South	May 1944	Goliad	8306	2
Mission Valley	May 1944	Victoria	8373	2
Las Mujeres	May 1944	Jim Hogg	5127	4
Burnell	June 1944	Karnes-Bee	6778	2
Goliad, North	June 1944	Goliad	2671	2
Green	June 1944	Karnes	6434	2
Brayton	June 1944	Nueces	7197	4
Welder	June 1944	Duval	2944	4
Pridham Lake	July 1944	Victoria	2959	2
Toro Creek	July 1944	Bee	3885	2
Jakalak	July 1944	Starr	1828	4
Charlotte	Aug. 1944	Atascosa	5000	1
Koontz	Aug. 1944	Victoria	4750	2
Dan Sullivan	Aug. 1944	Brooks	8505	4
Port Isabel	Aug. 1944	Cameron	5980	4
Richardson	Sept. 1944	Duval	1765	4
Sinton, West	Sept. 1944	San Patricio	5071	4
Clayton	Oct. 1944	Live Oak	6428	2
Hope	Oct. 1944	Lavaca	7710	2
Herbst	Oct. 1944	Duval	2821	4
McCaskill	Nov. 1944	Karnes	7404	2
Ragsdale	Nov. 1944	Victoria	4436	2
Anaqua	Dec. 1944	Victoria	2654	2
Falls City	Dec. 1944	Karnes	4651	2
Blanchard	Dec. 1944	Duval	3862	4
Almond	1944	Jim Wells	3222	4
Irving	Jan. 1945	Starr	3285	4
McDermott	Jan. 1945	Zapata	2609	4
Raymondville	Jan. 1945	Willacy	6346	4
Ricardo	Jan. 1945	Kleberg	7243	4
Leona River	Feb. 1945	Zavala	4212	1
Stockdale	Feb. 1945	Wilson	5148	1
Maedgen	Feb. 1945	San Patricio	4160	4
O'Neill	Feb. 1945	San Patricio	3375	4
Gormac	March 1945	Duval	2430	4
Mariposa	April 1945	Brooks	6654	4
Moody Ranch	May 1945	Jackson	8822	2
Word	May 1945	Lavaca	5862	2
Willman	May 1945	San Patricio	4545	4
Boyce	June 1945	Goliad	7640	2

June 1945–Feb. 1947

Yoakum	June 1945	DeWitt-Lavaca	8160	2
Flores	June 1945	Starr-Hidalgo	5840	4
Gyp Hill	June 1945	Brooks	4070	4
Rosalia	June 1945	Duval	3155	4
Staggs	June 1945	Webb	2800	4
Albrecht, West	July 1945	Live Oak	8160	2
Hinnant	July 1945	Live Oak	5772	2
Vienna	July 1945	Lavaca	8468	2
Riviera	July 1945	Kleberg	7453	4
Santa Maria	July 1945	Hidalgo	8854	4
Rachal	Aug. 1945	Brooks	6100	4
Hysaw	Sept. 1945	Karnes	3900	2
Theis	Sept. 1945	Bee	3565	2
Naylor	Sept. 1945	Jim Wells	3974	4
Goliad, West	Oct. 1945	Goliad	5010	2
La Sal Vieja	Oct. 1945	Willacy	8935	4
Berclair	Nov. 1945	Goliad-Bee	9905	2
Kay Creek	Nov. 1945	Victoria	2290	2
Sheriff	Nov. 1945	Calhoun	5710	2
Holland	Nov. 1945	Starr	2860	4
Parsons	Nov. 1945	Starr	3498	4
Marcado Creek	Dec. 1945	Victoria	4600	2
Borregas	Dec. 1945	Kleberg	5148	4
	100 million barrel field			
Refugio-Heard	1945	Refugio	6217	2
Ira	1945	Starr	3915	4
Jourdanton	Jan. 1946	Atascosa	7366	1
Muil	Feb. 1946	Atascosa	8987	1
Welder Ranch	Feb. 1946	Calhoun-Victoria	5898	2
Pita	March 1946	Brooks	7112	4
Poehler	April 1946	Goliad	8321	2
Sliva	April 1946	Bee	3020	2
Morales	May 1946	Jackson	4048	2
Cage Ranch	May 1946	Brooks	8022	4
Falfurrias	May 1946	Jim Wells-Kleberg-Brooks	7345	4
Wiegang	July 1946	Atascosa	3750	1
Hughes	Aug. 1946	Webb	1870	4
Clay, West	Sept. 1946	Live Oak	10121	2
Strauch	Oct. 1946	Bee	7800	2
Bishop	Nov. 1946	Nueces	6992	4
Round Lake	Nov. 1946	San Patricio	5440	4
Schwartz	Dec. 1946	Live Oak	4016	2
Farenthold	Dec. 1946	Nueces	5602	4
Lacy	Dec. 1946	Hidalgo-Cameron	7983	4
El Peyote	1946	Jim Hogg	3500	4
La Bahia	Jan. 1947	Goliad	5550	2
Meyersville	Jan. 1947	DeWitt-Victoria	8480	2
Wyrick	Jan. 1947	Refugio	4962	2
Cam	Jan. 1947	Duval	3319	4
Del Monte	Feb. 1947	Zavala	3303	1
Matagorda Bay	Feb. 1947	Calhoun	4966	2
Kennard	Feb. 1947	Starr	3844	4
McGill	Feb. 1947	Kennedy	7676	4

Roche	March 1947	Refugio	6300	2
Salem	March 1947	Victoria	4110	2
Cabeza Creek	April 1947	Goliad	7679	2
River	April 1947	Goliad	5528	2
Yoward	April 1947	Bee	7483	2
Dee Davenport	April 1947	Starr	4098	4
Haldeman, South	April 1947	Jim Wells	4494	4
Perez	April 1947	Webb	1265	4
Seventy-Six	April 1947	Duval	1322	4
Sommer	April 1947	Nueces	5982	4
Bloomington	May 1947	Victoria	4600	2
Harris	May 1947	Live Oak	7330	2
County Line	May 1947	Duval	2472	4
Mantor Briggs	May 1947	Nueces	4938	4
Watkins	May 1947	Webb	810	4
Joliet	June 1947	CaIdwell	2159	1
Little Alamo	June 1947	McMullen	2283	1
Caronician	June 1947	Live Oak	1281	2
Doss	June 1947	Hidalgo-Starr	3832	4
Edroy	June 1947	San Patricio	953	4
Joe Moss	June 1947	Zapata	715	4
Severiano	June 1947	Duval	2540	4
Squire	June 1947	Duval	3251	4
Green Lake	July 1947	Calhoun	5980	2
Halletsville	July 1947	Lavaca	7750	2
Minoak	July 1947	Bee	7560	2
Jay Simmons	July 1947	Starr	5026	4
Los Indios	July 1947	Hidalgo	6325	4
Brandt	Aug. 1947	Goliad	7708	2
Foester	Aug. 1947	Calhoun	7920	2
Heinzeville	Aug. 1947	Goliad	7560	2
Fulton Beach	Aug. 1947	Aransas	7133	4
Bear Creek	Sept. 1947	Medina	2240	1
Cecil	Sept. 1947	Hidalgo	4000	4
Rainbow Bend	Oct. 1947	LaSalle	5247	1
Pena	Oct. 1947	DeWitt	7512	2
Bullard	Oct. 1947	San Patrieio	6662	4
Nicholson	Oct. 1947	Webb	5574	4
Vidauri, West	Nov. 1947	Refugio	4816	2
Javelina	Nov. 1947	Hidalgo	5408	4
Charco	Dec. 1947	Goliad	4590	2
Jamie	Dec. 1947	Victoria	5102	2
New Angelita	Dec. 1947	San Patricio	5976	4
Murulla	1947	Duval	3920	4
Cologne	Feb. 1948	Victoria	2115	2
Holly	Feb. 1948	DeWitt	7575	2
Long Mott, East	Feb. 1948	Calhoun	7264	2
Sandy Creek	Feb. 1948	Jackson	4540	2
Andross	Feb. 1948	Duval	3804	4
Quien Sabe	Feb. 1948	Webb	2323	4
San Pablo	Feb. 1948	Jim Hogg	3838	4
Inez, South	March 1948	Victoria	516	2
Coastal	March 1948	Starr	5760	4

May 1948–May 1949

Gottschalt	May 1948	Goliad	7606	2
Marshall	May 1948	Goliad	8050	2
Lovia	May 1948	Duval	2660	4
Pantex	May 1948	Duval	2543	4
Sarita	May 1948	Kenedy-Kleberg	5894	4
Zaragosa	June 1948	Duval	2289	4
Beasley	July 1948	San Patricio-Bee	4900	2
Arrowhead	July 1948	Hidalgo	5967	4
El Panal	July 1948	Starr	4671	4
Copano Bay	Aug. 1948	Aransas	7143	4
DCRC-79	Aug. 1948	Duval	1454	4
Poquito Creek	Aug. 1948	Duval	2517	4 Trans. to Gruy F
Catarina	Sept. 1948	Dimmit	4888	1
Floresville, East	Sept. 1948	Wilson	2727	1
Blanco Creek	Sept. 1948	Refugio	6600	2
Melrose	Sept. 1948	Goliad	7845	2
Tom Lyne	Sept. 1948	Live Oak	9230	2
Borosa	Oct. 1948	Starr	4291	4
La Copita	Oct. 1948	Starr	4539	4
Palito Blanco	Oct. 1948	Jim Wells	4455	4
Semmes	Oct. 1949	Starr	2426	4
Vaello	Oct. 1948	Duval	4880	4
San Miguel Creek	Nov. 1948	McMullen	5406	1
Maxine	Nov. 1948	Live Oak	8820	2
Riverdale	Nov. 1948	Goliad	9178	2
Paso Real	Nov. 1948	Willacy	8585	4
Poteet	Dec. 1948	Atascosa	3567	1
Melbourne	Dec. 1948	Calhoun	8587	2
Speaks	Dec., 1948	Lavaca	1780	2
Nueces Bay	Dec. 1948	Nueces	3290	4
Salt Lake	Dec. 1948	Aransas	7180	4
Goliad	1948	Goliad	4254	2
Princess Louise	1948	San Patricio	5380	4
Karon, South	Jan. 1949	Live Oak	8180	2
Cody	Jan. 1949	Nueces	4323	4
Edinburg	Jan. 1949	Hidalgo	7563	4
Mercedes	Jan. 1949	Hidalgo	7266	4
Rita	Jan. 1949	Kenedy	7076	4
Santa Fe	Jan. 1949	Brooks	3040	4
Big Foot	Feb. 1949	Frio	3205	1
Breeden	Feb. 1949	Goliad	5086	2
Clarkwood, South	Feb. 1949	Nueces	7224	4
Gregg Wood	Feb. 1949	Starr	1314	4
Winter Garden	March 1949	Dimmit	2555	1
Fort Merrill	March 1949	Live Oak	4562	2
Rudman	March 1949	Bee	7010	2
Mustang Island	March 1949	Nueces	7433	4
Reynaga	March 1949	Starr	1130	4
Wolf Creek	April 1949	Guadalupe	2041	1
Bailey	April 1949	Nueces	5427	4
Cantu	April 1949	Duval	4674	4
Aldrete	May 1949	Refugio	6300	2
Hall Ranch	May 1949	Live Oak	7750	2

Maley	May 1949	Bee	7399	2
Daskam	May 1949	Hidalgo	4260	4
Flowella	May 1949	Brooks	7238	4
Spurs, South	June 1949	Live Oak	8314	2
Calallen	June 1949	Nueces-San Patricio	7212	4
Diego	June 1949	Duval	4730	4
Doc	June 1949	Jim Hogg	3742	4
Hidalgo, West	June 1949	Hidalgo	1938	4
Realitos, East	June 1949	Duval	4226	4
Pharr	July 1949	Hidalgo	9370	4
Jayeddie	Aug. 1949	Guadalupe	1970	1
Albrecht	Aug. 1949	Goliad	4512	2
Karon	Aug. 1949	Live Oak	7062	2
Donna	Aug. 1949	Hidalgo	6482	4
Laguna Larga	Aug. 1949	Kleberg	6497	4
London Gin	Aug. 1949	Nueces	4496	4
McCullough	Aug. 1949	Nueces	5538	4
Santa Teresa	Aug. 1949	Starr	4684	4
Viboras	Aug. 1949	Brooks	8110	4
	100 million barrel field			
Goree	Sept. 1949	Bee	7188	2
Cement	Sept. 1949	Nueces	6620	4
Saspamco	Oct. 1949	Wilson-Bexar	2833	1
Buena Vista	Oct. 1949	Starr	1700	4
Edlasater	Oct. 1949	Jim Hogg	2096	4
El Ebanito	Nov. 1949	Starr	4260	4
Mud Flats	Nov. 1949	Aransas-San Patricio	9244	4
Santos	Nov. 1949	Webb	2066	4
Bennview	Dec. 1949	Jackson	6153	2
Ben Shelton	Dec. 1949	Victoria	7420	2
Bedwell	Dec. 1949	Zapata	1301	4
Cox & Hamon	Dec. 1949	Duval	3572	4
Forty-Nine	Dec. 1949	Duval	1533	4
Hollow Tree	Dec. 1949	Jim Wells	2959	4
Orlee	Dec. 1949	Duval	1686	4
Russek	Jan. 1950	Lavaca	5284	2
Lou Ella	Jan. 1950	Jim Wells	5012	4
Portilla	Feb. 1950	San Patricio	7392	4
Virginia	Feb. 1950	Aransas	10505	4
Minnita	March 1950	Victoria	2952	2
El Ebanito	March 1950	Starr	6411	4
Hoffman	March 1950	Duval	2038	4
Swan Lake	April 1950	Jackson	6000	2
Cooksey	May 1950	Bexar	1558	1
Atlee	May 1950	Duval	5000	4
Gruy	May 1950	Duval	2610	4
La Parita	June 1950	Atascosa	1531	1
Triple -A-	June 1950	San Patricio	6652	4
Arneckeville	July 1950	DeWitt	7678	2
Billings	Aug. 1950	Webb	2389	4
Helen Gohlke	Aug. 1950	DeWitt	8096	2
Harvey	Aug. 1950	San Patricio	8104	4
Marks	Aug. 1950	Starr	2124	4

Sept. 1950–July 1953

Nixon	Sept. 1950	Gonzales	5839	1
Battle	Sept. 1950	Victoria	4756	2
Cottonwood Creek	Nov. 1950	DeWitt	7647	2
Hindes	Jan. 1951	Frio	5600	1
Pleasanton	Jan. 1951	Atascosa	8157	1
Francitas, North	March 1951	Jackson	8550	2
Pruitt	April 1951	Atascosa	4900	1
Parilla	April 1951	Duval	2200	4
Los Torritos	May 1951	Hidalgo	5848	4
Pharr, SW	May 1951	Hidalgo	8657	4
Higgins	June 1951	Wilson	5612	1
Hostetter	June 1951	McMullen	9680	1
Kyote	June 1951	Atascosa	3549	1
Red Fish Bay	June 1951	Nueces	7960	4
Magnolia Beach	Aug. 1951	Calhoun	8704	2
Stulting	Oct. 1951	Gonzales	2036	1
El Javali	Oct. 1951	Duval	2310	4
Viggo	Oct. 1951	Duval	3700	4
Captain Lucy	Dec. 1951	Jim Wells	5671	4
Tal Vez	Dec. 1951	Webb	1975	4
Cannan	Jan. 1952	Bee	6750	2
Westhoff Ranch	Jan. 1952	Jackson	6750	2
Little Rose	Jan. 1952	Jim Wells	5080	4
Gutierrez	Feb. 1952	Webb	1667	4
Acleto Creek	March 1952	Wilson	3068	1
Wherry & Green	March 1952	Atascosa	1710	1
Emma Haynes	March 1952	Goliad	7650	2
Crestonia	March 1952	Duval	5023	4
Howell	April 1952	Jim Wells	5310	4
Dinsmoor	May 1952	Atascosa	2060	1
Spartan	May 1952	San Patricio	8077	4
Christensen	June 1952	Live Oak	3700	2
Ewing	July 1952	San Patricio	6888	4
Tabasco	July 1952	Hidalgo	6082	4
Viboras	Aug. 1952	Brooks	8495	4
Alvarez	Sept. 1952	Starr	716	4
El Benadito	Sept. 1952	Starr	5391	4
Hodges	Sept. 1952	San Patricio	4914	4
Matze	Oct. 1952	Goliad	3214	2
Hailey	Nov. 1952	Live Oak	4255	2
Henze	Feb. 1953	DeWitt	4500	2
Ware	Feb. 1953	Starr	2266	4
Hynes Ranch	March 1953	Refugio	5940	2
Rand Morgan	March 1953	Jim Wells	5262	4
St. Joseph	March 1953	Webb	3085	4
De Spain	April 1953	Webb	2598	4
Mills Bennett	April 1953	Brooks	4625	4
Ransom Island	April 1953	Nueces	7657	4
Puerto Bay	May 1953	San Patricio	6746	4
Elaine	June 1953	Dimmit	3624	1
Barvo	June 1953	Goliad	4141	2
Steinmeyer	June 1953	Bee	4166	2
San Domingo	July 1953	Bee	2872	2

North Pasture	Aug. 1953	San Patricio	7659	4
Rowe	Aug. 1953	Jim Wells	5285	4
Mary Ellen O'Connor	Sept. 1953	Refugio	5856	2
Brelum	Oct. 1953	Duval	1956	4
Del Grullo	Nov. 1953	Kleberg	10603	4
Enos Cooper	Nov. 1953	San Patricio	7800	4
Janssen	Feb. 1954	Karnes	4085	2
Kellers Bay	Feb. 1954	Calhoun	8668	2
Moos	Feb. 1954	Webb	4109	4
Aloe	March 1954	Victoria	2645	2
Conchudo	March 1954	Webb	2327	4
Mary Bluntzer	March 1954	San Patricio	6114	4
Rockport, West	March 1954	Aransas	8284	4
Rodriquez	April 1954	LaSalle	412	1
Horn	May 1954	Atascosa	5262	1
Scotty	May 1954	Wilson	1610	1
Aguilares	May 1954	Webb	6723	4
Donna	May 1954	Hidalgo	6955	4
El Paistle	May 1954	Kenedy	6078	4
La Sara	June 1954	Willacy	6027	4
Crown, East	July 1954	Atascosa	4300	1
Cayo Del Oso	July 1954	Nueces	6647	4
Viola	July 1954	Nueces	4013	4
Alazan	August 1954 100 million barrel field	Kleberg	5748	4
McDaniel	Nov. 1954	Jackson	5860	2
Chevron	Nov. 1954	Kleberg	7760	4
Mudhill	Jan. 1955	Bee	3161	2
Het	Jan. 1955	San Patricio	2525	4
Prado	Jan. 1955	Jim Hogg	3732	4
Dilworth	Feb. 1955	McMullen	11170	1
Stacy	March 1955	Frio	4280	1
Patty	March 1955	Jim Hogg	3217	4
Leming	April 1955	Atascosa	2934	1
Linne	April 1955	Wilson	1362	1
Olivia	May 1955	Calhoun	8327	2
Jaboncillos Creek	May 1955	Duval	3100	4
Velma	June 1955	Atascosa	4635	1
Swinney Switch	June 1955	Live Oak	4322	2
Chess	June 1955	Willacy	6717	4
Santo Nino	June 1955	Webb	3754	4
Bowman	July 1955	Wilson	1065	1
Galba	July 1955	Atascosa	4770	1
May	July 1955	Kleberg	9320	4
Reyna	Aug. 1955	Starr	623	4
Stroman	Aug. 1955	Jim Hogg	3624	4
Braulia	Sept. 1955	Starr	4280	4
Dejay	Sept. 1955	Duval	2026	4
Elva	Sept. 1955	Duval	3518	4
Petrox	Sept. 1955	Duval	2024	4
Maude B. Traylor	Nov. 1955	Calhoun	8325	2
San Antonio Bay	Nov. 1955	Calhoun	7711	2
Thanksgiving	Nov. 1955	Duval	3286	4

Dec. 1955–Jan. 1959

Dunwolters	Dec. 1955	Karnes	3970	2
Rinehart	Jan. 1956	Bexar	1163	1
Alkek, North	Jan. 1956	Goliad	3350	2
Kaye	Feb. 1956	Wilson	2548	1
Roeder	Feb. 1956	DeWitt	9318	2
Hidalgo	April 1956	Hidalgo	6176	4
La Union	April 1956	Starr	2270	4
Navarro Cole	April 1956	Webb	1800	4
Bettis	May 1956	San Patricio	4852	4
Encino	May 1956	San Patricio	6390	4
Coughran	June 1956	Atascosa	6335	1
Embleton	July 1956	Jim Wells	5278	4
Fashing	Aug. 1956	Atascosa	10400	1
Johns	Aug. 1956	Duval	3954	4
Biel	Nov. 1956	Webb	1001	4
Knolle	Nov. 1956	Jim Wells	5093	4
Sacatosa	Dec. 1956	Maverick	1248	1
Paul White	Jan. 1957	Jim Wells	4624	4
Witte	Feb. 1957	Victoria	5847	2
Kens	March 1957	Guadalupe	2346	1
Ramada	April 1957	Nueces	3074	4
Stehle	April 1957	Jim Wells	5183	4
Waters	April 1957	Hidalgo	5970	4
Rincon, North	May 1957	Starr	7473	4
Touchstone	July 1957	Frio	3600	1
Bloomberg	July 1957	Hidalgo	3950	4
Viboras	July 1957	Brooks	8392	4
Indian Point	Aug. 1957	Nueces	9692	4
Mesquite Bonita	Aug. 1957	Duval	3109	4
Hornbuckle	Sept. 1957	Jackson	7209	2
Amargosa	Sept. 1957	Jim Wells	2746	4
San Carlos	Sept. 1957	Hidalgo	7559	4
La Huerta	Oct. 1957	Duval	3121	4
Pedernal	Oct. 1957	Starr	4589	4
Yturria	Oct. 1957	Starr	4913	4
Buckeye Knoll	Jan. 1958	Live Oak	3918	2
Washburn, North	March 1958	LaSalle	4718	1
Kentucky Mott	April 1958	Victoria	5884	2
Good Friday	April 1958	Duval	3562	4
Miller & Fox	April 1958	Jim Wells	3383	4
Torch	May 1958	Zavala	3006	1
Que Sera	May 1958	Webb	1129	4
Henry	June 1958	McMullen	3526	1
Lucille	July 1958	Live Oak	3237	2
Shepherd	July 1958	Hidalgo	2738	4
San Patricio	Aug. 1958	San Patricio	8086	4
Welhed	Aug. 1958	San Patricio	5847	4
Gabrysch	Sept. 1958	Jackson	6315	2
Kelly Ranch	Oct. 1958	Nueces	6269	4
Lonnie Glasscock	Nov. 1958	Victoria	3505	2
Alazan, North	Dec. 1958	Kleberg	7230	4
A.& H.	Jan. 1959	Duval	4142	4
Encinitas	Jan. 1959	Brooks	8035	4

Rowena, North	Jan. 1959	Jim Wells	4848	4
Hugh Fitzsimmons	March 1959	Dimmit	3750	1
Karen Beauchamp	March 1959	Goliad	2550	2
Rico	April 1959	Hidalgo	6924	4
Thirteen	June 1959	Dimmit	5804	1
Weaver & Olson	July 1959	Wilson	1460	1
Person	Aug. 1959	Karnes	10886	2
Buena Suerte	Aug. 1959	Duval	5345	4
Paisano Creek	Oct. 1959	Live Oak	3951	2
Malo Sueno	Oct. 1959	Webb	5550	4
Batesville, South	Dec. 1959	Zavala	4400	1
Renshaw	Dec. 1959	Atascosa	3213	1
El Toro, South	Feb. 1960	Jackson	5624	2
Presa De Oro	Feb. 1960	Webb	1078	4
Indio	April 1960	Zavala	4238	1
Milbur	April 1960	Milam	3475	1
Half Moon Reef	June 1960	Aransas	5823	4
C. A. Winn, East	Aug. 1960	Live Oak	4387	2
Arnold David	Aug. 1960	Nueces	6090	4
Panna Maria	Sept. 1960	Karnes	10864	2
Romeo	Nov. 1960	Jim Hogg	3587	4
Jake Hamon	Dec. 1960	McMullen	2728	1
Zapata	April 1961	Zapata	1050	4
Dinero	May 1961	Live Oak	5124	2
Laura Thompson	May 1961	Bee	5530	2
Monte Christo, South	June 1961	Hidalgo	5878	4
Bracero	Aug. 1961	Zavala	3816	1
Labus	Sept. 1961	Karnes	10740	2
Una West	Dec. 1961	Victoria	6500	2
Tres Encinos	Jan. 1962	Brooks	7156	4
Dubose	Feb. 1962	Gonzales	12135	1
Holdsworth	March 1962	Zavala	3801	1
San Roman	March 1962	Starr	4685	4
Spring Creek	April 1962	Live Oak	2526	2
Arnold Weldon	Aug. 1962	McMullen	3481	1
L. Ranch	Aug. 1962	Jackson	9064	2
Davy	Nov. 1962	Karnes	11001	2
Ella Borroum	Nov. 1962	Bee	6642	2
Doehrmann, North	Dec. 1962	DeWitt	8162	2
Wood High	Jan. 1963	Victoria	5567	2
Zimet	Jan. 1963	Webb	1202	4
Circle-A-	Feb. 1963	Goliad	5354	2
Papalote, East	Feb. 1963	Bee	3244	2
Zoller	March 1963	Calhoun	9131	2
McGregor	March 1963	Nueces	4014	4
Jack Frost	May 1963	LaSalle	1300	1
Loma Blanca	May 1963	Brooks	7480	4
Chupick	July 1963	Atascosa	5243	1
Thompsonville, NE.	Aug. 1963	Webb	5700	4
Texam, East	Sept. 1963	Live Oak	2260	2
Zarsky, South	Sept. 1963	Refugio	8264	2
McAllen Ranch	Sept. 1963	Hidalgo	12415	4
Glen Hummel	Oct. 1963	Wilson	2432	1

Nov. 1963-April 1969

Big Caesar	Nov. 1963	Kleberg	7315	4
Bayside	Feb. 1964	Refugio	8005	2
Palo Alto	March 1964	Victoria	6644	2
Texana	March 1964	Jackson	6618	2
Madero	March 1964	Kleberg	9712	4
Massingill, West	April 1964	Bee	5597	2
Canelo	April 1964	Kleberg	7276	4
Crowther	May 1964	McMullen	5997	1
Sixto	Sept. 1964	Starr	2247	4
Partido	Oct. 1964	Kleberg	5126	4
Hosek	Nov. 1964	Wilson	2392	1
Skipper	Nov. 1964	Brooks	9058	4
Austin Pierce	Dec. 1964	Gonzales	10856	1
Eloise	Jan. 1965	Victoria	6574	2
Romero Trap	Jan. 1965	Jim Wells	5131	4
Cousins	March 1965	McMullen	5744	1
Ginny, East	April 1965	San Patricio	7645	4
Hinde	April 1965	Starr	5420	4
Live Oak Lake	June 1965	Goliad	5005	2
Gloriana	Aug. 1965	Wilson	1630	1
Chemcel	Sep. 1965	Nueces	7445	4
Escondido	Oct. 1965	Jim Hogg	4854	4
Dragoon Creek	Nov. 1965	McMullen	1022	1
Saner Ranch	Nov. 1965	Maverick	1460	1
Wheeler-Mag	Nov. 1965	McMullen	1193	1
Houdman	Dec. 1965	Live Oak	2356	2
Rio-Sheerin	Jan. 1966	Starr	778	4
Cabazos	Feb. 1966	Kenedy	5072	4
Neely, East	Feb. 1966	Duval	1156	4
Wipif	April 1966	Maverick	2115	1
Monson, Southwest	May 1966	Karnes	6273	2
Corpus Channel, NW	June 1966	Nueces	10418	4
Phil Power	July 1966	Refugio	5208	2
Concordia, West	April 1967	Nueces	5176	4
Shell Point	April 1967	Aransas	6935	4
Mosquito Point	May 1967	Calhoun	9185	2
Rancho Viejo	May 1967	Jim Hogg	4853	4
Bennview, East	June 1967	Jackson	6564	2
Burgentine Lake, SW	June 1967	Aransas	7683	4
Encinal Channel	Aug. 1967	Nueces	10200	4
Carancahua Creek	March 68	Jackson	9657	2
Julian	March 1968	Kenedy	8062	4
Portland, West	March 1968	San Patricio	9096	4
Mikeska, North	June 1968	Live Oak	9091	2
Panther Reef	June 1968	Calhoun	8705	2
Estes Cove	June 1968	Aransas	7807	4
Rattlesnake Point	Aug. 1968	Aransas	8860	4
Orcones	Oct. 1968	Duval	5300	4
Las Pintas	Nov. 1968	Webb	7113	4
Fitzpatrick	Jan. 1969	Maverick	2020	1
Indian Creek	Jan. 1969	Jackson	9610	2
Devils Water Hole	March 1969	McMullen	6990	1
Big Wells, Northeast	April 1969	Zavala	5435	1

Balfour	April 1969	Dewitt	3294	2
Judd City	June 1969	Starr	5969	4
W. P. Holloway	July 1969	Atascosa	6486	1
Palafox	Sept. 1969	Webb	3058	4
Cherokee	Oct. 1969	Milam	3050	1
M & F	Oct. 1969	Zapata	5270	4
Berkow	Feb. 1970	Bee	4920	2
Big Wells	April 1970	Dimmit	5425	1
Don Woody	Aug. 1970	Starr	3994	4
McCampbell	Aug. 1970	San Patricio	12221	4
Pita Island	Aug. 1970	Nueces	8942	4
Bun, North	Sept. 1970	Maverick	2347	1
Good Luck	Sept. 1970	Dimmit	5849	1
Hammond	Sep. 1970	Zavala	3259	1
King Ranch, West	Dec. 1970	Willacy	5720	4
Storey	May 1971	LaSalle	5256	1
Duke	Sept. 1971	Atascosa	4714	1
Pena Creek	Sept. 1971	Dimmit	3400	1
Kincaid-Winn	Oct. 1971	Maverick	7739	1
Bolland Ranch	Aug. 1972	Live Oak	4525	2
Elpar	Sept. 1972	Jim Wells	5494	4
Bias Uribe	Oct. 1972	Zapata	1726	4
El Puerto	Jan. 1973	Starr	856	4
Valentine	Feb. 1973	Lavaca	9154	2
Freddy Hutt	March 1973	McMullen	4975	1
Las Tiendas	March 1973	Webb	6813	4
Shary	April 1973	Hidalgo	7270	4
Bulloh	May 1973	Milam	2012	1
Boory	July 1973	Cameron	5790	4
Laguna Larga	July 1973	Nueces	8682	4
Stets	Aug. 1973	Duval	3212	4
Stumberg Ranch	Sept. 1973	Dimmit	4630	1
New Zapata	Jan. 1974	Zapata	1774	4
Clark Ranch	June 1974	Webb	7935	4
Saint Thirteen	July 1974	Dimmit	3585	1
Bernsen	Aug. 1974	Jim Wells	5007	4
Elsa, South	Sept. 1974	Hidalgo	7808	4
Jasper-Webb	Nov. 1974	Webb	3576	4
Zone 21-B	Dec. 1974	Kleberg	7100	4
Rochelle	Feb. 1975	Maverick	2347	1
Rocky Creek, East	Feb. 1975	Dimmit	2250	1
Laredo, South	March 1975	Webb	5742	4
Ray Point	July 1975	Live Oak	6284	2
Spinach	Aug. 1975	Zavala	3954	1
Randolph	Sep. 1975	Zavala	3767	1
Sutil	Oct. 1975	Wilson	6297	1
Christine, Southeast	Jan. 76	Atascosa	5238	1
Pawelek	Feb. 76	Wilson	7238	1
Monteola	Feb. 1976	Bee	13439	2
Dilley	March 1976	Frio	7554	1
Mag	March 1976	Gonzales	7484	1
Goose Island	March 1976	Aransas	7104	4
Patty Huff	April 1976	Atascosa	5340	1

June 1976-July 1980

Wheeler	June 1976	McMullen	9920	1
Dos Arroyos	July 1976	Zavala	5074	1
Camaron Ranch	Aug. 1976	LaSalle	2235	1
Finley-Webb	Aug. 1976	Webb	4768	4
Pilgrim	Sept. 1976	Gonzales	8027	1
Beeville, West	Sep. 1976	Bee	3927	2
Menking	Sept. 1976	Lavaca	8977	2
Nine Mile Point, W.	Oct. 1976	Aransas	11780	4
Cosmopolitan	Nov. 1976	Gonzales	6907	1
Palmer	Nov. 1976	Wilson	2871	1
Block 901	Jan. 1977	Nueces	8715	4
Cost	March 1977	Gonzales	7296	1
Cebolla	Aug. 1977	Zavala	3660	1
Winston	Aug. 1977	Gonzales	7376	1
Gaffney, Southwest	Aug. 1977	Goliad	5336	2
Escondido Creek	Aug. 1977	Kleberg	7346	4
Smithville	Sept. 1977	Bastrop	6580	1
Sherman Offshore, SW	Oct. 1977	Calhoun	3017	2
Pilosa, East	May 1977	Zavala	2574	1
Mien, North	Jan. 1978	Jim Hogg	2910	4
Dos Payasos, East	March 1978	Webb	7206	4
linde	March 1978	Starr	7054	4
Padre, North	March 1978	Kleberg	8782	4
Tar Baby	April 1978	Hidalgo	15532	4
Donna Buchanan	May 1978	Hidalgo	7831	4
Potrero Lopena, SW	June 1978	Kenedy	7412	4
Sweet Home	July 1978	Lavaca	6576	2
Mustang Isl. Blk. 881-L	July 1978	Kleberg	14526	4
Black Diamond	Nov. 1978	Jim Hogg	4552	4
Hinojosa, North	Nov. 1978	Kleberg	6021	4
Borchers, North	Dec. 1978	Lavaca	8018	2
Clear	Dec. 1978	Aransas	6212	4
San Diego, Southwest	Dec. 1978	Duval	5908	4
Cooke	April 1979	LaSalle	4118	1
Grass Island	April 1979	Calhoun	4940	2
Smiley, Southeast	Aug. 1979	Gonzaies	8515	1
Muffin	Aug. 1979	Kenedy	9462	4
Blakeway	Sept. 1979	Dimmit	4546	1
Stumberg	Sept. 1979	Dimmit	4195	1
Black Diamond	Sept. 1979	Jim Hogg	4734	4
Jennings Ranch	Aug. 1979	Zapata	9112	4
Margarita	Aug. 1979	Zapata	6843	4
Ike Pryor	Nov. 1979	Zavala	2636	1
Flowella	Nov. 1979	Brooks	11216	4
Leona	Jan. 1980	Zavala	3224	1
Peach Creek	Jan. 1980	Gonzales	7869	1
Pruske	Jan. 1980	Wilson	7553	1
Patton	Feb. 1980	Karnes	8101	2
Alligator Slough	March 1980	Zavala	3691	1
Randado Ranch	May 1980	Zapata	5918	4
Star Brite, West	May 1980	Duval	5847	4
Gonzales	June 1980	Gonzales	8790	1
Katie Welder	July 1980	Calhoun	7654	2

Little New York	Aug. 1980	Gonzales	1954	1
Derrick Bacon	Sept. 1980	Wilson	6500	1
Independence, West	Oct. 1980	Zavala	3857	1
J.Y.F.	Oct. 1980	Wilson	7707	1
Marcelina Creek	Oct. 1980	Wilson	8060	1
Milano	Oct. 1980	Milam	4250	1
Fentress	Nov. 1980	Caldwell	1788	1
Christian	Dec. 1980	Gonzales	7295	1
Los Cuates Ranch	Dec. 1980	Jim Hogg	4372	4
Roma, East	Jan. 1981	Starr	3936	4
Pecan Gap	April 1981	Milam	5185	1
Giddings	May 1981	Bastrop	7480	1
Laura	May 1981	Calhoun	7464	2
Devils Run	June 1981	Refugio	5254	2
Block 1064W	June 1981	Willacy	5930	4
Bronco-Medina	Aug. 1981	Medina	482	1
Holman	Sept. 1981	Live Oak	4476	2
Hasse	Oct. 1981	Wilson	5950	1
Jetero	Oct. 1981	Zavala	4061	1
Gato Creek, S. E.	Nov. 1981	Webb	9901	4
A.W.P.	Dec. 1981	McMullen	9800	1
Cindy Ann	Dec. 1981	Wilson	6049	1
Ashford	Jan. 1982	Lavaca	5248	2
Stedman Island	Jan. 1982	Nueces	8808	4
Montemayor	Feb. 1982	Duval	5516	4
Winn-Dulce	March 1982	Zavala	2790	1
Block 772-L	March 1982	Nueces	10838	4
Dilworth Ranch	May 1982	McMullen	5524	1
Clareville	May 1982	Bee	4448	2
Bartell Pass	June 1982	Aransas	10780	4
Pandora, Northwest	Aug. 1982	Wilson	5054	1
Sophia	Aug. 1982	McMullen	5661	1
Pawnee, South	Sept. 1982	Bee	6103	2
Monte Christo	Oct. 1982	Hidalgo	5940	4
Rheinlander	Nov. 1982	Caldwell	1615	1
Salinas	Nov. 1982	Starr	2018	4
Big Mule	Jan. 1983	McMullen	5556	1
Matagorda Block 629	Jan. 1983	Calhoun	4864	2
Circle-B-	April 1983	Goliad	4977	2
Crass	Aug. 1983	Refugio	5508	2
Hi-Luk	Aug. 1983	San Patrieio	5658	4
Tn Bar, North	Sept. 1983	LaSalle	8096	1
Dry Hollow	Oct. 1983	Lavaca	2598	2
Woods Cemetery	Nov. 1983	Dewitt	6551	2
Cooke Ranch	March 1984	LaSalle	7823	1
TW	March 1984	Jackson	7846	2
Kathleen Ann	June 1984	Wilson	5999	1
Los Rubios, East	July 1984	Duval	5540	4
Venada	July 1984	Webb	9604	4
Jackson Ranch	Nov. 1984	Live Oak	5376	2
Bina, Northeast	Dec. 1984	Kleberg	7624	4
Dixie	Jan. 1985	Milam	3142	1
Cecily	April 1985	Dewitt	4952	2

July 1985–May 1999

Laughlin-Kibbe	July 1985	Jim Wells	5466	4
Guadalupe Bend	July 1985	Guadalupe	2179	1
Lacal	Sept. 1985	Willacy	8370	4
Speary	Dec. 1985	Karnes	9432	2
Medico	Sept. 1986	Jim Hogg	3088	4
W.C. Finch	Dec. 1986	Milam	1930	1
Gmen	Jan. 1987	Nueces	10344	4
Roen	Jan. 1987	Webb	16259	4
Dolores Creek, South	May 1987	Webb	8057	4
Alvarado	July 1987	Starr	8140	4
Kelly Lynne	Sept. 1987	Wilson	4815	1
Reymet	Oct. 1987	San Patricio	7702	4
Cotton Row	Nov. 1987	Hidalgo	12929	4
J.L.B.	Dec. 1987	Karnes	3263	2
Conn Brown Harbor	Dec. 1987	San Patricio	8055	4
Doughty	Feb. 1988	Nueces	196765	4
Eschberger	Oct. 1988	Nueces	11803	4
Martin Ranch	Nov. 1988	Duval	5458	4
Southern Comfort	March 1989	Calhoun	9114	2
Hubberd	April 1989	Webb	6612	4
Kitty Burns	July 1989	Live Oak	3637	2
Suemaur	Oct. 1989	Staff	4446	4
Two M	Nov. 1989	Maverick	2470	1
Sawgrass	Dec. 1989	Victoria	2076	2
Bennett Ranch	April 1990	Duval	5395	4
Bueno Sueno	April 1990	Webb	5580	4
Cuthbertson	June 1990	Dewitt	8053	2
First Shot	Aug. 1990	Gonzales	11369	1
Matagorda Blk. 562-L	Aug. 1990	Calhoun	6606	2
Ruthie H.	Nov. 1990	Duval	7035	4
Matagorda Isl. Bik. 721-L	Dec. 1990	Aransas	6976	4
Siene Bean	Oct. 1992	Refugio	5773	2
El Anzuelo	Feb. 1993	Kenedy	8929	4
Sublime, South	Aug. 1993	Lavaca	2494	2
Holyfield Fan Man Sand	Nov. 1993	Lavaca	4848	2
Tocquigny	April 1994	Dimmit	1935	1
Travis Higgins	July 1994	Maverick	4694	1
Reven	July 1994	Jackson	4355	2
Fandango, West	Aug. 1994	Zapata	14080	4
El Infernillo	Nov. 1994	Kleberg	8097	4
Alte Hunde	Dec. 1994	Zapata	6490	4
Cortez	Jan. 1995	Starr	1367	4
Gasper	April 1995	Wilson	2660	1
Beccero Creek	April 1995	Webb	6490	4
EDR	Aug. 1995	Duval	4259	4
Amore	March 1996	Hidalgo	5940	4
Tordilla-NC	Oct. 1996	Willacy	9328	4
Mondo Sueno	Jan. 1997	San Patricio	5650	4
Brazil, Southwest	Sept. 1997	McMullen	9772	1
Prue Ranch	Nov. 1997	Frio	3001	1
El Mesquitito	March 1998	Zapata	10220	4
SVD	March 1999	Duval	3387	4
Ann Mag, East	May 1999	Brooks	13434	4

Oct. 1999-March 2003

Roleta	Oct. 1999	Zapata	7792	4
Ranura	Nov. 1999	Kenedy	9979	4
Agula Creek	Dec. 1999	Victoria	8466	2
Molina	Jan. 2000	Bee	5181	2
Ortiz Frio	Jan. 2000	Dewitt	3035	2
Mustang Is!. Bik. 883	March 2000	Nueces	11390	4
CD Nabors	Jan. 2001	Starr	7466	4
Comanche-Halsell	Feb. 2002	Maverick	6731	1
Guajolote Suerte	March 2002	Calhoun	5797	2
G.O.M. St. 904	April 2002	Nueces	10506	4
Heintz	Oct. 2002	Starr	8520	4
MI 632	Feb. 2003	Calhoun	2004	2
MI 631L	March 2003	Calhoun	5436	2

Compiled by Ray Govett, Ph. D.,
Consulting Geologist & Engineer

Endnotes

The Early Days

1. James A. Clark, *Chronological History of the Petroleum and Natural Gas Industries*, (Houston: Clark Book Co., 1963), 1-4. Clark attributes the information about the great flood to *This Fascinating Oil Business*, by Max Ball, and says that Ball took it from a Babylonian tale inscribed on a tablet about 4,000 years ago.

2. Wayne Gard, *The First 100 Years of Texas Oil and Gas*, (Dallas: Texas Mid-Continent Oil and Gas Association), 2; Clark 13–14.

3. Nancy Heard, "First Oil in Texas," *South Texas Blowout*, (Alice, Texas: *The Corpus Christi Caller-Times*, 1951), 11; Larry Smallwood, "Oil Boom Continues: Early South Texas Oil, Piedras Pintas Revisited," *Traveler*, May 1999, 6; *Laredo Times*, 6 July 1901; *Corpus Christi Caller*, 27 April 1952.

4. Papers of family of William A. Tinney (hereinafter referred to as Tinney Papers); Heard, *Blowout*, 13; *Corpus Christi Caller*, 27 April 1952.

5. Tinney Papers; Smallwood 8; *Corpus Christi Caller*, 27 April 1952, and 24 May1952.

6. Heard, *Blowout*, 50.

7. "Texas' First Oilman: Tinny (sic) Started Drilling in 1899," *Traveler*, November 2002, 4; *New Handbook of Texas*, (Austin: Texas State Historical Association, 1996), 1:486; Tinney Papers; Heard, *Blowout*, 50.

8. Frank X. Tolbert, "The Story of Lyne Taliaferro (Tol) Barret," *Texas & Oil: 100 Years of Growth*, Texas Mid-Continent Oil & Gas Association, May 1966, 1; C.A. Warner, *Texas Oil and Gas Since 1543*, (Houston: Gulf Publishing Co., 1939), 6-7, 113; Rachel Bluntzer Hebert, *The Forgotten Colony: San Patricio de Hibernia*, (Burnet, Texas: Eakin Press, 1981), 285.

9. Tolbert 7, 11, 12–14.

10. Gard 7-8; Warner, 16; *Texas Almanac and State Industrial Guide*, (Dallas: *Dallas Morning News*, 1931), 194.

11. Gard 8; Warner 17.

12. Warner 19.

13. Roy Grimes, *300 Years in Victoria County*, (Victoria, Texas: The Victoria Advocate Publishing Co.), 394.

14. Walter Rundell, Jr., *Early Texas Oil: A Photographic History 1866–1936*, (College Station: Texas A&M University Press, 1977), 23; Warner 115; Gard 9-11; Bill Walraven telephone interview with Paul Ledvina, Exxon/Mobil Co. Archivist, Fairfax, Va., February 2003.

15. Keith Guthrie, *The History of San Patricio County*, (Austin: San Patricio County Historical Commission, 1986), 74.

16. Warner 281; Coleman McCampbell, *Texas Seaport: the Story of the Growth of Corpus Christi*, (New York, New York: Exposition Press, Inc., 1952), 6; Bill Walraven, "Natural gas was only a nuisance," *Corpus Christi Caller*, May 7, 1977, B1.

17. "First Drilling In This Section Was Started in 1902," *Corpus Christi Caller*, 24 Oct. 1937; *Corpus Christi, A Guide*, (Corpus Christi: Corpus Christi Chamber of Commerce, 1942) 170, 172.

17. Nancy Heard, "Blowouts Old Stuff in South Texas," *South Texas Blowout*, 18.

18. *The History of Nueces County*, (Jenkins Publishing Company, 1972), 143.

19. *Corpus Christi Caller*, 9 Nov. 1922; 10 Nov. 1922; 3 Dec. 1936; 11 Dec. 1936.

20. Walraven, "Natural gas was only a nuisance"; Mody C. Boatright, *Folklore of the Oil Industry*, (Dallas: Southern Methodist University Press, 1963), 114.

21. *Corpus Christi Caller*, 4 Sept. 1936.

22. *San Antonio Express*, 21 Nov. 1936.

23. *The History of Nueces County*, 144.

24. Ralph Storm, interviewed by Bill Walraven; Walraven, "Natural gas was only a nuisance."

25. O.G. McClain interview on Corpus Christi Geological Society Oil Video 4; O.G. McClain letter to Owen R. Hopkins, 27 June 1998.

26. Walraven, "Natural gas was only a nuisance."

27. Clyde T. Reed, "Search for Water Brings Area's First Gas Well," *The The Corpus Christi Caller-Times*, 23 April 1939; "First Drilling in This Section Was Started in 1902."

28. Warner 281, 286; *The History of Nueces County*, 141.

29. *San Patricio County News*, 23 April 1915.

30. Nancy Heard, "Old Oil Boom Days Weren't So Good," *San Antonio Express*, 7 Nov. 1965; "South Texas Has Plenty More Oil Not Found, 80-Year-Old Killam Believes," *San Antonio Express*, 25 April 1954; "Pescadito Kid Recalls Rough, Ready Oil Days," *San Antonio Express*, 21 Jan. 1968; "Laredo," "Indian Territory," Tape # 183, W. A. Owens interview of O.W. Killam, 7 May 1956, from the collections of the Center for American History, the University of Texas at Austin; *New Handbook of Texas*, 3: 1094; "Wildcatting in the Laredo Area," "Doodlebugs," Tape # 192, Mody C. Boatright interview of O. W. Killam, 5 Sept. 1956, from the collections of the Center for American History, the University of Texas at Austin; Pamphlet from Killam's office;

Michael Black, Mirando City: *A New Town in a New Oil Field*, (Laredo, Texas; Laredo Publishing Company, 1972), 73, 42-43; Radcliffe Killam, unnumbered videotape of the Corpus Christi Geological Society, 25 Oct. 2001; Diana Davids Olien and Roger M. Olien, *Oil in Texas, The Gusher Age, 1895-1945*, (Austin: The University of Texas Press, 2002), 109.

31. Kilgore Collection, Special Collections and Archives, Bell Library, Texas A&M Univesity-Corpus Christi.

32. Owens interview of O.W. Killam; Radcliffe Killam, interviewed by Bill Walraven in Killam's Laredo office, 28 Jan. 2003.

33. Riley Froh, *Edgar B. Davis, Wildcatter Extraordinary*, (Luling, Texas; The Luling Foundation, 1994), 8, 11–15, 17, 19, 24–25, 27–28, 41, 46–48, 99, 143, 245; Riley Froh, *Edgar B. Davis and Sequences in Business Capitalism: From Shoes to Rubber to Oil*, (New York, N. Y. : Garland, 1993), 8; W. Carlton "Tubby" Weaver, interviewed by Bill Walraven in a series of meetings and telephone calls, 2002–2004; Woolsey manuscript on exhibit in Central Texas Oil Patch Museum, Luling, Texas; *The Luling Signal*, 14 May 1924.

34. Gerald Lynch, *Roughnecks, Drillers, and Toolpushers*, (Austin: University of Texas Press), 40–42.

35. Lynch 44; Froh, *Wildcatter*, 41–42, 55-56, 62–73, 85–87.

36. 144. George Hawn, interviewed by Bill Walraven in Hawn's Corpus Christi office, 25 Nov. 2002.

37. "Wingy Smith," *Historic Legends of Western Oil*, 4:28, "Last of the Wildcatters," 4:5–6, Special Collections and Archives, Bell Library, Texas A&M University-Corpus Christi.

38. Dana Blankenhorn, "Hugh Roy Cullen: Oil gifts still flow from King of the Texas Wildcatters," *Houston Business Journal*, 17 Sept. 1984, 1B, 13B, 15B; Ed Kilmon and Theon Wright, *Hugh Roy Cullen: A Story of American Opportunity*, (New York: Prentice-Hall, Inc., 1954), 137–143, 176–181, 185–189; *National Oil Scouts Association 1937 Yearbook*, 122–123; Warner 307; Clark 181; *Texas Almanac 2003-2004*, 567; "Cullen Dead at 76," *Houston Chronicle*, 5 July 1957; "Mr. Cullen Dies—and a City Mourns Its Loss—Rites are Tomorrow," *Houston Press*, 5 July 1957.

39. Guthrie, *History of San Patricio County*. 75, 77; Bill and Marjorie Walraven, *Empresarios' Children: The Welders of Texas*, published by the Rob and Bessie Welder Wildlife Foundation, (Corpus Christi, Texas: Javelina Press, 2000), 202.

40. "First Nueces County Gas Producer Comes in on Dunn Property in 1922," *Corpus Christi Caller*, 26 Nov. 1933; Olien and Olien, *Oil in Texas, The Gusher Age*, 114–115; *The History of Nueces County*, 142; Anne Dodson, "W. Armstrong Price, He recalls the early days when oil was just being discovered in South Texas," *The Corpus Christi Caller-Times*, 17 July 1977; W. Armstrong Price, "Pioneer Oil Co.'s No. 1 Meaney Gas Blowout Well," article prepared for the Historical Markers Subcommittee of Nueces County Historical Survey Committee, copy in possession of Bill Walraven.

41. George Hogan, "Flat Tire Helped in Finding Gas," *South Texas Blowout*, 21; Warner 290.

42. Anne Dodson, "First Nueces County oil strike big news in 1930," *Corpus Christi Caller*, 17 July 1977, 5G; Bill Walraven, "At 96, area's first geologist still hard at his studies," *Corpus Christi Caller*, 11 April 1985; Anne Dodson, "W. Armstrong Price, He recalls the early days . . . ," *Corpus Christi Times*; Grady Phelps, "W. Armstrong Price not retiring type, *Corpus Christi Times*, 3 Feb. 1976; O. G. McClain, oil videotape 4 of the Corpus Christi Geological Society; Anne Dodson, "South Texas geology pioneer Price dies at 98," *Corpus Christi Caller*, 3 Nov. 1987.

A Time of Growth

1. *The Independent Petroleum Monthly*, March 1961.

2. Maston Nixon, "The Coming of Southern Alkali Corporation to Corpus Christi," Kilgore Collection, Box 30, F. 30.9, Special Collections and Archives, Bell Library, Texas A&M University-Corpus Christi; Maston Nixon, Biography Form, *The Corpus Christi Caller-Times*; "Businessman Maston Nixon dies at 70," *The Corpus Christi Caller-Times*, 3 April 1966; "Bluff Business Section Started by Nixon Ahead of Oil Boom Here," *The Corpus Christi Caller-Times*, 26 Sept. 1954.

3. Nixon Biography Form.

4. "The Nixon Story: Legal Difficulties Lead To Founding Own Oil Co.," *The San Angelo Standard-Times*, 24 Dec. 1951, 2.

5. "Bluff Business Section Started"; *Independent Petroleum Monthly*; letter from George M. Placke, president of the Corpus Christi Area Heritage Society, to W. Carlton Weaver, 3 May 2002; "Businessman Maston Nixon Dies at 70."

6. O. G. McClain, oil videotape of the Corpus Christi Geological Society.

7. Corpus Christi Oilman: "Warren of Renwar Named Chief Roughneck for '58, *Corpus Christi Caller*, 27 Oct. 1958; Guy Ira Warren, Independent Oil Operator, Investments, Ranching, *Outstanding Men of Texas*, 208; Biography-Guy I. Warren, on file in the library of *The Corpus Christi Caller-Times*; Frank La Roe, "Guy Warren Sees City As Tidelands Center," *Corpus Christi Caller-Times*, 12 Oct. 1952; Nancy Heard, "Italians Welcomed AMG, Colonel Warren Believes,"

Corpus Christi Times, 11 Feb. 1946; "Corpus Christi Gains Capital, Growth, Leaders From Oil," *Corpus Christi Caller*, 14 Oct. 1956.

8. "Corpus Christi Gains Capital"

9. "Sam E. Wilson, Jr." *Texas Edition: Men of Achievement*, Evelyn Miller Crowell, Ed., (Dallas: John Moranz Association, 1940), 87.

10. Author's 2002–2004 interviews of Tubby Weaver.

11. *Texas Edition: Men of Achievement*, 86; Tubby Weaver, interviewed by author; Maston Nixon, "The Conception of the Multi-Story Nixon Building at the Corner of Broadway and Leopard, December 1962," Maston Nixon, Envelope # 1, library of *The Corpus Christi Caller-Times.*

12. *History of Nueces County*, 145–146.

13. James A. Clark, *Three Stars for the Colonel*, (New York: Random House, 1954), 68; "Railroad Commission," *New Handbook of Texas*, 5: 409; 2: 341; "Deal of the Century," *AAPG Explorer, Special Issue: A Century,* 39; Leonard Nikoloric, "Battles of the RRC," *Corpus Christi Magazine*, May 1985, 30; "Conservation ends chaotic development, insures future supplies," *Humble Way*, 15:1 May–June 1959, 18; Lawrence Goodwyn, *Texas Oil, American Dreams*, (Austin: Texas State Historical Association, 1996), 50.

14. "Culberson, Olin Wellborn Nichols," *Handbook of Texas*, 2: 436.

15. Tom Lea, *The King Ranch*, (King Ranch, 1957), 502.

16. Lea 504–506.

17. Charles J. V. Murphy, "Treasures in Oil and Cattle," *Fortune*, 80:2, 1 Aug. 1969, 111.

18. *These Forty Years*, Humble Oil and Refining Co., June 1957; *Humble Way*, June 1957; Olien and Olien, *Oil in Texas*, 63-64; John Cypher, *Bob Kleberg and the King Ranch*, (Austin: University of Texas Press, 1995), 17; Mary Fritz, "A Giant Who 'Went Blithely," *AAPG Explorer, A Special Issue*, 38; *New Handbook of Texas*, 5: 118.

19. Fritz 39.

20. *Handbook of Texas* 5, 318–319; Fritz, 38–39; "In and Out at Harbor Island," *Humble Way*, 13:2, July-August 1957, 4; Cypher 17; Lea 612; "Oil on the King Ranch," *The Humble Way*, 9:2, July-August 1953, 9.

21. Vonda Davis, "Humble Discovery: King Ranch Oil Find Has 21st Birthday," *The Corpus Christi Caller-Times*, 11 Oct. 1966; "North Alazan: 4-in-1 Oil Well," *Humble Way*, 15:2, July-August 1959, 6; "Laying the Big Line," *Humble Way*, 14:6, March-April 1959, 24.

22. Claude V. D'Unger, "Is end near for oil and gas operations on the island?" *Corpus Christi Caller-Times*, 21 Jan. 2005; "They're Still Making History on Party Six," *Search*, 1:3, 11–12; "Getting Off One More Shot," *Search*, 1:3, 15-16.

23. "One More Shot," 15–16.

24. Dorothy Abbott McCoy, *Oil, Mud and Guts, Birth of a Texas Town*, (Brownsville, Texas: Dorothy Abbott McCoy, 1977), 2-5, 100.

25. James Mosier, interviewed by author at Mosier's Corpus Christi home,

26. McCoy 128–129; Mike Cox, *Texas Ranger Tales, Stories That Need Telling*, (Plano, Texas: Republic of Texas Press, 1997), 209–210; Author's interview of Radcliffe Killam; Olien and Olien, *Oil in Texas*, 211; Janet R. Edwards, "Life at the Oil Camp: Anything But Crude," *Texas Highways*, March 1991, 18–20; videotaped interview with Dick White, Corpus Christi Geological Society.

27. Edwards, 21.

28. Wallace Smith, "Jack Pratt Promoted First Area Oil Field," *Alice Daily Echo*, 1964.

29. Author's interviews of Tubby Weaver; Tubby Weaver, Corpus Christi Geological Society videotaped and audio interviews.

30. Eugene Deaver, interviewed by author at Deaver's Corpus Christi residence, August 2003; Jack Graham, interviewed by author at Graham's office, August 2003; *Legends of Texas Oil*; "A minute biography of . . . Tom Graham," *Corpus Christi Times*, 4 July 1941.

31. John Crutchfield, videotaped interview, Corpus Christi Geological Society.

32. *Entrepreneurship in Oil: Mr. Paul R. Haas and the Prado Oil and Gas Company*, (Austin: Interviews conducted by Larry Segrest, sponsored by the Moody Foundation, Oil Business History Project, The University of Texas at Austin, February,1971), 35; video interview of John Crutchfield, Corpus Christi Geological Society.

33. Carole Kneeland, "It takes brains—and a little luck—to find oil," *Corpus Christi Times*, 1 June 1978.

34. Letter from Lois J. Green to W. Carlton "Tubby" Weaver, 12 April 2004; "Arnold O. Morgan, local oilman, dies," *Corpus Christi Caller*, 29 May 1981; "Oilman Morgan is dead," *Corpus Christi Times*, 29 May 1981; Kneeland.

35. Olien and Olien, *Oil in Texas*, 232; *Corpus Christi Caller*, 21 June 1942; U. S. Merchant Ships Sunk or Damaged in World War II, November 23, 2004, <http://www.usmm.org/shipsunkdamaged.html>; Early Deane, "The Dope Bucket," *Corpus Christi Caller*, 27 Jan. 1942, 5.

36. Nancy Heard, *The Corpus Christi Caller-Times*, 2 Jan. 1944; Chester Wheless, videotaped interview, Corpus Christi Geological Society.

37. Olien and Olien, *Oil in Texas*, 232-233.

38. Olien and Olien, *Oil in Texas*, 234; United Carbon Co., Box 30, F. 30.9, Kilgore Collection, Special Collections and Archives,

Bell Library, Texas A&M University-Corpus Christi; author's interviews of Tubby Weaver.

39. Bruce Scrafford, "Developments in South Texas in 1944," *Bulletin of the American Association of Petroleum Geologists*, 29:6 (June 1945), 777; 23 Nov. 2004, <http://www.tsha.utexas.edu/handbook/online/articles/view/BB/dob8.html>; "Pipelines and LNG Importer Serving New England and New York," Northeast Gas Association New England Natural Gas Market Update, 30 May 2003.

40. O. G. McClain, Corpus Christi Geological Society videotaped interview.

New Horizons

1. O. G. McClain-Oil videotape 6, Corpus Christi Geological Society.

2. "Oil and Gas Industry," *New Handbook of Texas*, 4: 1126.

3. "An Empire of Gas Has To Expand Fast," *Business Week*, 28 Jan. 1956; 1959 Texas Eastern Gas Transmission Co. Financial Report; William Vrana, interviewed by author at Vrana's Corpus Christi home, 2002. William Vrana, "The Rise and Demise of a Grand Conglomerate Company, 1943–1999."

4. Louis Beecherl, "Pioneers of Texas Oil," interviews from the Center for American History, the University of Texas at Austin.

5. Jess P. Roach, "Millard Holland 'Bulger' Major, 1919–1996," *AAPG Bulletin*, April 1997, 674–675; M.H. Major, "It Was Fun While It Lasted," (Corpus Christi, Texas, 26 Feb. 1992), 71–123.

6. 36. James C. Freeman, "Trials, Tribulations and Luck of a Petroleum Geologist," May 1996.

7. Vrana, "Rise and Demise; "As J. I. Case Sowed, So Shall It Reap," *Business Week*, 21 July 1990, 27; author's interview of Vrana.

8. George Hawn, interviewed by author at Hawn's Corpus Christi office, 25 Nov. 2002.

9. Radcliffe Killam, Corpus Christi Geological Society's videotaped interviews; Radcliffe Killam, interviewed by author at Killam's Laredo, Texas, office, 28 Jan. 2003; John G. Hurd, interviewed in his office in San Antonio 4 June 1992 by Lawrence Goodwyn and Barbara Griffith, TIPRO Archives, The Center for American History, the University of Texas at Austin.

10. W. Ray Brown, "Grand Jury Indicts Collier for Perjury," *Corpus Christi Caller*, 2 Aug. 1942, 1A; Lee Durst, Corpus Christi Geological Society videotaped interview, Tape No. 970-80, Quadrant Productions Video, 24 Dec. 2003.

11. Jon Spradley, interviewed by author at Spradley's Corpus Christi office, 10 Sept. 2004.

12. O.G. McClain, Corpus Christi Geological Society videotaped interview, Quadrant Productions videos 1-2 tape 4, 3, 2, 5; O.G. McClain letter to Owen R. Hopkins, chief geologist, Suemaur Exploration & Production, 27 June 1998.

13. Clark, *Three Stars for the Colonel*, 205, 209–210, 217, 233, 231; "Culberson, Olin Wellborn Nichols," *New Handbook of Texas*, 2: 436; "Railroad Commission," *New Handbook of Texas*, 5: 410; Jim Herring, interviewed by author at Herring's home in Rockport, Texas, 13 Oct. 2004; Ernest O. Thompson, Statement to the House Ways and Means Committee, U. S. House of Representatives, Washington, D. C., Appendix A, Clark, *Three Stars for the Colonel*, 251; "Thompson, Ernest Othmer," *New Handbook of Texas*, 6: 471; "Oil and Gas Industry," New Handbook of Texas, 4:1128; Leonard Nikoloric, "Battles of the RRC," *Corpus Christi Magazine*, May 1985, 31.

14. Jerome "Jerry" O"Brien, Corpus Christi Geological Society videotaped interview, Quadrant Productions; Anne Gruner Schlumberger, *The Schlumberger Adventure: Two Brothers Who Pioneered in Petroleum Technology*, (New York, N. Y.: Arco Publishing Co., 1982), 48; *The History of Nueces County*, (Jenkins Publishing Co, 1972); Debbie Dorsett, "Jerry O'Brien," *Bulletin of the South Texas Geological Society*, 41:8, April 2001, 18–20; Jerry O'Brien interviewed by author in O'Brien's San Antonio office, 30 Oct. 2002; Donald and Bonnie Sue Jacobs, "Petroleum Pioneers: Jerry O'Brien: 70 Years in the Oil Patch," *South Texas, The Newsletter*, April/Early Summer, 2001, 29; "Texas Oil History," *San Antonio Express-News*, 7 Nov. 1965; Carmina Danini, "One of the last WWI vets dies," *San Antonio Express-News*, 16 Nov. 2003; Weatherston obituary, *San Antonio Express-News*, 16 Nov. 2003; "A Salute," South Texas Geological Society, 18 Oct. 2001.

16. Slick, Thomas Baker, Jr., The Handbook of Texas Online, 28 Nov. 2004, <http://www.tsha.utexas.edu/handbook/online/articles/view/SS/fsl7.html>; "Looking Into the Future: Dreaming Realist From Yale Conceived Research Center," Tom Slick File, Library File, *San Antonio Express-News*; Ray Miles, *King of the Wildcatters, The Life and Times of Tom Slick*, (College Station: Texas A&M University Press, 1996), 122; Biographical Sketch of Thomas Baker Slick, 20 Oct. 1948, Library, *San Antonio Express-News*; Memorandum, Tom Slick, Jr., Files, *San Antonio Express-News*; Cindy Tumiel, "S. A. science giants at loggerheads," *San Antonio Express-News*, 29 Sept. 2003.

17. Bonnie Weise, "An Interview with Louis H. Haring, Jr.," *Bulletin of the South Texas Geological Society*, 43:9, May 2003; Louis J. Haring, Jr., manuscript in possession of the author; Louis H. Haring,

interviewed by the author at Petroleum Club in San Antonio, Texas, 4 June 2003.

18. Wilford Stapp, "Geology Is My Love, Music is My Passion," *Bulletin of the South Texas Geological Society*, 15:9, May 2000, 7–13; Wilford Lee Stapp Biographical Summary, *Bulletin of the South Texas Geological Society*, 15:9, May 2000, 8; Wilford Stapp, interviewed by the author at the Petroleum Club in San Antonio, Texas, 4 June 2003.

19. Jay Endicott, recorded audiotaped interview.

20. "Chuck" Forney, interviewed by author 29 Feb. 2004.

21. J. Frank La Roe, "Wyatt has one of city's largest independent firms," *Corpus Christi Caller*, 1956; Jane Wolfe, *Blood Rich: When Oil Billions, High Fashion, and Royal Intimacies Are Not Enough*, (Boston: Little, Brown, and Co., 1993), 52, 103–109, 118 ; Lee Durst, interviewed by the author 24 Dec. 2003; Spencer Pearson, "Wyatt's 'Hardly Able Oil' hardly ended there," *Corpus Christi Caller-Times*, 14 July 1986; Jay Endicott, recorded audiotaped interview; Lynn Pentony, "Coastal States grew to giant-size in 10 years," *Corpus Christi Caller-Times*, 3 Nov. 1965; Beecherl interview; *New Handbook of Texas*, 5, 409–410; "Wyatt to resign as Coastal chairman," *Corpus Christi Caller-Times*, 26 March 1997; "Wyatt plans to hand title to James Paul," *Corpus Christi Caller-Times*; Anne Pearson, "Making Money and Enemies: Wyatt renowned for scraps," *Houston Chronicle*, 14 April 1991.

22. Dick Conolly, interviewed by author at Town & Country Restaurant, Corpus Christi, October 2003.

23. William C. Johnson, videotaped interview at Corpus Christi Country Club, 10 June 2004.

24. Hewitt Fox, Corpus Christi Geological Society videotaped interview; W. Rinehart Miller and Hewitt B. Fox, *The Oil Game*, (Corpus Christi, Texas).

25. "Bill" Miller (2), interviewed by the author.

26. Daniel Pedrotti, videotaped interview at Corpus Christi Country Club, 10 June 2004.

27. Video presentation of TRT Holdings.

28. Wallace Graham, interviewed by the author 3 Aug. 2004 in San Antonio and 14 Aug. 2004 at Corpus Christi home of Robert Graham. Robert Graham, interviewed by author at Graham's Corpus Christi home 14 Aug. 2004.

29. George Tanner, interviewed by author at Tanner's Corpus Christi office, September 2004.

30. William T. Vogt, Jr., interviewed by author at Vogt's Corpus Christi office, September 2004.

31. Richard C. Wilshusen, interviewed by author at Wilshusen's Corpus Christi office, October 2004.

32. Jerry Clark, interviewed by author at Clark's Corpus Christi office, 24 Sept. 2004.

33. Lawrence Hoover, interviewed by author at Hoover's Corpus Christi office, 3 Sept. 2004.

34. Cassette tape interview of Al Harris courtesy of the Corpus Christi Geological Society.

35. Ray Govett, unpublished manuscript, 28 Sept. 2004.

36. "South Texas oil barons didn't get rich overnight, *Corpus Christi Caller-Times*, 17 July 1977; Brian O'Brien, telephone interviews by author, September 2004, 11 Oct. 2004.

37. Lindon Curry, videotaped interview at Corpus Christi Country Club, 10 June 2004.

38. Robert Buschman, interviewed by author at Buschman's San Antonio office, 30 Dec. 2003.

39. John David Scott, *Mr. Independent Hank Harkins: his climb from roughneck to oil magnate!*, Limited printing, John David Scott, 1985, 8; *Men of Achievement in Texas*, Garland A. Smith, Ed., (Austin: Garland A. Smith, 1973), 146.

40. Lucien Flournoy, interviewed by the author in Flournoy's office in Alice, Texas, 19 Nov. 2002; "Lucien Flournoy," *Men of Achievement in Texas*, 66; Dan A. Hughes, interviewed by the author in Hughes's office in Beeville, Texas, 21 March 2003; Darla Morgan, "Success gives Alice oilman power—behind the throne," *Corpus Christi Caller-Times*, 3 Nov. 1985; Mary Lee Grant, "Oil man Lucien Flournoy has wrested his wealth from the ground," *Corpus Christi Caller-Times*, 15 Aug. 1999; Eleanor Mortensen, Around the City, *Corpus Christi Times*, 4 Jan. 1983.

41. William "Bill" Carl, interviewed by author in Carl's home near Beeville, Texas.

42. Dan A. Hughes, interviewed by the author in Hughes's office in Beeville, Texas; Hughes Company brochure; Texas Top Gas Producers, 27 Nov. 2004, <http://www.lasser.com/data/ttop_100-_gas.shtml>.

The Move Offshore

1. Lucille Glasscock, *A Texas Wildcatter: A Fascinating Saga of Oi*l, (San Antonio: The Naylor Company, 1952), 6–11, 15–22, 27–32, 36–38, 43–44, 49–57, 103; Genny McNamara, "A Time of Boom and Bust," *The Corpus Christi Caller-Times*, 1 July 1962.

2. Glassscock, 110–113; C.G. Glasscock-Tidelands Oil Co. Annual Report

1957, 6–12, 18; C.G. Glasscock-Tidelands Oil Co. Annual Report 1958, 5; Ralph Storm, interviewed by author at Storm's Corpus Christi home.

3. Annual Report 1958, 2–5; "James C. Storm," *Men of Achievement in Texas*, 30; Ralph Storm interview; Célika Storm, interviewed by author in Mrs. Storm's Corpus Christi office; "Storm Brothers of Corpus Christi Pit Ingenuity Against Oil Doldrums," *The Corpus Christi Caller-Times*, 16 Aug. 1964, 19; Nancy Heard, "Phillips' McMullen Wildcat Reaches 3rd Deepest Mark," *San Antonio Express-News*, 12 June 1965, 5-D; "GM Diesels Drill to 24,420 Feet," *Power News*, Stewart & Stevenson Companies, 5:2, 1; "2 Storm Drilling Companies Sold," *Corpus Christi Times*, 2 July 1968; Industry Pioneers Hall of Fame Plaque, Offshore Energy Center, Galveston, Texas; Erik T. Borgen, "'Jim' Storm, 'Mr. Offshore Drilling,' Thrives on 60 Hours Weeks, Scorns Retirement Plans," *Business & Energy International*, September 1975, 6; "Port Mourns Chairman's Death: a Tribute to James C. Storm," *Port of Corpus Christi Channels*, Fall 1991.

4. Laurence McNeil, Corpus Christi Geological Society videotaped interview; Paul R. Haas, Corpus Christi Geological Society videotaped interview; *Entrepreneurship in Oil: Mr. Paul R. Haas and the Prado Oil and Gas Company*, (Austin: interviews conducted by Larry Segrest, sponsored by the Moody Foundation, Oil Business History Project, The University of Texas at Austin, February, 1971), 6–7, 2–4, 157; "La Gloria Personnel Deemed Tops in Civic Achievements," *Corpus Christi Caller*, 14 Oct. 1956; "Wilson Funeral 4 p.m. Tomorrow," *Corpus Christi Times*, 21 Feb. 1947.

5. *Entrepreneurship in Oil*, Laurence McNeil, video interview; *Entrepreneurship in Oil*, 27.

6. Chester Wheless, Corpus Christi Geological Society videotaped interview.

7. *Entrepreneurship in Oil*, 24, 29–32, 36-37, 40-43, 73; video interview with Laurence McNeil; John David Scott, *El Gordo... El Magnifico*, (Corpus Christi Exploration Company, 1981), 64.

8. Laurence McNeil, interviewed by author at McNeil's Corpus Christi office, 12 Sept. 2003.

9. Scott, 20–24, 48, 121, 139–140, 50, 112; William Volk, Corpus Christi Geological Society videotaped interview.

10. Scott, 45, 21, 68, 51, 128, 48–49, 145, 40, 52–56; author's interview of McNeil.

11. Jim Herring, interviewed by author at Herring's home in Rockport, Texas, 13 Oct. 2004; author's interview of McNeil; Paul Haas video interview; *Texas Business 1980;* Scott, 58.

12. J. Frank LaRoe, "30,000 miles-2 months, Turnbull Inspects Firm's Foreign Rig Operations," *Corpus Christi Caller*; Barta, Carolyn, *Bill Clements, Texan to His Toenails*, (Austin: Eakin Press, 1996); J. Frank LaRoe, "Residents' Drilling Firm Gets $50 million Contract," *The Corpus Christi Caller-Times*; Transocean, 27 Nov. 2004, <http://www.deepwater.com/History.cfm>.

13. "Another Day That Lives in Infamy," *AAPG Explorer, Special Issue: A Century*, 60; Santa Barbara Wildlife Care Network, 28 Nov. 2004, <http://www.silcom.com/~sbwcn/spill.shtml>; Randy Bissell, interviewed by author in Bissell's Corpus Christi office, May 2004.

Notes to page 276-311

Refinery Row

1. *Texas Oil and Gas Since 1543*, 309; Unpublished manuscript, Southwestern Oil and Refining Co. history.

2. Kate Robbins, CITGO Petroleum Corporation.

3. Michael Arndt, "Corpus Christi refinery buyout fits world trend," from the *Chicago Tribune*, published in *The Corpus Christi Caller-Times*, 9 Oct. 1988; CITGO Petroleum Corp., About CITGO, Company History, 29 Oct. 2003, <http://www.citgo.com/AboutCITGO/CompanyHistory.jsp>; *Corpus Christi Caller-Times*, 23 Feb. 1992.

4. Carl Newlin, interviewed by author, November 2003; Southwestern Oil & Refining Co. history, unpublished manuscript in possession of the author.

5. About Valero, 21 April 2004, <http://www.valero.com/conl.php@p=History+of+Valero>; Tom Whitehurst, Jr., "Celebration here greets the founder of Valero," *Caller-Times*, 14 June 2001; Brad Olson, "Valero earns employee-friendly stripes, again," *Corpus Christi Caller-Times*, 13 Jan. 2005.

6. West, Dr. Robert V., Jr., Tesoro Petroleum Corp., biographical summary in library file, *San Antonio Express-News*; "Tesoro In Top 500," *San Antonio Express*, 2 June 1974; Tesoro Today, A Profile, 30 May 2003, 'Who We Are," <http://www.tesoropetroleum.com/whoweare.html>.

7. "El Paso reports $1.93 billion loss for 2003," *Corpus Christi Caller-Times*, 1 Oct. 2004; <http://www.valero.com/conl.php@p=History+of+Valero>.

8. *Corpus Christi Caller-Times*, October 31, 2003; "Howell Corp. okays refinery sale, *Corpus Christi Caller-Times*, 5 Sept. 1981; "Coastal States Gas Grew to Giant-Size in 10 Years, *Corpus Christi Caller*, 3 Nov. 1965; "Wyatt to resign as Coastal

chairman," *Corpus Christi Caller-Times*, 26 March 1997; ; "an OLD WORLD INDUSTRY in a whole NEW WORLD," Valero Energy Corporation 2003 Summary Annual Report; Bill Day, "Valero signs deal for Aruba refinery, *San Anonio Express-News*, 5 Feb. 2004; Valero 2003 Annual Report; "Valero to announce $8 billion deal," *San Antonio Express-News*, 25 April 2005.

9. Jan Jarboe Russell, "Fueled by anger and ego, veteran oil man jumps back in game," *San Antonio Express-News*, 13 June 2004.

A Changing World

1. William E. "Bill" Carl, author's interview; William E. Carl interview at the TIPRO Annual Convention in Houston, June 6, 1993. Tape in TIPRO Archives at the Center for American History, The University of Texas, Austin; author's interview of Dan Hughes.

2. Author's interview of Larry McNeil; author's interview of Bill Carl; author's interview of Randy Bissell.

3. Author's interview of George Tanner; Chester Wheless, Corpus Christi Geological Society videotaped interview, Quadrant Productions; William D. English presentation at meeting of the Corpus Christi League of Women Voters, Corpus Christi Seaman's Center, 15 July 2004.

4. Weldon Oliver, interviewed by author at the Central Texas Oil Patch Museum in Luling, Texas, 12 March 2003; W. E. Findley, interviewed by author in Findley's office in Alice, Texas, October 2004.

5. Author's interview of Findley.

6. Floyd Nix, interviewed by author in Nix's Corpus Christi office, 2004; Olien and Olien, *Oil in Texas, The Gusher Age*, 105; Warner 283; the *Corpus Christi Caller;* George H. Fancher, with Robert L. Whiting and James H. Cretsinger, *The Oil Resources of Texas*, (Texas Petroleum Research Committee, 1954), 149.

7. Author's interview of Jon Spradley.

8. Author's telephone interview of Brian O'Brien; <http://www.statesman.com/news/content/auto/epaper/eddition s/sunday/news_f35e54B6d32a313300c5.html>

9. Radcliffe Killam, Corpus Christi Geological Society videotaped interview; David Killam, Corpus Christi Geological Society videotaped interview.

10. Author's interview of Wilford Stapp.

11. O. G. McClain, Corpus Christi Geological Society videotaped interview; author's interview of Richard Wilshusen.

12. James C. 'Jake' Venable, interviewed by author at Pathfinder Energy Services, Inc., 19 Oct. 2004.

13. Cecilia Venable, interviewed by author at her home and at the Corpus Christi Public Library, November 2004; author's interview with Bill Miller (2).

14. Author's interview of Jerry Clark; Dan Pedrotti, Corpus Christi Geological Society videotaped interview.

15. Tom Davidson, interviewed by author in Davidson's office in Boerne, Texas, 21 July 2004.16. Author's interview of William Carl; author's interview of Larry McNeil; author's interview of Bill Findley; author's interviews of Richard Wilshusen.

17. Author's interview of Randy Bissell; author's telephone interviews of Brian O'Brien.

Bibliography

Interviews

Beecherl, Louis. Pioneers of Texas Oil. Interviews from the Center for American History, the University of Texas at Austin.

Bissell, Randy. Interviewed by author at Bissell's Corpus Christi office. May 2004.

Boatright, Mody C. "Wildcatting in the Laredo Area." "Doodlebugs," Interview of O. W. Killam. 5 Sept. 1956. Pioneers of Texas Oil. Tape # 192. From the collections of the Center for American History, the University of Texas at Austin.

Buschman, Robert. Interviewed by author at his office in San Antonio, Texas. 30 Dec. 2003.

Carl, William E. Interview at the TIPRO Annual Convention in Houston. 6 June 1993. Tape in TIPRO Archives at the Center for American History, the University of Texas at Austin.

_____________. Corpus Christi Geological Society videotaped interview. Quadrant Productions Video. 10 June 2004 at the Corpus Christi Country Club.

_____________. Interviewed by author at Carl's home near Beeville, Texas. 7 Feb. 2003.

Clark, Jerry. Interviewed by author at Clark's Corpus Christi office. 24 Sept. 2004.

Conolly, Dick. Interviewed by author at Town & Country Restaurant, Corpus Christi. October 2003.

Crutchfield, John. Corpus Christi Geological Society videotaped interview. Quadrant Productions Video. 22 Aug. 2001 at the Corpus Christi Town Club.

Curry, Landon. Interviewed and videotaped at the Corpus Christi Country Club. Quadrant Productions Video. 10 June 2004.

Davidson, Tom. Interviewed by author at Davidson's office in Boerne, Texas, July 2004.

Deaver, Eugene. Interviewed by author at assisted living facility in Calallen, Corpus Christi, Texas. August 2003.

Durst, Lee. Corpus Christi Geological Society videotaped interview. Tape No. 970-80, Quadrant Productions Video. 5 Nov. 2003 at 4802 Ocean Drive, Corpus Christi.

Endicott, Jay. Corpus Christi Geological Society videotaped interview. Quadrant Productions Video. 4 Feb. 2004 at 4146 Harry St.,Corpus Christi.

___________. Self-recorded audiotaped interview.

"Entrepreneurship in Oil, Mr. Paul R. Haas and the Prado Oil and Gas Company." Austin: Interviews conducted by Larry Segrest. Sponsored by the Moody Foundation. Oil Business History Project. The University of Texas at Austin. February 1971.

Findley, W. E. Interviewed by author in Findley's office in Alice, Texas. Fall 2004.

Flournoy, Lucien. Interviewed by author in Flournoy's office in Alice. Texas. 19 Nov. 2002.

Forney, "Chuck." Corpus Christi Geological Society videotaped interview. Quadrant Productions Video. 4 Feb. 2004 at 4146 Harry St., Corpus Christi.

______________. Interviewed by author.

Fox, Hewitt. Corpus Christi Geological Society videotaped interview. Quadrant Productions Video. 4 Feb. 2004 at 4146 Harry St., Corpus Christi.

Fullbright, Bob. Interviewed by author at Fullbright's office in Hebbronville, Texas. 23 Oct. 2002.

Graham, Jack. Interviewed by author at Graham's Corpus Christi office. August 2003.

Graham, Robert. Interviewed by author at Graham's Corpus Christi home 14 Aug. 2004.

Graham, Wallace.Interviewed by author in San Antonio 3 Aug. 2004 and at Robert Graham's Corpus Christi home 14 Aug. 2004.

Haas, Paul R. Corpus Christi Geological Society videotaped interview. Quadrant Productions Video. 27 June 2001 at the 600 Building, Corpus Christi.

Haring, Louis H. Interviewed by author at the Petroleum Club in San Antonio, Texas. 4 June 2003.

Harris, Albert T. "Al." Taped interview courtesy of the Corpus Christi Geological Society.

Hawn, George. Interviewed by author at Hawn's Corpus Christi office. 25 Nov. 2002.

Herring, Jim. Interviewed by author at Herring's home in Rockport, Texas. 13 Oct. 2004.

Hill, David "Tex." Interviewed by author in San Antonio. 5 June 2003.

Hoover, Lawrence. Interviewed by author at Hoover's Corpus Christi office. 3 Sept. 2004.

Hurd, John G. Interview in his office in San Antonio 4 June 1992 by Lawrence Goodwyn and Barbara Griffith. TIPRO Archives. The Center for American History, the University of Texas at Austin.

Hughes, Dan A. Interviewed by author in Hughes's office in Beeville, Texas. 21 March 2003.

Johnson, William C. Interviewed and videotaped at Corpus Christi Country Club. Quadrant Productions Video. 10 June 2004.

Killam, David. Corpus Christi Geological Society videotaped interview. Quadrant Productions Video. 25 October 2001at the San Antonio Petroleum Club.

____________. Interviewed by author at Country Club in Laredo, Texas. 28 Jan. 2003.

Killam, Radcliffe. Corpus Christi Geological Society videotaped interview. Quadrant Productions Video. 25 Oct. 2001 at the San Antonio Petroleum Club.

____________. Interviewed by author at Killam office in Laredo, Texas. 28 Jan. 2003.

"Laredo," "Indian Territory." O. W. Killam interview by W. A. Owens. Pioneers in Texas Oil. Tape # 183. 7 May 1956. From the collections of the Center for American History, the University of Texas at Austin.

Ledvina, Paul. Exxon/Mobil Co. Archivist interviewed by author by telephone from Ledvina's office in Fairfax, Va. February 2003.

McClain, O. G. Corpus Christi Geological Society videotaped interview. Tapes 4, 6. Quadrant Productions Video. 27 June 2001 at the 600 Building, Corpus Christi.

McNeil, Lawrence. Corpus Christi Geological Society videotaped interview. Quadrant Productions Video. 22 Aug. 2001 at the Corpus Christi Town Club..

__________________. Interviewed by author at McNeil's Corpus Christi office. 12 Sept. 2003.

Miller, "Bill" (2). Interviewed by author.

Miller, William B. Corpus Christi Geological Society videotaped interview. Quadrant Productions Video. 5 Nov. 2003 at 4802 Ocean Drive, Corpus Christi.

Mosier, James. Interviewed by author at Mosier's Corpus Christi home. September 2002.

Newlin, Carl. Corpus Christi Geological Society videotaped interview. Quadrant Productions Video. 5 Nov. 2003 at 4802 Ocean Drive.

___________. Interviewed by author at Newlin's Corpus Christi home. November 2003.

Nix, Floyd. Interviewed by author at Nix's Corpus Christi office. 2004.

O'Brien, Brian. Interviewed by author by telephone from O'Brien's Houston office. September 2004. 11 Oct. 2004.

O'Brien, Jerome "Jerry." Corpus Christi Geological Society videotaped interview. Quadrant Productions Video. 25 October 2001 at the San Antonio Petroleum Club.

____________________. Interviewed by author in O'Brien's office in San Antonio, Texas. 30 October 2002.

O'Connor, Karen. Corpus Christi Geological Society videotaped interview. Quadrant Productions Video. 10 June 2004 at the Corpus Christi Country Club.

Oliver, Weldon. Interviewed by author at the Central Texas Oil Patch Museum in Luling, Texas. 11 March 2003.

Pedrotti, Daniel. Videotaped interview at Corpus Christi Country Club. 10 June 2004.

Popejoy, W. L. Interviewed by author by telephone. 2 Oct. 2004.

Spradley, Jon. Interviewed by author at Spradley's Corpus Christi office. 10 Sept. 2004.

Stapp, Wilford Lee. Interviewed by author at the Petroleum Club in San Antonio, Texas. 4 June 2003.

Storm, Celika (Mrs. James Storm). Interviewed by author at her Corpus Christi office.

Storm, Ralph. Interviewed by author at Storm's Corpus Christi home.

Tanner, George. Interviewed by author at Tanner's Corpus Christi office, September 2004.

Venable, Cecilia. Interviewed by author at Venable home. 8 Nov. 2004.

Venable, James C. "Jake." Interviewed by author at Pathfinder Energy Services, Inc. 19 Oct. 2004.

Vogt, William T., Jr. Interviewed by author at Vogt's Corpus Christi office, September 2004.

Volk, William. Corpus Christi Geological Society videotaped interview. Quadrant Productions Video. 4 Feb. 2004 at 4146 Harry St., Corpus Christi.

Vrana, William "Bill." Corpus Christi Geological Society videotaped interview. Quadrant Productions Video. 4 Feb. 2004 at 4146 Harry St., Corpus Christi.

__________________. Interviewed by author at Vrana's Corpus Christi home. 2002.

Weaver, William Carlton "Tubby." Corpus Christi Geological Society videotaped interviews. Quadrant Productions Video. 28 June 2001 at the 600 Building, Corpus Christi.

__________________________. Interviewed by author in a series of meetings and telephone interviews from 2002–2004.

Wheless, Chester. Corpus Christi Geological Society videotaped interview. Quadrant Productions Video. 22 Aug. 2001 at the Corpus Christi Town Club.

White. Dick. Corpus Christi Geological Society videotaped interview. Quadrant Productions Video. 28 June 2001 at the 600 Building, Corpus Christi.

Wilshusen, Richard C. Interviewed by author at Wilshusen's Corpus Christi office. 1 Oct. 2004.

Books

Barta, Carolyn. *Bill Clements. Texan to His Toenails.* Austin: Eakin Press. 1996.

Black, Michael. *Mirando City: A New Town in a New Oil Field.* Laredo, Texas; Laredo Publishing Company, 1972.

Boatright, Mody C. *Folklore of the Oil Industry.* Dallas: Southern Methodist University Press, 1963.

Boone, Laura Phipps. *The Petroleum Dictionary.* Norman, Oklahoma: University of Oklahoma Press, 1952.

Clark, James A. *Chronological History of the Petroleum and Natural Gas Industries.* Houston: Clark Book Co.,1963.

__________. *Three Stars for the Colonel.* New York: Random House. 1954.

Corpus Christi, A Guide. Corpus Christi: Corpus Christi Chamber of Commerce, 1942.

Cox, Mike. *Texas Ranger Tales. Stories That Need Telling.* Plano: Republic of Texas Press, 1997.

Cypher, John. *Bob Kleberg and the King Ranch.* Austin: University of Texas Press. 1995.

Fancher, George H., with Robert L. Whiting and James H. Cretsinger. *The Oil Resources of Texas.* Texas Petroleum Research Committee, 1954.

Froh, Riley. *Edgar B. Davis and Sequences in Business Capitalism: From Shoes to Rubber to Oil.* New York, N. Y.: Garland. 1993.

__________. *Edgar B. Davis. Wildcatter Extraordinary.* Luling. Texas; The Luling Foundation. 1994.

Gard, Wayne. *The First 100 Years of Texas Oil and Gas.* Dallas: Texas Mid-Continent Oil and Gas Association.

Glasscock, Lucille. *A Texas Wildcatter: A Fascinating Saga of Oil.* San Antonio: The Naylor Company. 1952.

Goodwyn, Lawrence. *Texas Oil. American Dreams.* Austin: Texas State Historical Association. 1996.

Grimes, Roy. *300 Years in Victoria County.* Victoria. Texas: The Victoria Advocate Publishing Co.

Guthrie, Keith. *The History of San Patricio County.* Austin: San Patricio County Historical Commission. 1986.

Rachel Bluntzer Hebert. *The Forgotten Colony: San Patricio de Hibernia.* Burnet. Texas: Eakin Press. 1981.

The Historical Encyclopedia of Texas. Ellis A. Davis, Ed. Texas Historical Society, 1937.

The History of Nueces County. Jenkins Publishing Company. 1972.

Lea, Tom. *The King Ranch.* King Ranch. 1957.

Men of Achievement in Texas. Garland A. Smith, Ed. Austin: Garland A. Smith. 1973.

Lynch, Gerald. *Roughnecks. Drillers. and Toolpushers.* Austin: University of Texas Press.

McCampbell, Coleman. *Texas Seaport: The Story of the Growth of Corpus Christi.* New York, N. Y.: Exposition Press., Inc. 1952.

McCoy, Dorothy Abbott. *Oil. Mud and Guts. Birth of a Texas Town.* Brownsville, Texas: Dorothy Abbott McCoy. 1977.

Miles, Ray. *King of the Wildcatters. The Life and Times of Tom Slick.* College Station: Texas A&M University Press. 1996.

Miller, W. Rinehart, and Hewitt B. Fox. *The Oil Game.* Corpus Christi, Texas.

New Handbook of Texas. Austin: Texas State Historical Association. 1996. Volumes 2, 4, 5, and 6.

Olien, Diana Davids, and Roger M. Olien. *Oil in Texas. The Gusher Age.* 1895-1945. Austin: The University of Texas Press. 2002.

Rundell, Walter, Jr. *Early Texas Oil: A Photographic History 1866–1936.* College Station: Texas A&M University Press. 1977.

Texas Edition: Men of Achievement. Evelyn Miller Crowell, Ed. Dallas: John Moranz Association. 1940.

Schlumberger, Anne Gruner. *The Schlumberger Adventure: Two Brothers Who Pioneered in Petroleum Technology.* New York. N. Y.: Arco Publishing Co. 1982.

Scott, John David. *El Gordo... El Magnifico.* Corpus Christi Exploration Company. 1981.

__________. *Mr. Independent Hank Harkins: his climb from roughneck to oil magnate!* Limited printing. John David Scott. 1985.

These Forty Years. Humble Oil and Refining Co. June 1957.

Tolbert, Frank X. "The Story of Lyne Taliaferro (Tol) Barret." *Texas & Oil: 100 Years of Growth.* Texas Mid-Continent Oil & Gas Association. May 1966.

Walraven, Bill. *Corpus Christi: The History of a Texas Seaport.* Woodland Hills, Calif.: Windsor Publications, Inc. 1982.

__________ and Marjorie K. Walraven. *Empresarios' Children: The Welders of Texas.* Published by the Rob and Bessie Welder Wildlife Foundation. Sinton, Texas: Javelina Press. 2000.

Warner, C. A. *Texas Oil and Gas Since 1543.* Houston: Gulf Publishing Co. 1939.

Wolfe, Jane. *Blood Rich: When Oil Billions, High Fashion, and Royal Intimacies Are Not Enough.* Boston. Little, Brown and Co. 1993.

Journal Articles

American Oil & Gas Reporter (April 2003 and April 2004).

"an OLD WORLD INDUSTRY in a whole NEW WORLD." Valero Energy Corporation 2003 Summary Annual Report.

"Another Day That Lives in Infamy." *AAPG Explorer. Special Issue: A Century*. 60.

"As J. I. Case Sowed. So Shall It Reap." *Business Week* (21 July 1990): 27.

Blankenhorn, Dana. "Hugh Roy Cullen: Oil gifts still flow from King of the Texas Wildcatters." *Houston Business Journal* (17 Sept. 1984): 1B, 13B, 15B.

Borgen, Erik T. "'Jim' Storm. 'Mr. Offshore Drilling.' Thrives on 60 Hours Weeks, Scorns Retirement Plans." *Business & Energy International* (September 1975).

C.G. Glasscock-Tidelands Oil Co. Annual Report 1957.

C.G. Glasscock-Tidelands Oil Co. Annual Report 1958.

"Conservation ends chaotic development, insures future supplies." *Humble Way* 15 (1 May–June 1959): 18.

Dan A. Hughes Company Brochure.

"Deal of the Century." *AAPG Explorer. Special Issue: A Century*. 39.

Dorsett, Debbie. "Jerry O'Brien." *Bulletin of the South Texas Geological Society* 41 (8 April 200): 18.

Edwards, Janet R. "Life at the Oil Camp: Anything But Crude." *Texas Highways* (March 1991).

"An Empire of Gas Has To Expand Fast." *Business Week* (28 Jan. 1956).

Fritz, Mary. "A Giant Who 'Went Blithely." *AAPG Explorer*: 38.

"Getting Off One More Shot." *Search* 1, no. 3. 15-16.

"GM Diesels Drill to 24,420 Feet." *Power News*. Stewart & Stevenson Companies 5, no. 2: 1.

Heard, Nancy. "Blowouts Old Stuff in South Texas." *South Texas Blowout. The Corpus Christi Caller-Times* (1951).

______________. "First Oil in Texas." *South Texas Blowout*

Hogan, George. "Flat Tire Helped in Finding Gas." *South Texas Blowout*: 21.

Humble Way (June 1957).

"In and Out at Harbor Island." *Humble Way* 13, no. 2 (July-August 1957): 4.

The Independent Petroleum Monthly (March 1961).

Jacobs, Donald and Bonnie Sue Jacobs. "Petroleum Pioneers: Jerry O'Brien: 70 Years in the Oil Patch." *South Texas. The Newsletter* (April/Early Summer 2001).

Koch Reflections. Koch Refining Co. (1953).

"Last of the Wildcatters." *Historic Legends of Western Oil*. Special Collections and Archives. Bell Library. Texas A&M University-Corpus Christi. 4, no. 5-6: 28.

"Laying the Big Line." *Humble Way* 14, no. 6 (March-April 1959): 24.

Murphy, Charles J. V. "Treasures in Oil and Cattle." *Fortune* 80, no. 2 (1 Aug. 1969): 111.

National Oil Scouts Association 1937 Yearbook.

Nikoloric, Leonard. "Battles of the RRC." *Corpus Christi Magazine* (May 1985): 30.

"North Alazan: 4-in-1 Oil Well." *Humble Way* 15, no. 2 (July-August 1959): 6.

Oil & Gas Journal (20 Aug. 1936).

"Oil on the King Ranch." *The Humble Way* 9, no. 2 (July-August 1953): 9.

"Pipelines and LNG Importer Serving New England and New York." Northeast Gas Association New England Natural Gas Market Update. (30 May 2003).

"Port Mourns Chairman's Death: a Tribute to James C. Storm." *Port of Corpus Christi Channels* (Fall 1991).

Roach, Jess P. "Millard Holland 'Bulger' Major, 1919–1996." Bulletin of the American Association of Petroleum Geologists (April 1997): 674–675.

Scrafford, Bruce. "Sam E. Wilson, Jr., Developments in South Texas in 1944." *Bulletin of the American Association of Petroleum Geologists* 29 no. 6 (June 1945): 777.

Smallwood, Larry. "Oil Boom Continues: Early South Texas Oil. Piedras Pintas Revisited." *Traveler* (May 1999).

Stapp, Wilford. "Geology Is My Love, Music is My Passion." *Bulletin of the South Texas Geological Society* 15 no. 9 (May 2000).

*Texas Almanac and State Industrial Guide. Dallas Morning News (*1931).

Texas Eastern Gas Transmission Co. Financial Report 1959.

"Texas' First Oilman: Tinny (sic) Started Drilling in 1899." *Traveler* (November 2002).

"They're Still Making History on Party Six." *Search* 1, no. 3.

"They Still Drill with Cable Tools," *The Humble Way* 4, no. 3 (Sept.-Oct. 1948), 24–27.

Weise, Bonnie. "An Interview with Louis H. Haring. Jr." *Bulletin of the South Texas Geological Society* 43 no. 9 (May 2003).

Correspondence

Green, Lois J. Letter to W. Carlton "Tubby" Weaver. 12 April 2004.

McClain, O. G. Letter to Owen R. Hopkins, chief geologist, Suemaur Exploration & Production. 27 June 1998.

Placke, George M., president of the Corpus Christi Area Heritage Society. Letter to W. Carlton "Tubby" Weaver. 3 May 2002.

Monographs

Freeman, James C. "Trials, Tribulations and Luck of a Petroleum Geologist." May 1996. Monograph in possession of the author.

Major, M. H. "It Was Fun While It Lasted." Corpus Christi, Texas. 26 Feb. 1992. Monograph in possession of the author.

Nixon, Maston. "The Coming of Southern Alkali Corporation to Corpus Christi." Kilgore Collection. Box 30. F. 30.9. Special Collections and Archives, Bell Library. Texas A&M University-Corpus Christi.

______________. "The Conception of the Multi-Story Nixon Building at the Corner of Broadway and Leopard. December 1962." Maston Nixon. Envelope # 1. The Corpus Christi Caller-Times Library.

Price, W. Armstrong. "Pioneer Oil Co.'s No. 1 Meaney Gas Blowout Well." Article prepared for the Historical Markers Subcommittee of Nueces County Historical Survey Committee and in possession of the author.

"A Salute." South Texas Geological Society. 18 Oct. 2001.

Vrana, William. "The Rise and Demise of a Grand Conglomerate Company 1943–1999." Article in possession of the author.

Newspapers

Alice Daily Echo.

Chicago Tribune.

Corpus Christi Caller.

The Corpus Christi Caller-Times.

Corpus Christi Times.

Houston Chronicle.

Laredo Times.

The Luling Signal.

The San Angelo Standard-Times.

San Antonio Express.

San Antonio Express-News.

Unpublished Sources

Biographical Sketch of Thomas Baker Slick. 20 Oct. 1948. Library. *San Antonio Express-News.*

English, William D. Presentation at meeting of the Corpus Christi League of Women Voters. Corpus Christi Seaman's Center. 15 July 2004.

Govett, Ray. Prepared statement. 28 Sept. 2004.

Haring. Louis J., Jr. Unpublished manuscript in the possession of the author.

Industry Pioneers Hall of Fame Plaque. Offshore Energy Center, Galveston, Texas.

Slick, Tom, Jr. Files. Memorandum, *San Antonio Express-News.*

Southwestern Oil and Refining Co. history. Article in possession of the author.

Tinney, William A. Family Papers.

TRT Holdings video presentation.

United Carbon Co. Box 30. F. 30.9. Kilgore Collection. Special Collections and Archives. Bell Library. Texas A&M University-Corpus Christi.

Warren, Guy I. Biographical File. *The Corpus Christi Caller-Times.*

West, Dr. Robert V., Jr. Tesoro Petroleum Corporation. Biographical summary in Library File. *San Antonio Express-News.*

Woolsey, Vernon. Manuscript on exhibit in Central Texas Oil Patch Museum. Luling, Texas.

Government Documents

Ernest O. Thompson, Statement to the House Ways and Means Committee, U. S. House of Representatives, Washington, D. C., Appendix A, Clark, *Three Stars for the Colonel*, 251.

Internet Sources

About Valero. 21 April 2004. <http://www.valero.com/conl.php@p=History+of+Valero>.

About CITGO. Company History. 29 Oct. 2003. <http://www.citgo.com/AboutCITGO/CompanyHistory.jsp.html>.

Fortune Archives. <http:.www.fortune.com/fortune/2000/05/29/ten.html>.

Handbook of Texas Online. 27 November 2004. <http://www.tsha.utexas.edu/handbook/online/articles/view/BB/dob8.html>.

Robbins, Kate. <KROBBINcitgo.com>. CITGO Petroleum Corporation. E-mail of 11 Nov. 2003.

Santa Barbara Wildlife Care Network. 28 Nov. 2004, <http://www.silcom.com/~sbwcn/spill.shtm>.

Tesoro Today: A Profile. 'Who We Are." 30 May 2003. <http://www.tesoropetroleum.com/whoweare>.

Texas Top Gas Producers. 27 Nov. 2004. <http://www.lasser.com/data/top_100_gas.shtml>.

Transocean. 27 Nov. 2004. <http://www.deepwater.com/History.cfm>.

U. S. Merchant Ships Sunk or Damaged in World War II. 27 November 2004. <http://www.usmm.org/shipsunkdamaged.html>.

Photo Credits

The shadow of his camera in the foreground symbolizes the priceless pictorial legacy that 'Doc' McGregor left to Corpus Christi and South Texas.

Page 2 — 'Doc' McGregor Collection, Corpus Christi Museum of Science and History (hereinafter cited as McGregor Collection

Page 3 — (Left) McGregor Collection; (right) Sammy Gold Photography

Page 6 — McGregor Collection

Page 11 — Courtesy of Radcliffe Killam

Page 12 — Courtesy of George Hawn

Page 13 — McGregor Collection

Page 14 — McGregor Collection

Page 15 — (Top left) Corpus Christi Public Library; (top right) McGregor Collection; (bottom) Central Texas Oil Patch Museum, Luling, Texas

Page 16	McGregor Collection
Page 17	(Left, top and bottom) Courtesy of Kilgore Postcard Collection, Special Collections & Archives, Texas A&M University-Corpus Christi; (right) Bill Walraven photo
Page 18	(Top) McGregor Collection; (bottom) Corpus Christi Public Library
Page 19	(Top right) Sinton Public Library; (bottom) McGregor Collection
Page 20	(Top) Bill Walraven photo; (bottom) C.G. Glasscock-Tidelands Oil Company, 1957 Annual Report, Courtesy of Ralph Storm
Page 21	Courtesy of Carl Newlin
Pages 22–23	McGregor Collection
Page 24	(Top) Bill Walraven photo; (bottom) McGregor Collection
Page 25	Sammy Gold Photography
Pages 26–27	McGregor Collection
Pages 28–29	Tubby Weaver photo courtesy of Tubby Weaver; others by Bill Walraven
Pages 30-31	McGregor Collection
Page 32	Bill Walraven photo
Page 35	McGregor Collection
Page 36	Corpus Christi Public Library
Page 37	McGregor Collection
Page 38	(Left) Corpus Christi Public Library; (right top and bottom) McGregor Collection
Page 41	Courtesy of Bill Tinney
Page 42	*South Texas Blowout*, Kilgore Collection, Special Collections & Archives, Texas A&M University-Corpus Christi
Page 43	Bill Walraven photo
Page 44	Courtesy of Bob Fulbright
Page 46	McGregor Collection
Page 48	Corpus Christi Public Library
Page 49	McGregor Collection
Page 50	Corpus Christi Public Library
Page 51	Corpus Christi Public Library
Pages 52–59	McGregor Collection
Page 60	(Top) Corpus Christi Public Library; (bottom) McGregor Collection
Pages 62–63	Courtesy of Radcliffe Killam
Pages 64–65	Courtesy of Bob Fullbright
Pages 66–68	Courtesy of Radcliffe Killam
Page 70	(Top) Courtesy of Bob Fullbright; (bottom) Courtesy of Radcliffe Killam
Page 71	Courtesy of Radcliffe Killam
Pages 72–80	Central Texas Oil Patch Museum, Luling, Texas
Page 81	(Top) Central Texas Oil Patch Museum; (bottom) Bill Walraven photo
Pages 82–83	Central Texas Oil Patch Museum
Page 84	Bill Walraven photo
Page 85	Courtesy of George Hawn
Pages 87–88	McGregor Collection
Page 91	Photograph of Hugh Roy Cullen, University of Houston Historical Photographs Collection. Courtesy of Special Collections, University of Houston Libraries
Page 92	Photograph of Hugh Roy Cullen, University of Houston Historical Photographs Collection. Courtesy of Special Collections, University of Houston Libraries
Page 93	*South Texas Blowout*
Page 94–96	McGregor Collection
Page 98–99	Corpus Christi Geological Society
Pages 100–101	Courtesy of the *Oil & Gas Journal*
Pages 102–103	McGregor Collection
Page 104	(Top) C. C. Public Library; (bottom) McGregor Collection
Page 105	Corpus Christi Public Library
Pages 106–111	Courtesy of the *Oil & Gas Journal*
Page 113	Courtesy of *The Corpus Christi Caller-Times*
Page 114	McGregor Collection
Page 118	Courtesy of *The Corpus Christi Caller-Times*
Page 119	(Top) *Humble Way*, Courtesy of Exxon Corp.; (bottom) Central Texas Oil Patch Museum
Page 120	Archives, Texas A&M University-Kingsville
Pages 123–127	*Search* magazine, courtesy of Exxon Corporation
Page 128	(Top) Sinton Public Library; (bottom) James Mosier
Page 129	*Freer's Oil-O-Rama, 30 Years of Progress*, Kilgore Collection, Special Collections and Archives, Texas A&M University-Corpus Christi
Pages 130–131	McGregor Collection
Page 132	Courtesy of James Mosier
Page 133	(Top) McGregor Collection; (bottom) *Freer's Oil-O-Rama, 30 Years of Progress*, Kilgore Collection, Special Collections and Archives, Texas A&M University-Corpus Christi
Pages 134–137	McGregor Collection
Page 138	Courtesy of Tubby Weaver
Page 139	McGregor Collection
Page 140	Courtesy of Tubby Weaver
Pages 141–144	McGregor Collection
Page 145	Courtesy of Tubby Weaver
Page 146	Courtesy of Jack Graham
Page 147	Courtesy of *The Corpus Christi Caller-Times*
Page 148	Courtesy of Jack Graham
Page 149	McGregor Collection
Page 150	(Top) Bill Walraven photo; (bottom and page 139) McGregor Collection
Page 152	*South Texas Oil Directory*, 1973–1974
Pages 153–157	McGregor Collection
Page 159	*South Texas Blowout*
Page 160	Sammy Gold Photography
Pages 161–162	McGregor Collection

Pages 164-271

Page 164	Courtesy of *Business Week*
Page 166–168	McGregor Collection
Page 169	Corpus Christi Geological Society, 1964–65 Directory
Page 170	Courtesy of Bob Fullbright
Page 171	Corpus Christi Geological Society, 1964–65 Directory
Page 172	McGregor Collection
Page 173	Courtesy of George Hawn
Page 174-175	Courtesy of Radcliffe Killam
Pages 176–180	McGregor Collection
Page 183	(Top) Bill Walraven photo; (Bottom) Courtesy of O. G. McClain
Pages 184–185	McGregor Collection
Page 186	Corpus Christi Geological Society Directory, 1964–65, Courtesy of the Corpus Christi Geological Society
Page 187–188	McGregor Collection
Page 189	Courtesy of James Herring
Page 192	(Top) Courtesy of Jerome O'Brien; (bottom) McGregor Collection
Page 194	McGregor Collection
Page 195	(Right) *South Texas Blowout*
Page 198	South Texas Geological Society Directory, 1961–1962, Courtesy of the Corpus Christi Geological Society
Page 199	UT Institute of Texan Cultures at San Antonio, No. L-3384
Page 200	UT Institute of Texan Cultures at San Antonio, No. Z-2380-A
Page 201	McGregor Collection
Page 202	South Texas Geological Society Directory, 1961–1962, Courtesy of the Corpus Christi Geological Society
Pages 203–205	McGregor Collection
Page 207	Sammy Gold Photography
Page 208	McGregor Collection
Page 209	South Texas Geological Society Directory, 1961–1962, Courtesy of the Corpus Christi Geological Society
Page 210	McGregor Collection
Page 212	(Left) Courtesy of O. G. McClain; (right) McGregor Collection
Page 213	Corpus Christi Geological Society Directory, 1960–1961, Courtesy of the Corpus Christi Geological Society
Page 214	McGregor Collection
Page 217	All-American Wildcatters, Courtesy of Dan Hughes
Page 218	McGregor Collection
Pages 220	McGregor Collection
Page 223	Courtesy of Stewart & Stevenson
Pages 224–226	McGregor Collection
Page 228	Courtesy of Wallace and Robert Graham
Page 229	McGregor Collection
Pages 230–231	South Texas Oil & Gas Professional Directory, 1985, Courtesy of the Corpus Christi Geological Society
Page 232	South Texas Oil & Gas Professional Directory, 1985, Courtesy of the Corpus Christi Geological Society
Page 233	Bill Walraven photo
Page 234	Courtesy of Wallace and Robert Graham
Pages 236–237	(Left) Courtessy of Wallace and Robert Graham; (right) Courtesy of the Rob and Bessie Welder Wildlife Foundation
Page 240	South Texas Oil & Gas Professional Directory, 1985, Courtesy of the Corpus Christi Geological Society
Page 241	Archives and Special Collections, Texas A&M University-Kingsville
Page 242	Courtesy of Carl Newlin
Page 243	McGregor Collection
Page 244	South Texas Oil & Gas Professional Directory, 1985, Courtesy of the Corpus Christi Geological Society
Page 246	All-American Wildcatters, Courtesy of Dan Hughes
Page 247	McGregor Collection
Page 248	South Texas Oil & Gas Professional Directory, 1985, Courtesy of the Corpus Christi Geological Society
Page 250	Courtesy of Jay Endicott
Page 251	South Texas Oil & Gas Professional Directory, 1985, Courtesy of the Corpus Chisti Geological Society
Page 253	Alice Public Library
Page 254	South Texas Oil Directory, 1973–1974, Courtesy of the Corpus Christi Geological Society
Pages 256–258	Courtesy of Lucien Flournoy
Page 259	All-American Wildcatters, Courtesy of Dan Hughes
Page 260	South Texas Oil Directory, 1973–1974
Page 261	(Top)**A**ll-American Wildcatters, Courtesy of Dan Hughes; (bottom) Courtesy of Ralph Storm
Page 263	McGregor Collection
Page 264	Courtesy of Joe McCullough
Page 265	All-American Wildcatters, Courtesy of Dan Hughes
Page 266	Bill Walraven photo
Page 268	Courtesy of Tom Davidson
Page 269	Sammy Gold Photography
Page 270	(Top) McGregor Collection; (Bottom) Courtesy of Ralph Storm
Page 271	From *A Texas Wildcatter, A Fascinating Saga of Oil,* courtesy of the Glasscock family

Page 272	(Top) Courtesy of Ralph Storm; (bottom) McGregor Collection
Pages 274–275	Sammy Gold Photography
Page 276	Courtesy of Ralph Storm
Page 277	Courtesy of Mrs. Celika Storm
Pages 278–279	Courtesy of Ralph Storm
Page 280	Courtesy of Paul Haas
Pages 281	Courtesy of O. G. McClain
Pages 282	(Top) South Texas Oil Directory, Courtesy of the Corpus Christi Geological Society; (bottom) McGregor Collection
Page 283	Courtesy of Texas A&M University-Kingsville
Page 285	McGregor Collection
Page 288–290	Courtesy of Corpus Christi Exploration Company
Page 292	Courtesy of Ralph Storm
Page 293	Courtesy of Corpus Christi Exploration Company
Page 295	Bill Walraven photo
Page 296	Courtesy of David Crow, Southeastern Drilling Co.
Page 297	(Top) Courtesy of David Crow, Southeastern Drilling Co.; (bottom) McGregor Collection
Page 298	Bill Walraven photo
Page 300	*Humble Way*, Courtesy of Exxon Corp.
Page 302	Courtesy of Carl Newlin
Page 303	Corpus Christi Public Library
Page 304	McGregor Collection
Page 306	Courtesy of the *Oil & Gas Journal*
Page 307	McGregor Collection
Pages 308–309	Courtesy of Carl Newlin
Page 311	McGregor Collection
Page 312	Courtesy of Bill Findley, FESCO
Pages 313–314	Bill Walraven photos
Page 319	Artists' rendition by Payne & Rowlett, Courtesy of Bill English, Cheniere LNG, Inc.
Pages 320–321	Courtesy of Weldon Oliver, International Services, Inc.
Pages 322–323	Courtesy of Bill Findley, FESCO
Page 325	McGregor Collection
Page 326	Bill Walraven photo
Page 327	Corpus Christi Geological Society Directory, 1967–1968
Page 331	Bill Walraven photo
Page 332	(Top) McGregor Collection; (bottom) Bill Walraven photo
Page 333	Bill Walraven photos
Page 334	John C. Davis drawing
Page 337	(Top) Bill Walraven photo; (bottom) Courtesy of Tom Davidson
Page 338	Bill Walraven photo
Page 342	Courtesy of the Corpus Christi Geological Society
Page 362	McGregor Collection

Index

A

B

E

F

G

H

J

K

L

M

O

S

T

U

V

W

Y

Z

This book was published by the Corpus Christi Geological Society and printed by Taylor Publishing Company of Dallas, Texas, using Adobe InDesign CS software. Cover and dust jacket illustrations by John C. Davis, Jr., Dallas, Texas

Headlines are Futura Medium and Futura Extra Bold Condensed, and body type is 11-point and 10-point Minion Condensed and Minion Condensed Italic. Photographs were edited using Extensis software and/or Adobe Photoshop CS.